ALGEBRAIC GEOMETRY

ALGEBRAIC GEOMETRY

Proceedings of the Conference at Berlin
9–15 March 1988

Edited by

H. KURKE

Sektion Mathematik, Berlin, Germany

and

J. H. M. STEENBRINK

Mathematical Institute, Nijmegen, The Netherlands

Reprinted from

COMPOSITIO MATHEMATICA

Volume 76, Nos 1 & 2, 1990

KLUWER ACADEMIC PUBLISHERS

DORDRECHT / BOSTON / LONDON

ISBN 0-7923-0934-0

Published by Kluwer Academic Publishers,
P.O. Box 17, 3300 AA Dordrecht, The Netherlands.

Kluwer Academic Publishers incorporates
the publishing programmes of
D. Reidel, Martinus Nijhoff, Dr W. Junk and MTP Press.

Sold and distributed in the U.S.A. and Canada by
Kluwer Academic Publishers,
101 Philip Drive, Norwell, MA 02061, U.S.A.

In all other countries, sold and distributed by
Kluwer Academic Publishers Group,
P.O. Box 322, 3300 AH Dordrecht, The Netherlands.

Printed on acid-free paper

COMPOSITIO MATHEMATICA

Volume 76, Nos 1 & 2, October 1990

Special Issue

ALGEBRAIC GEOMETRY

Proceedings of the Conference at Berlin, 9–15 March 1988

Compositio Mathematica **76:** vii, 1990.

Preface

The Conference on Algebraic Geometry, held in Berlin 9–15 March 1988, was organised by the Sektion Mathematik of the Humboldt-Universität. The organising committee consisted of H. Kurke, W. Kleinert, G. Pfister and M. Roczen. The Conference is one in a series organised by the Humboldt-Universität at regular intervals of two or three years, with the purpose of providing a meeting place for mathematicians from eastern and western countries.

The present volume contains elaborations of part of the lectures presented at the Conference and some articles on related subjects. All papers were subject to the regular refereeing procedure of *Compositio Mathematica,* and H. Kurke acted as a guest editor of this journal.

The papers focus on actual themes in algebraic geometry and singularity theory, such as vector bundles, arithmetical algebraic geometry, intersection theory, moduli and Hodge theory.

We are grateful to all those who, by their hospitality, their presence at the Conference, their support or their written contributions, have made this Conference to a success.

The editors

Compositio Mathematica **76**: viii, 1990.

List of participants

R. Achilles, *Halle*
K. Altmann, *Berlin*
Y. Alwadi, *Berlin*
S. Aouira, *Berlin*
A. Aramova, *Sofia*
L. Badescu, *Bucharest*
C. Banica, *Bucharest*
F. Bardelli, *Pavia*
G. Barthel, *Konstanz*
K. Behnke, *Hamburg*
A. Campillo, *Valladolid*
J. Castellanos, *Madrid*
J. Coanda, *Bucharest*
W. Decker, *Kaiserslautern*
F. Delgado, *Valladolid*
B. Dgheim, *Berlin*
A. Dimca, *Bucharest*
K. Drechsler, *Halle*
A. Durfee, *Bonn*
D. Eisenbud, *Waltham*
H. Esnault, *Bonn*
J. Eßer, *Bonn*
J. Feustel, *Berlin*
Th. Fiedler, *Berlin*
T. Fimmel, *Berlin*
H. Flenner, *Göttingen*
J. Franke, *Berlin*
M. Furushima, *Bonn*
F. Gaeta, *Madrid*
H. Gräbe, *Erfurt*
G.-M. Greuel, *Kaiserslautern*
K. Haberland, *Jena*
Z. Hajto, *Kraków*
H. Hauser, *Innsbruck*
F. Herrlich, *Bochum*
L. Hille, *Berlin*
L. Hoa, *Halle*
Th. Höfer, *Bonn*
R. Holzapfel, *Berlin*
B. Hunt, *W. Lafayette*
D. Huybrecht, *Berlin*
F. Ischebeck, *Münster*
J. Jurkiewicz, *Warschau*
C. Kahn, *Waltham*
G. Janelidze, *Tbilisi*
G. Kempf, *Baltimore*
W. Kleinert, *Berlin*
S. Kloth, *Köthen*
B. Kreußler, *Berlin*
H. Kurke, *Berlin*
A. Landia, *Tbilisi*
H. Lange, *Erlangen*
O. Laudal, *Oslo*
A. Lipkowski, *Beograd*
V. Lomadse, *Tbilisi*
N. Manolache, *Bucharest*
F. Marko, *Bratislava*
B. Martin, *Berlin*
H. Meltzer, *Berlin*
T. Muhammed, *Halle*
G. Müller, *Mainz*
W. Müller, *Berlin*
U. Nagel, *Halle*
N. Nakayama, *Bonn*
Y. Namikawa, *Bonn*
A. Némethi, *Bucharest*
V. Palamodov, *Moscow*
A. Parisinski, *Gdansk*
G. Pfister, *Berlin*
D. Popescu, *Bucharest*
P. Pragacz, *Torún*
H. Reimann, *Berlin*
M. Roczen, *Berlin*
H. Roloff, *Erfurt*
D. Rothe, *Berlin*
P. Schenzel, *Halle*
H. Schönemann, *Berlin*
O. Schröder, *Halle*
T. Shiota, *Princeton*
Th. Siebert, *Berlin*
O. Song, *Berlin*
J. Spies, *Berlin*
J. Steenbrink, *Nijmegen*
M. Szurek, *Warsaw*
J. Tschinkel, *Berlin*
N. Tu Cuong, *Hanoi*
E. Viehweg, *Bonn*
W. Vogel, *Halle*
U. Vollmer, *Jena*
E. Zink, *Berlin*

Compositio Mathematica **76**: 1–5, 1990.

2-Cocycles and Azumaya algebras under birational transformations of algebraic schemes

F.A. BOGOMOLOV[1] & A.N. LANDIA[2]

[1]*Steklov Mathematical Institute of the Academy of Sciences of USSR, Vavilov Street 42, Moscow 117333, USSR;* [2]*Mathematical Institute of the Academy of Sciences of Georgian SSR, Z. Rukhadze Street 1, Tbilisi 380093, USSR*

Received 17 November 1988; accepted in revised form 23 November 1989

The basic question whether the injection $\mathrm{Br}(X) \to H^2(X, \mathcal{O}_X^*)_{\mathrm{tors}}$ is an isomorphism arose at the very definition of the Brauer group of an algebraic scheme X. Positive answers are known in the following cases:

1. the topological Brauer group $\mathrm{Br}(X_{\mathrm{top}}) \cong H^2(X, \mathcal{O}_{\mathrm{top}}^*)_{\mathrm{tors}} \cong H^3(X, \mathbb{Z})_{\mathrm{tors}}$ (J.-P. Serre); in the etale (algebraic) case the isomorphism is proved for
2. smooth projective surfaces (A. Grothendieck);
3. abelian varieties;
4. the union of two affine schemes (R. Hoobler, O. Gabber).

The first author has formulated a birational variant of the basic question, while considering the unramified Brauer group in [1]. The group $\mathrm{Br}_V(K(X)) = \cap \mathrm{Br}(A_v) \subseteq \mathrm{Br}(K(X))$ (intersection taken over all discrete valuation subrings A_v of the rational function field $K(X)$) is isomorphic to $H^2(\tilde{X}, \mathcal{O}^*)$, where $\tilde{X}$ is a nonsingular projective model of X, i.e. a nonsingular projective variety birationally equivalent to X.

QUESTION. Given a cocycle class $\gamma \in H^2(\tilde{X}, \mathcal{O}^*)$, is it possible to find a nonsingular projective model $\bar{X}$ such that γ is represented by a $\mathbb{P}^n$-bundle (i.e. by an Azumaya algebra) on $\bar{X}$?

The case where X is a nonsingular projective model of V/G, with G a γ-minimal group and V a faithful representation of G, was considered in [2]. O. Gabber in his letter to Bogomolov (12.1.1988) has given an affirmative answer to the question in the case of general algebraic spaces. In this paper we give a simple version of his proof for algebraic schemes.

Let X be a scheme, $\gamma \in H^2(X, \mathcal{O}^*)$, $\{U_i\}$ an affine cover of X. Then the restriction of γ to each U_i is represented by an Azumaya algebra A_i. If we would have isomorphisms $A_{i|U_i \cap U_j} \cong A_{j|U_i \cap U_j}$, we could glue the sheaves $\{A_j\}$ and get an

Azumaya algebra on X, representing γ. But we have isomorphisms $A_{i|U_i \cap U_j} \otimes \mathrm{End}(E_{ij}) \cong A_{j|U_i \cap U_j} \otimes \mathrm{End}(E_{ji})$ for certain vector bundles E_{ij}, E_{ji} on $U_i \cap U_j$.

THEOREM. *Let X be a noetherian scheme, $\gamma \in H^2(X, \mathcal{O}_X^*)$. There exists a proper birational morphism $\alpha: \tilde{X} \to X$ such that $\alpha^*(\gamma)$ is represented by an Azumaya algebra on $\tilde{X}$.*

Proof. It is enough to consider X which are connected. Suppose that $\{U_i\}$ is an affine open cover of X and that γ is non-trivial on at least one U_i. We will construct an Azumaya algebra on a birational model of X by an inductive process which involves adjoining one by one proper preimages of the subsets U_i and, by an appropriate birational change of the scheme and Azumaya algebra obtained, extending the new algebra to the union. We start with some affine open subset U_0 and an Azumaya algebra A_0 on it.

Now suppose by induction that we already have an Azumaya algebra $\bar{A}_k$ on the scheme X_k, a Zariski-open subset of the scheme $\bar{X}_k$, equipped with a proper birational map $\bar{\alpha}_k: \bar{X}_k \to X$ such that $X_k = \bar{\alpha}_k^{-1}(U_0 \cup \cdots \cup U_k)$. Let U_{k+1} intersect $U_0 \cup \cdots \cup U_k$ and $\bar{U}_{k+1} = \bar{\alpha}^{-1}(U_{k+1})$. Suppose that on U_{k+1}, γ is represented by the Azumaya algebra A_{k+1}. In the same vein as above we have an isomorphism

$$\bar{A}_{k|X_k \cap \bar{U}_{k+1}} \otimes \mathrm{End}(E_{k,k+1}) \cong \bar{\alpha}_k^*(A_{k+1})|_{X_k \cap \bar{U}_{k+1}} \otimes \mathrm{End}(E_{k+1,k})$$

and we need to extend $E_{k,k+1}$ to X_k and $E_{k+1,k}$ to $\bar{U}_{k+1}$ from their intersection. After this we will change $\bar{A}_k$ and $\bar{\alpha}_k^*(A_{k+1})$ by the other representatives $\bar{A}_k \otimes \mathrm{End}(E_{k,k+1})$, $\bar{\alpha}_k^*(A_{k+1}) \otimes \mathrm{End}(E_{k+1,k})$ of the same Brauer classes and glue these Azumaya algebras, hence the proof.

First, extend both sheaves E as coherent sheaves. This can be done by the following

LEMMA. *Let X be a noetherian scheme, $U \subseteq X$ a Zariski-open subset, E a coherent sheaf on U. Then there exists a coherent sheaf E' on X such that $E'|_U \cong E$.* This is Ex. II.5.15 in [4].

Note that we can assume that in our inductive process we add neighborhoods U_{k+1} of no more than one irreducible component (or an intersection of irreducible components) of X, different from those contained in X_k. Thus we assume $X_k \cap U_{k+1}$ to be connected and the rank of E to be constant on $X_k \cap U_{k+1}$, hence E' will be locally generated by n elements, where n is the rank of E.

LEMMA (see [3], Lemma 3.5). *Let X be a noetherian scheme, E a coherent sheaf on X, locally free outside a Zariski closed subset Z on X. Then there exists a coherent sheaf I of ideals on X such that the support of $\mathcal{O}_X/I$ is Z with the following* *property. Let $\alpha: \tilde{X} \to X$ be the blowing up of X with center I, then the sheaf $\bar{\alpha}(E) := $ the quotient of $\alpha^*(E)$ by the subsheaf of sections with support in $\alpha^{-1}(Z)$, is locally free on $\tilde{X}$.*

Proof. The proof consists of two parts. First: to reduce the number of local generators to get this number constant on the connected components of X (the minima are the values of the (local) rank function of E). Second, to force the kernel of the (local) presentations $\mathcal{O}_V^m \to E|_V \to 0$ to vanish for all neighborhoods from some cover $\{V\}$. Both parts are proved by indicating the suitable coherent sheaves of ideals and blowing up X with respect to these sheaves. Let $\mathcal{O}_V^m \xrightarrow{f} E|_V - 0$ be a local presentation of E. Then $\mathrm{Ker}(f)$ is generated by all relations $\Sigma_{i=1}^m c_i a_i = 0$ where $\{a_i\}$ stand for the free basis of $\mathcal{O}_V^m$. The coherent sheaf of ideals in the first case is the sheaf defined locally as the ideal I_V in $\mathcal{O}_V$ generated by all c_i such that $\Sigma_{i=1}^m c_i a_i \in \mathrm{Ker}(f)$ and in the second case as $J_X = \mathrm{Ann}(\mathrm{Ker}(f))$. As the number of generators is constant in the case we are interested in, we give the details only for the second part of the proof and refer to [3] for the first.

Let $\alpha: X' \to X$ be the blowing up of X with respect to J_X and let $\bar{\alpha}(E)$ be as in the statement of the Lemma. Let

$$0 \to (\mathrm{Ker}(\bar{f}))|_{V'} \to \mathcal{O}_{V'}^m \xrightarrow{\bar{f}} \bar{\alpha}(E)|_{V'} \to 0$$

be the local presentation of $\bar{\alpha}(E)$. We have $\alpha^{-1}(\mathrm{Ann}(f)) \subseteq \mathrm{Ann}(\mathrm{Ker}(\bar{f}))$. Let $\mathfrak{p} \in Z'$, $V' = \mathrm{Spec}(A')$ an affine neighborhood of $\mathfrak{p}$ in X' and let $\Sigma_{i=1}^m c_i a_i \in \mathrm{Ker}(\bar{f})|_{V'}$ map to a nonzero element in $\mathrm{Ker}(\bar{f})_{\mathfrak{p}}$. Denote by γ a generator of the invertible sheaf $\alpha^{-1}(\mathrm{Ann}(f))$ on $V'' = \mathrm{Spec}(A'') \subseteq V'$ for suitable A''. It is clear that there exists for given $\mathfrak{p}$ and V'' a finite sequence of open affine neighborhoods $V_1'', \ldots, V_s''$ such that $X' \backslash Z' = V_1''$, $V'' = V_s''$ and $V_j'' \cap V_{j+1}'' \neq \emptyset$ for $j = 1, \ldots, s-1$. So suppose $V' \cap (X' \backslash Z') \neq \emptyset$ and $\mathfrak{q} \in V'' \cap (X \backslash Z)$. Then $(c_i)_{\mathfrak{q}} = 0$ for $i = 1, \ldots, m$ and $\mathfrak{q} \in \mathrm{Spec}(A_\gamma'')$ hence $\gamma^k c_i = 0$ for $i = 1, \ldots, m$ for some k. Since γ is not a zero divisor, we conclude that $c_i = 0$ for $i = 1, \ldots, m$. Thus (maybe after considering a finite sequence of points $\mathfrak{q}_1, \ldots, \mathfrak{q}_s$) we prove that $(\mathrm{Ker}(f))_{\mathfrak{p}}$ is trivial for every $\mathfrak{p} \in X'$. □

In this way we glue the two sheaves $\bar{A}_k$ and A_{k+1} and get an Azumaya algebra on $\tilde{X}_k \cup \tilde{U}_{k+1}$. As the scheme X is quasi-compact, we obtain an Azumaya algebra on $\tilde{X}$ after a finite number of such steps.

Now we have to show that this process can be done in such a way that the class [A] of the Azumaya algebra A constructed in this way is equal to $\bar{\alpha}^*(\gamma)$. Again this goes by induction on k. We have $X_{k+1} = U \cup V$ with $U = \bar{\alpha}_{k+1}^{-1}(U_0 \cup \cdots \cup U_k)$ and $V = \bar{\alpha}_{k+1}^{-1}(U_{k+1})$. We have the exact sequence

$$H^1(U \cap V, \mathcal{O}^*) \to H^2(X_{k+1}, \mathcal{O}^*) \to H^2(U, \mathcal{O}^*) \oplus H^2(V, \mathcal{O}^*)$$

and by induction hypothesis, $\bar{\alpha}^*_{k+1}(\gamma) - [\bar{A}_{k+1}]$ maps to zero in $H^2(U, \mathcal{O}^*) \oplus H^2(V, \mathcal{O}^*)$ so it comes from $\beta \in H^1(U \cap V, \mathcal{O}^*)$. By blowing up X_{k+1} we may assume that β is represented by a line bundle which extends to U. Then β maps to zero in $H^2(X_{k+1}, \mathcal{O}^*)$, hence $\bar{\alpha}^*_{k+1}(\gamma) - [\bar{A}_{k+1}] = 0$. □

Note that we need not bother about the compatibility of isomorphisms, because at each step we choose a new isomorphism between the Azumaya algebra A on $U_1 \cup \cdots \cup U_j$ from the preceding step and A_k on U_k, modulo $\mathrm{End}(E)$, $\mathrm{End}(E_k)$.

COROLLARY 1. *Let G be a finite group, V a faithful complex representation of G. Then there exists a nonsingular projective model X of V/G such that* $\mathrm{Br}(X) = H^2(X, \mathcal{O}^*)$.

Proof. The group $H^2(X, \mathcal{O}^*)$ is a birational invariant of nonsingular projective varieties and is isomorphic to $H^2(G, \mathbb{Q}/\mathbb{Z})$ if X is a model of V/G (see [1]). It remains to recall that the group $H^2(G, \mathbb{Q}/\mathbb{Z})$ is finite. □

COROLLARY 2. *Let X be a noetherian scheme over $\mathbb{C}$, Z a closed subscheme of X and $\gamma \in H^2_Z(X, \mathcal{O}^*)$. Then there exists a proper morphism $\alpha: X' \to X$ which is an isomorphism above $X \backslash Z$ and maps γ to zero in $H^2_{\alpha^{-1}(Z)}(X', \mathcal{O}^*)$.*

Proof. First, let's have $\alpha(\gamma)$ map to zero in $H^2(X, \mathcal{O}^*)$. To do this, desingularize X by $X' \to X$. Then in the following exact sequence (in etale cohomology), β will be injective:

$$\begin{array}{ccccccc} H^1(X'\backslash Z', \mathcal{O}^*) & \to & H^2_{Z'}(X', \mathcal{O}^*) & \to & H^2(X', \mathcal{O}^*) & \xrightarrow{\beta} & H^2(X'\backslash Z', \mathcal{O}^*) \\ & & & & \uparrow & & \\ & & & & H^2_Z(X, \mathcal{O}^*) & & \end{array}$$

The injectivity is due to the injectivity of $H^2(X', \mathcal{O}^*) \to H^2(K(X'), \mathcal{O}^*)$ for a nonsingular irreducible scheme X'.

Now γ comes from $\gamma' \in H^1(X'\backslash Z', \mathcal{O}^*) = \mathrm{Pic}(X'\backslash Z')$. It is obvious that Picard elements lift to Picard elements by the blowing ups from the theorem. Thus from the diagram

$$\begin{array}{ccccc} H^1(X'', \mathcal{O}^*) & \to & H^1(X''\backslash Z'', \mathcal{O}^*) & \to & H^2_{Z''}(X'', \mathcal{O}^*) \\ & & \uparrow & & \uparrow \\ & & H^1(X'\backslash Z', \mathcal{O}^*) & \to & H^2_{Z'}(X', \mathcal{O}^*) \end{array}$$

we conclude that γ becomes trivial on Z'' by $X'' \to X'$ which extends γ' to X''. □

Now let us return to the problem of an isomorphism $\mathrm{Br}(X) \to H^2(X, \mathcal{O}^*)$ for nonsingular quasi-projective varieties. The theorem reduces the general problem to the following

QUESTION. Let X' be a blowing up of a nonsingular variety X along a smooth subvariety S and let A' be an Azumaya algebra on X'. Does there exist an Azumaya algebra A on X such that its inverse image on X' is equivalent to A'?

In case the restriction of A' to the pre-image of S is trivial, the question reduces to the one, whether a vector bundle on this preimage can be extended to X as a vector bundle. For example, if $\dim(X) = 2$ then S is a point and its proper preimage is a $\mathbb{P}^1$ with self-intersection -1. Since the map $\mathrm{Pic}(X') \to \mathrm{Pic}(\mathbb{P}^1)$ is surjective, any vector bundle on $\mathbb{P}^1$ can be extended to X'.

Therefore we obtain a simple proof of the basic theorem in the case $\dim(X) = 2$ using the birational theorem.

In the case of $\dim(X) = 3$ the same procedure reduces the basic problem to the analogous problem of extending vector bundles from $\mathbb{P}^2$ and ruled surfaces to a variety of dimension three.

References

[1] Bogomolov, F.A., Brauer group of quotients by linear representations. *Izv. Akad. Nauk. USSR, Ser. Mat.* 51 (1987) 485–516.

[2] Landia, A.N., Brauer group of projective models of quotients by finite groups. Dep. in GRUZNIITI 25.12.1987, no. 373-Г87.

[3] Moishezon, B.G., An algebraic analog of compact complex spaces with sufficiently large field of meromorphic functions I. *Izv. Akad. Nauk. USSR, Ser. Mat.* 33 (1969) 174–238.

[4] Hartshorne, R., Algebraic Geometry. Graduate Texts in Math. 52, Springer Verlag, Berlin etc. 1977.

Compositio Mathematica **76**: 7–17, 1990.

Monads and cohomology modules of rank 2 vector bundles

WOLFRAM DECKER*

Universität Kaiserslautern, Fachbereich Mathematik, Erwin-Schrödinger-Straße, 6750 Kaiserslautern, FRG

Received 19 August 1988; accepted 20 July 1989

Introduction

Monads are a useful tool to construct and study rank 2 vector bundles on the complex projective space $\mathbb{P}_n, n \geqslant 2$ (compare [O-S-S]). Horrocks' technique of eliminating cohomology [Ho 2] represents a given rank 2 vector bundle $\mathscr{E}$ as the cohomology of a monad

$$(M(\mathscr{E})) \quad \mathscr{A} \xrightarrow{\varphi} \mathscr{B} \xrightarrow{\psi} \mathscr{C}$$

as follows.

First eliminate the graded $S = \mathbb{C}[z_0, \ldots, z_n]$-module $H^1\mathscr{E}(*) = \oplus_{m \in \mathbb{Z}} H^1(\mathbb{P}_n, \mathscr{E}(m))$ by the universal extension

$$0 \to \mathscr{E} \to \mathscr{Q} \to \tilde{L}_0 \to 0,$$

where

$$L_0 \to H^1\mathscr{E}(*) \to 0.$$

is given by a minimal system of generators ($\sim$ stands for sheafification).

If $n = 2$ take this extension as a monad with $\mathscr{A} = 0$.

If $n \geqslant 3$ eliminate dually $H^{n-1}\mathscr{E}(*)$ by the universal extension

$$0 \to \tilde{L}_0^{\vee}(c_1) \to \mathscr{K} \to \mathscr{E} \to 0$$

(where $c_1 = c_1(\mathscr{E})$ is the first Chern-class). Then notice, that the two extensions

* Partially supported by the DAAD.

can be completed to the display

$$\begin{array}{ccccccccc}
 & & & & 0 & & 0 & & \\
 & & & & \downarrow & & \downarrow & & \\
0 & \longrightarrow & \tilde{L}_0^\vee(c_1) & \longrightarrow & \mathscr{K} & \longrightarrow & \mathscr{E} & \longrightarrow & 0 \\
 & & \| & & \downarrow & & \downarrow & & \\
0 & \longrightarrow & \tilde{L}_0^\vee(c_1) & \xrightarrow{\varphi} & \mathscr{B} & \longrightarrow & \mathscr{Q} & \longrightarrow & 0 \\
 & & & & \downarrow{\scriptstyle\psi} & & \downarrow & & \\
 & & & & \tilde{L}_0 & = & \tilde{L}_0 & & \\
 & & & & \downarrow & & \downarrow & & \\
 & & & & 0 & & 0 & &
\end{array}$$

of a monad

$$\tilde{L}_0^\vee(c_1) \xrightarrow{\varphi} \mathscr{B} \xrightarrow{\psi} \tilde{L}_0$$

for $\mathscr{E}$.

To get a better understanding for $\mathscr{B}$, φ and ψ consider first the case $n = 2, 3$. Then $\mathscr{B}$ is a direct sum of line bundles by Horrocks' splitting criterion [Ho 1]. Taking cohomology we obtain a free presentation

$$B \xrightarrow{\psi} L_0 \longrightarrow H^1\mathscr{E}(*) \longrightarrow 0$$

with $B = H^0\mathscr{B}(*)$. The crucial point is that this is minimal [Ra]. Moreover, if $n = 3$, then B is self-dual [Ra]: $B^\vee(c_1) \simeq B$. We will see below that up to isomorphism φ is the dual map of ψ.

Let us summarize and slightly generalize. Consider an arbitrary graded S-module N of finite length with minimal free resolution (m.f.r. for short)

$$0 \longrightarrow L_{n+1} \xrightarrow{\alpha_n} L_n \longrightarrow \cdots \longrightarrow L_1 \xrightarrow{\alpha_0} L_0 \longrightarrow N \longrightarrow 0.$$

If $n = 2$ then $N \simeq H^1\mathscr{E}(*)$ for some rank 2 vector bundle $\mathscr{E}$ on $\mathbb{P}_2$ iff $\operatorname{rk} L_1 = \operatorname{rk} L_0 + 2$ (compare [Ra]). In this case $\mathscr{E}$ is uniquely determined as $\widetilde{\ker \alpha_0}$:

$$\begin{array}{ccccccccccc}
0 & \longrightarrow & \tilde{L}_0^\vee(c_1) & \xrightarrow{\alpha_0^\vee(c_1)} & \tilde{L}_1^\vee(c_1) & \xrightarrow{\alpha_1} & \tilde{L}_1 & \xrightarrow{\alpha_0} & \tilde{L}_0 & \longrightarrow & 0. \\
 & & & & & \searrow & & \nearrow & & & \\
 & & & & & & \mathscr{E} & & & & \\
 & & & & & \nearrow & & \searrow & & & \\
 & & & & 0 & & & & 0 & &
\end{array}$$

(This sequence is self-dual by Serre-duality [Ho 1, 5.2], since $\mathscr{E}^\vee(c_1) \simeq \mathscr{E}$).

For $n = 3$ there is an analogous result. Answering Problem 10 of Hartshorne's list [Ha] we prove:

PROPOSITION 1. *N is the first cohomology module of some rank 2 vector bundle on $\mathbb{P}_3$ iff*

(1) $\operatorname{rk} L_1 = 2 \operatorname{rk} L_0 + 2$ *and*
(2) *there exists an isomorphism* $\Phi: L_1^\vee(c_1) \xrightarrow{\simeq} L_1$ *for some* $c_1 \in \mathbb{Z}$ *such that* $\alpha_0 \circ \Phi \circ \alpha_0^\vee(c_1) = 0$.

In this case any Φ satisfying (2) *defines a monad*

$$(M_\Phi) \quad \tilde{L}_0^\vee(c_1) \xrightarrow{\Phi\circ\alpha_0^\vee(c_1)} \tilde{L}_1 \xrightarrow{\alpha_0} \tilde{L}_0$$

and $\mathscr{E}$ is a 2-bundle on $\mathbb{P}_3$ with $H^1\mathscr{E}(*) \simeq N$ (and $c_1 = c_1(\mathscr{E})$) iff $(M(\mathscr{E})) \simeq (M_\Phi)$ for some Φ.

To complete the picture let us mention a result of Hartshorne and Rao (not yet published). If $N \simeq H^1\mathscr{E}(*)$ as above then $L_0^\vee(c_1) \xrightarrow{\varphi} L_1$ is part of a minimal system of generators for $\ker \alpha_0$. In other words: There exists a splitting

$$0 \to L_4 \to L_3 \to L_2' \oplus L_0^\vee(c_1) \to L_1 \to L_0 \to H^1\mathscr{E}(*) \to 0$$

inducing the monad

$$(M(\mathscr{E})) \quad \tilde{L}_0^\vee(c_1) \to \tilde{L}_1 \to \tilde{L}_0$$

and the m.f.r.

$$0 \to \tilde{L}_4 \to \tilde{L}_3 \to \tilde{L}_2' \to \mathscr{E} \to 0$$

resp.

For $n \geqslant 4$ there is essentially only one indecomposable 2-bundle known on $\mathbb{P}_n$: The Horrocks-Mumford-bundle $\mathscr{F}$ on $\mathbb{P}_4$ with Chern-classes $c_1 = -1$, $c_2 = 4$. We prove:

PROPOSITION 2. *The* m.f.r. *of $H^2\mathscr{F}(*)$ decomposes as*

$$\begin{array}{c} 0 \to H_2 \xrightarrow{\beta_1} H_1 \xrightarrow{\begin{pmatrix}\beta_0' \\ \beta_0''\end{pmatrix}} L_0^\vee(c_1) \oplus L_1 \xrightarrow{\begin{pmatrix}0 & \alpha_0 \\ \alpha_0^\vee(c_1) & *\end{pmatrix}} L_0 \oplus L_1^\vee(c_1) \to \\ \searrow B \nearrow \quad\quad 0 \nearrow \quad \searrow 0 \\ \to H_1^\vee(c_1) \to H_2^\vee(c_1) \to H^2\mathscr{F}(*) \to 0 \end{array}$$

with $B = H^0\mathscr{B}(*)$, *inducing the monad*

$$(M(\mathscr{F}))\quad \tilde{L}_0^\vee(c_1) \to \mathscr{B} \to \tilde{L}_0$$

and the minimal free presentation

$$L_1 \xrightarrow{\alpha_0} L_0 \longrightarrow H^1\mathscr{F}(*) \longrightarrow 0.$$

The corresponding m.f.r. *decomposes as*

$$0 \to L_5 \to L_4 \to L_3' \oplus H_2 \xrightarrow{\begin{pmatrix} * & 0 \\ * & \beta_1 \end{pmatrix}} L_2' \oplus H_1 \xrightarrow{(* \ \ \beta_0'')} L_1 \xrightarrow{\alpha_0} L_0 \to$$
$$\to H^1\mathscr{F}(*) \to 0$$

inducing the m.f.r.

$$0 \to \tilde{L}_5 \to \tilde{L}_4 \to \tilde{L}_3' \to \tilde{L}_2' \to \mathscr{F} \to 0.$$

$(M(\mathscr{F}))$ is the monad given in [H-M]. Using its display we can compute the above m.f.r.'s explicitly. Especially we reobtain the equations of the abelian surfaces in $\mathbb{P}_4$ ([Ma 1], [Ma 2]).

Of course we may deduce from $\mathscr{F}$ some more bundles by pulling it back under finite morphisms $\pi\colon \mathbb{P}_4 \to \mathbb{P}_4$. The above result also holds for the bundles $\pi^*\mathscr{F}$ with $(M(\pi^*\mathscr{F})) = \pi^*(M(\mathscr{F}))$.

There is some evidence (but so far no complete proof), that a splitting as in Proposition 2 occurs for every indecomposable 2-bundle on $\mathbb{P}_4$. This suggests a new construction principle for such bundles by constructing their H^2-module first.

Proof of Proposition 1

Let $n = 3$ and N be a graded S-module of finite length with m.f.r.

$$0 \to L_4 \to L_3 \to L_2 \to L_1 \xrightarrow{\alpha_0} L_0 \to N \to 0.$$

Suppose first that $N \simeq H^1\mathscr{E}(*)$ for some 2-bundle $\mathscr{E}$ on $\mathbb{P}_3$ (with first Chern-class c_1). As seen in the introduction, Horrocks' construction leads to a monad

$$(M(\mathscr{E}))\quad \tilde{L}_0^\vee(c_1) \xrightarrow{\varphi} \tilde{L}_1 \xrightarrow{\alpha_0} \tilde{L}_0$$

for $\mathscr{E}$. The dual sequence

$$\tilde{L}_0^\vee(c_1) \xrightarrow{\alpha_0^\vee(c_1)} \tilde{L}_1^\vee(c_1) \xrightarrow{\varphi^\vee(c_1)} \tilde{L}_0$$

is a monad for $\mathscr{E}^\vee(c_1) \simeq \mathscr{E}$. The induced presentation of N has to be isomorphic to that one given by the m.f.r.:

$$\begin{array}{ccccccc} L_1^\vee(c_1) & \xrightarrow{\varphi^\vee(c_1)} & L_0 & \to & N & \to & 0 \\ {\scriptstyle \Phi^\vee(c_1)}\downarrow \wr\| & & \wr\| & & \wr\| & & \\ L_1 & \xrightarrow{\alpha_0} & L_0 & \to & N & \to & 0. \end{array}$$

Dualizing gives (2) since $\alpha_0 \circ \varphi = 0$ and thus also a monad (M_Φ) for $\mathscr{E}$, isomorphic to $(M(\mathscr{E}))$ (replace φ by $\Phi \circ \alpha_0^\vee(c_1)$).

Conversely if N satisfies (2), we obtain a monad (M_Φ) by sheafification. (Since $\tilde{N} = 0$, α_0 is a bundle epimorphism. Dually $\alpha_0^\vee(c_1)$ is a bundle monomorphism.) Let $\mathscr{E}$ be the cohomology bundle of (M_Φ). Then $H^1\mathscr{E}(*) \simeq N$. $\mathscr{E}$ has rank 2, if N satisfies (1). □

REMARK 1. (i) Let $N \simeq H^1\mathscr{E}(*)$ as above with induced splitting

$$0 \to L_4 \to L_3 \to L_2' \oplus L_0^\vee(c_1) \to L_1 \to L_0 \to H^1\mathscr{E}(*) \to 0$$

as in the introduction. Recall that $\mathscr{E}$ is stable iff $H^0(\mathbb{P}_3, \mathscr{E}(m)) = 0$ for $m \leqslant -c_1/2$. Thus $\mathscr{E}$ is stable iff L_2' has no direct summand $S(m)$ with $m \geqslant c_1/2$. Notice that this condition only depends on N.

(ii) If N satisfies (1) and has only one generator, then (2) is obviously equivalent to the symmetry condition $L_1^\vee(c_1) \simeq L_1$. Thus [Ra, 3.1] is a special case of Proposition 1.

EXAMPLES. (i) The well-known Null correlation bundles are by definition the bundles corresponding to the S-module $\mathbb{C}$. Consider the Koszul-presentation

$$4S \xrightarrow{\alpha_0} S(1) \to \mathbb{C} \to 0, \quad \alpha_0 = (z_0, z_1, z_2, z_3).$$

The isomorphisms $4S \xrightarrow{\Phi} 4S$ with $\alpha_0 \circ \Phi \circ \alpha_0^\vee(c_1) = 0$ are precisely the 4×4 skew symmetric matrices with nonzero determinant. Two such matrices give isomorphic bundles iff they differ by a scalar (use [O-S-S, II, Corollary 1 to 4.1.3]). The moduli space of Null correlation bundles is thus isomorphic to $\mathbb{P}_5 \backslash \mathbb{G}$, where $\mathbb{G}$ is the Plucker embedded Grassmanian of lines in $\mathbb{P}_3$.

Unlike the case $n = 2$ the bundle is not uniquely determined by the module.

(ii) The S-module

$$6S \xrightarrow{\alpha_0} S(1) \oplus S(2) \to N \to 0, \; \alpha_0 = \begin{pmatrix} z_0 & z_1 & z_2 & z_3 & 0 & 0 \\ 0 & 0 & z_0^2 & z_1^2 & z_2^2 & z_3^2 \end{pmatrix}$$

satisfies (1) and the symmetry condition $L_1^\vee \simeq L_1$, i.e. the necessary conditions of [Ra]. But N does not satisfy (2).

Cohomology modules of the Horrocks-Mumford-bundle $\mathscr{F}$

We first recall the construction of $\mathscr{F}$ [H-M]. Let

$$V = \mathrm{Map}(\mathbb{Z}_5, \mathbb{C})$$

be the vector space of complex valued functions on $\mathbb{Z}_5$. Denote by

$$H \subset N \subset \mathrm{SL}(5, \mathbb{C})$$

the Heisenberg group and its normalizer in $\mathrm{SL}(5, \mathbb{C})$ resp.

Let $V_0 = V, V_1, V_2, V_3$ and

$$W = \mathrm{Hom}_H(V_1, \Lambda^2 V)$$

be defined as in [H-M]. The V_i are irreducible representations of H and N of degree 5. W is an irreducible representation of N/H of degree 2. It is unimodular, so it comes up with an invariant skew symmetric pairing.

Let $\mathbb{P}_4 = \mathbb{P}(V)$ be the projective space of lines in V. The Koszul-complex on $\mathbb{P}(V)$ is the exact sequence

$$(K) \quad \cdots \to \mathcal{O}(-1) \otimes \Lambda^2 V \to \mathcal{O} \otimes \Lambda^3 V \to \cdots,$$

$$\begin{array}{ccc} & \searrow \quad \nearrow & \\ & (\Lambda^2 \mathscr{T})(-3) & \\ & \nearrow \quad \searrow & \\ 0 & & 0 \end{array}$$

obtained by exterior multiplication with the tautological subbundle

$$\mathcal{O}(-1) \to V \otimes \mathcal{O}.$$

The exterior product provides (K) with a self-duality (with values in $\mathcal{O}(-1) \otimes \Lambda^5 V \simeq \mathcal{O}(-1)$).

This induces the natural pairings

$$(\Lambda^i\mathcal{T})(-i-1)\otimes(\Lambda^{4-i}\mathcal{T})(i-5)\xrightarrow{\wedge}(\Lambda^4\mathcal{T})(-6)\simeq\mathcal{O}(-1)$$

and is compatible with the action of SL(5, $\mathbb{C}$).

It can be extended to $(K)\otimes W$ by tensoring with the invariant form, then being compatible with the action of N.

As in the proof of [H-M, Lemma 2.4] it follows, that $(K)\otimes W$ decomposes as

$$\begin{array}{ccccc} & \mathcal{O}(-1)\otimes V_1 & \begin{pmatrix} 0 & \alpha_0 \\ \alpha_0^\vee(-1) & \star \end{pmatrix} & \mathcal{O}\otimes V_3 & \\ \cdots \to & \oplus & \xrightarrow{\hspace{6em}} & \oplus & \to \cdots, \\ & \mathcal{O}(-1)\otimes V_1\otimes U & & \mathcal{O}\otimes V_3\otimes U & \\ & & \searrow \qquad \nearrow & & \\ & & (\Lambda^2 T)(-3)\otimes W & & \\ & \nearrow & & \searrow & \\ & 0 & & 0 & \end{array}$$

given by the splitting into irreducible N-modules. Moreover the induced

$$\mathcal{O}(-1)\otimes V_1\to(\Lambda^2\mathcal{T})(-3)\otimes W\to\mathcal{O}\otimes V_3$$

is the self-dual Horrocks-Mumford-monad, whose cohomology is $\mathcal{F}$ (normalized such that $c_1\mathcal{F}=-1$).

To proof Proposition 2 consider the display

$$\begin{array}{ccccccc} & & & 0 & & 0 & \\ & & & \downarrow & & \downarrow & \\ 0\longrightarrow & \mathcal{O}(-1)\otimes V_1 & \longrightarrow & \mathcal{K} & \longrightarrow & \mathcal{F} & \longrightarrow 0 \\ & \| & & \downarrow & & \downarrow & \\ 0\longrightarrow & \mathcal{O}(-1)\otimes V_1 & \to & (\Lambda^2\mathcal{T})(-3)\otimes W & \longrightarrow & \mathcal{Q} & \longrightarrow 0 \\ & & & \downarrow & & \downarrow & \\ & & & \mathcal{O}\otimes V_3 & = & \mathcal{O}\otimes V_3 & \\ & & & \downarrow & & \downarrow & \\ & & & 0 & & 0 & \end{array}$$

It first follows that $H^2\mathcal{F}(*)=W$ is a vector space, sitting in degree -2 (compare [H-M]). Its m.f.r. is the Koszul-complex obtained from $(K)\otimes W$ by taking global sections. So it decomposes, inducing the presentation

$$S(-1) \otimes V_1 \otimes U \xrightarrow{\alpha_0} S \otimes V_3 \longrightarrow H^1\mathscr{F}(*) \longrightarrow 0$$

and the Horrocks-Mumford-monad. But this is just

$$(M(\mathscr{F})) \quad \mathcal{O}(-1) \otimes V_1 \to \mathscr{B} \to \mathcal{O} \otimes V_3:$$

apply e.g. [O-S-S, II, Corollary 1 to 4.1.3] (notice that $H^0\mathscr{F} = 0$ implies $H^0\mathscr{B} = H^0\mathscr{B}^*(-1) = 0$ by construction of $(M(\mathscr{F}))$).

It remains to show that α_0 is minimal and that the corresponding m.f.r. of $H^1\mathscr{F}(*)$ decomposes, inducing the m.f.r., say,

$$0 \longrightarrow F_3 \xrightarrow{\gamma_2} F_2 \xrightarrow{\gamma_1} F_1 \xrightarrow{\gamma_0} F_0 \longrightarrow F \longrightarrow 0$$

of $F = H^0\mathscr{F}(*)$.

From the second row of the display we obtain the m.f.r. of $Q = H^0\mathscr{Q}(*)$:

$$\begin{array}{ccccccccc}
 & & & & 0 & & 0 & & \\
 & & & & \downarrow & & \downarrow & & \\
 & & & & S(-3) \otimes W & = & S(-3) \otimes W & & \\
 & & & & \downarrow{\scriptstyle \beta_1} & & \downarrow{\scriptstyle \beta_1} & & \\
 & & & & S(-2) \otimes V \otimes W & = & S(-2) \otimes V \otimes W & & \\
 & & & & \downarrow{\scriptstyle \beta_0} & & \downarrow{\scriptstyle \beta_0''} & & \\
0 & \to & S(-1) \otimes V_1 & \to & (S(-1) \otimes V_1) \oplus (S(-1) \otimes V_1 \otimes U) & \to & S(-1) \otimes V_1 \otimes U & \to & 0 \\
 & & \| & & \downarrow & & \downarrow & & \\
0 & \to & S(-1) \otimes V_1 & \longrightarrow & H^0(\Lambda^2\mathscr{T})(* - 3) \otimes W & \longrightarrow & Q & \longrightarrow & 0 \\
 & & & & \downarrow & & \downarrow & & \\
 & & & & 0 & & 0 & &
\end{array}$$

The third column of the display gives rise to the commutative diagram

$$
\begin{array}{ccccccccccc}
 & & 0 & & & & & & & & \\
 & & \downarrow & & & & & & & & \\
 & & F_3 & & 0 & & & & & & \\
 & & {\scriptstyle\gamma_2}\downarrow & & \downarrow & & & & & & \\
 & & F_2 & \xrightarrow{\alpha_3'} & S(-3)\otimes W & & & & & & \\
 & & {\scriptstyle\gamma_1}\downarrow & & \downarrow{\scriptstyle\beta_1} & & & & & & \\
 & & F_1 & \xrightarrow{-\alpha_2'} & S(-2)\otimes V\otimes W & & & & & & \\
 & & {\scriptstyle\gamma_0}\downarrow & & \downarrow{\scriptstyle\beta_0''} & & & & & & \\
 & & F_0 & \xrightarrow{\alpha_1'} & S(-1)\otimes V_1\otimes U & \xrightarrow{\alpha_0} & S\otimes V_3 & \longrightarrow & H^1\mathscr{F}(*) & \longrightarrow & 0 \\
 & & \downarrow & & \downarrow & & \| & & \| & & \\
0 & \longrightarrow & F & \longrightarrow & Q & \longrightarrow & S\otimes V_3 & \longrightarrow & H^1\mathscr{F}(*) & \longrightarrow & 0 \\
 & & \downarrow & & \downarrow & & & & & & \\
 & & 0 & & 0 & & & & & &
\end{array}
$$

with exact columns and bottom row.

The induced

$$
0\to F_3 \xrightarrow{\gamma_2} F_2 \xrightarrow{\begin{pmatrix}\gamma_1\\ \alpha_3'\end{pmatrix}} F_1\oplus(S(-3)\otimes W) \xrightarrow{\begin{pmatrix}\gamma_0 & 0\\ \alpha_2' & \beta_1\end{pmatrix}} F_0\oplus(S(-2)\otimes V\otimes W) \xrightarrow{(\alpha_1' \;\; \beta_0'')}
$$

$$
\to S(-1)\otimes V_1\otimes U \xrightarrow{\alpha_0} S\otimes V_3 \to H^1\mathscr{F}(*)\to 0
$$

is exact and it is minimal, iff $\alpha_1', \alpha_2', \alpha_3'$ have no entries in $\mathbb{C}\backslash\{0\}$. But since $H^0\mathscr{F}(1)=0$ [H-M], these maps have only entries in degrees $\geqslant 1$. □

REMARK 2 (i) Let us describe $(M(\mathscr{F}))$ more explicitly by choosing convenient bases of $V_1, V_3 = V_1^*, W$ and forgetting the N-module structure (compare the proof of [H-M, Lemma 2.5].)

Choose the basis $e_0, \ldots, e_4$ of $V = \mathrm{Map}(\mathbb{Z}_5, \mathbb{C})$ given by $e_i(j) = \delta_{ij}$ and its dual basis $z_0, \ldots, z_4 \in V^*$.

Define

$$A = (a_{ij})_{\substack{0 \leqslant i \leqslant 4 \\ 0 \leqslant j \leqslant 1}} \quad \text{by}$$

$$\left.\begin{aligned} a_{i0} &= e_{i+2} \wedge e_{i+3} \\ a_{i1} &= e_{i+1} \wedge e_{i+4} \end{aligned}\right\} \quad i \bmod 5.$$

Then w_0, w_1, given by $w_j(e_i) = a_{ij}$ is a basis of W. Identifying $W \simeq \mathbb{C}^2$, the invariant form on W becomes the standard symplectic form $Q = \left(\begin{smallmatrix} 0 & 1 \\ -1 & 0 \end{smallmatrix}\right)$ on $\mathbb{C}^2$.

We thus may rewrite $(M(\mathscr{F}))$ as

$$(M(\mathscr{F})) \quad 5\mathcal{O}(-1) \xrightarrow{Q \cdot {}^tA} 2(\Lambda^2 \mathscr{T})(-3) \xrightarrow{A} 5\mathcal{O},$$

the matrices operating by exterior multiplication.

(ii) From the explicit form of $(M(\mathscr{F}))$ we can compute α_0 explicitly. Choose a convenient basis of $\Lambda^2 V \otimes W = (V_1 \oplus U)$. Then

$$15S(-1) \xrightarrow{\alpha_0} 5S \longrightarrow H^1\mathscr{F}(*) \longrightarrow 0$$

is the matrix

$$\alpha_0 = \begin{pmatrix} 0 & z_3 & 0 & 0 & z_2 & 0 & 0 & z_1 & z_4 & 0 & z_0 & 0 & 0 & 0 & 0 \\ z_3 & 0 & z_4 & 0 & 0 & 0 & 0 & 0 & z_2 & z_0 & 0 & z_1 & 0 & 0 & 0 \\ 0 & z_4 & 0 & z_0 & 0 & z_1 & 0 & 0 & 0 & z_3 & 0 & 0 & z_2 & 0 & 0 \\ 0 & 0 & z_0 & 0 & z_1 & z_4 & z_2 & 0 & 0 & 0 & 0 & 0 & 0 & z_3 & 0 \\ z_2 & 0 & 0 & z_1 & 0 & 0 & z_0 & z_3 & 0 & 0 & 0 & 0 & 0 & 0 & z_4 \end{pmatrix}.$$

Resolving it (use e.g. [B-S]), we obtain the m.f.r. of $H^1\mathscr{F}(*)$. Its shape is

$$0 \to 2S(-8) \to 20S(-6) \to 35S(-5) \oplus 2S(-3) \to (15S(-4) \oplus 4S(-3)) \oplus 10S(-2)$$
$$\to 15S(-1) \to 5S \to H^1\mathscr{F}(*) \to 0.$$

(iii) Consider the induced m.f.r. of F and its dual

$$\cdots \to 35S(-5) \xrightarrow{\gamma_0} 15S(-4) \oplus 4S(-3) \xrightarrow{\Gamma} 15S(3) \oplus 4S(2) \xrightarrow{{}^t\gamma_0} 35S(4) \to \cdots$$

$$\begin{array}{ccccc} & \searrow & & \nearrow & \\ & & F \simeq F^{\vee}(-1) & & \\ & \nearrow & & \searrow & \\ 0 & & & & 0\,. \end{array}$$

Γ can be computed by resolving ${}^t\gamma_0$ (use again [B-S]). We thus obtain explicit bases for the spaces of sections $H^0\mathscr{F}(m)$. Especially we get the equations of the zero-schemes of sections of $\mathscr{F}(3)$, including the abelian surfaces in $\mathbb{P}_4$.

(iv) Let $\pi: \mathbb{P}_4 \to \mathbb{P}_4$ be a finite morphism and d^4 its degree. Then $\pi^*\mathscr{F}$ is a stable 2-bundle with Chern-classes $c_1 = -d$, $c_2 = 4d^2$. Proposition 2 and the above remarks also hold for $\pi^*\mathscr{F}$: Replace (K) by $\pi^*(K)$, $(M(\mathscr{F}))$ by $\pi^*(M(\mathscr{F})) = (M(\pi^*\mathscr{F}))$ and $z_0, \ldots, z_4$ in α_0 by $f_0, \ldots, f_4$, where $f_0, \ldots, f_4$ are the forms of degree d defining π.

Acknowledgement

I'd like to thank R. Hartshorne for helpful discussions.

References

[B-S] D. Bayer and M. Stillman: Macaulay, a computer algebra system for algebraic geometry.

[Ha] R. Hartshorne: Algebraic vector bundles on projective spaces: A problem list. *Topology* 18 (1979) 117–228.

[H-M] G. Horrocks and D. Mumford: A rank 2 vector bundle on $\mathbb{P}^4$ with 15,000 symmetries. *Topology* 12 (1973) 63–81.

[Ho 1] G. Horrocks: Vector bundles on the punctured spectrum of a local ring. *Proc. London Math. Soc.* 14 (1964) 689–713.

[Ho 2] G. Horrocks: Construction of bundles on $\mathbb{P}^n$. In: *Les équations de Yang-Mills*, A. Douady, J.-L. Verdier, séminaire E.N.S. 1977–1978. *Astérisque* 71–72 (1980) 197–203.

[Ma 1] N. Manolache: Syzygies of abelian surfaces embedded in $\mathbb{P}^4$. *J. reine angew. Math.* 384 (1988) 180–191.

[Ma 2] N. Manolache: The equations of the abelian surfaces embedded in $\mathbb{P}^4(\mathbb{C})$. Preprint, Bukarest, 1988.

[O-S-S] C. Okonek, M. Schneider and H. Spindler: Vector bundles on complex projective spaces. Boston, Basel, Stuttgart: Birkhäuser 1980.

[Ra] P. Rao: A note on cohomology modules of rank two bundles. *J. Algebra* 86, (1984) 23–34.

Compositio Mathematica **76**: 19–47, 1990.

On the Milnor fibrations of weighted homogeneous polynomials

ALEXANDRU DIMCA
Department of Mathematics, INCREST, Bd. Pacii 220, RO-79622 Bucharest, Romania; Current address: The Institute for Advanced Study, Princeton, NJ 08540, USA

Received 6 October 1988; accepted 20 July 1989

Let $\mathbf{w} = (w_0, \ldots, w_n)$ be a set of integer positive weights and denote by S the polynomial ring $\mathbb{C}[x_0, \ldots, x_n]$ graded by the conditions $\deg(x_i) = w_i$. For any graded object M we denote by M_k the homogeneous component of M of degree k. Let $f \in S_N$ be a weighted homogeneous polynomial of degree N.

The Milnor fibration of f is the locally trivial fibration $f\colon \mathbb{C}^{n+1} \setminus f^{-1}(0) \to \mathbb{C} \setminus \{0\}$, with typical fiber $F = f^{-1}(1)$ and geometric monodromy $h\colon F \to F$, $h(x) = (t^{w_0}x_0, \ldots, t^{w_n}x_n)$ for $t = \exp(2\pi i/N)$. Since $h^N = 1$, it follows that the (complex) monodromy operator $h^*\colon H^{\cdot}(F) \to H^{\cdot}(F)$ is diagonalizable and has eigenvalues in the group $G = \{t^a; a = 0, \ldots, N-1\}$ of the N-roots of unity.

We denote by $H^{\cdot}(F)_a$ the eigenspace corresponding to the eigenvalue t^{-a}, for $a = 0, \ldots, N-1$.

When f has an isolated singularity at the origin, the only nontrivial cohomology group $H^k(F)$ is for $k = n$ and the dimensions $\dim H^n(F)_a$ are known by the work of Brieskorn [2]. But as soon as f has a nonisolated singularity, it seems that even the Betti numbers $b_k(F)$ are known only in some special cases, see for instance [9], [14], [17], [22], [25].

The first main result of our paper is an *explicit formula* for the cohomology groups $H^k(F)$ and for the eigenspaces $H^k(F)_a$. Let $\Omega^{\cdot}$ be the complex of global algebraic differential forms on $\mathbb{C}^{n+1}$, graded by the convention $\deg(u\, dx_{i_1} \wedge \cdots \wedge dx_{i_k}) = p + w_{i_1} + \cdots + w_{i_k}$ for $u \in S_p$. We introduce a new differential on $\Omega^{\cdot}$, namely $D_f(\omega) = d\omega - (|\omega|/N)\, df \wedge \omega$, for $\omega \in \Omega^k_p$ with $|\omega| = p$ the degree of ω and d the usual exterior differential, similar to Dolgachev [8], p. 61.

For $a = 0, \ldots, N-1$ we denote by $\Omega^{\cdot}_{(a)}$ the subcomplex in $\Omega^{\cdot}$ given by $\bigoplus_{s \geq 0} \Omega^{\cdot}_{-a+sN}$.

To a D_f-closed form $\omega \in \Omega^{k+1}$ we can associate the element $\delta(\omega) = [i^* \Delta(\omega)]$ in the de Rham cohomology group $H^k(F)$, where Δ is the contraction with the Euler vector field (as in [12], p. 467 in the homogeneous case and [8], p. 43 in the weighted homogeneous case) and $i\colon F \to \mathbb{C}^{n+1}$ denotes the inclusion.

THEOREM A. *The maps* $\delta\colon H^{k+1}(\Omega^{\cdot}, D_f) \to \tilde{H}^k(F)$ *and* $\delta\colon H^{k+1}(\Omega^{\cdot}_{(a)}, D_f) \to \tilde{H}^k(F)_a$ *are isomorphisms for any* $k \geq 0$, $a = 0, \ldots, N-1$, *with* $\tilde{H}$ *denoting reduced cohomology.*

The proof of this Theorem depends on a comparison between spectral sequences naturally associated to the two sides of these equalities see (1.8).

Our second main theme is that these spectral sequences can be used to perform *explicit computations* and to derive interesting *numerical formulas*, in spite of the fact that the E_1-term has infinitely many nonzero entries and that degeneration at the E_2-term happens only in special cases (see (3.10) and (3.11) below).

The eigenspaces $H^{\cdot}(F)_0$ are particularly interesting. If $\mathbf{P} = \mathbf{P}(\mathbf{w})$ denotes the weighted projective space Proj(S), V the hypersurface $f = 0$ in $\mathbf{P}$ and $U = \mathbf{P} \setminus V$ the complement, then there is a natural identification $H^{\cdot}(F)_0 = H^{\cdot}(U)$. We conjecture an inclusion between the filtration on $H^{\cdot}(F)_0$ induced by the spectral sequence mentioned above and the (mixed) *Hodge filtration* on $H^{\cdot}(U)$, having a substantial consequence for explicit computations and extending to the singular case an important result of Griffiths [12], see (2.7.ii) below.

To prove the analogous result for these filtrations on the whole $H^{\cdot}(F)$, we establish first some subtle properties of the *Poincaré residue operator*

$R: H^{\cdot}(\mathbb{C}^{n+1} \setminus F) \to H^{\cdot - 1}(F)$ (see (1.6), (1.20), (1.21) and (2.6)) which may be useful in their own.

Note that the Betti numbers $b_k(V)$ are completely determined by $b_k(U)$ and hence one can get by our method at least upper bounds for all $b_k(V)$ as well as the exact value of the top interesting one (i.e. $b_{n+m-1}(V)$ where $m = \dim f^{-1}(0)_{\text{sing}}$) in a *finite* number of steps see (2.8).

Then we specialize to the case when f has a one-dimensional singular locus, a situation already studied (without the weighted homogeneity assumption) by N. Yomdin and, more recently and more completely, by D. Siersma, R. Pellikaan, D. van Straten, T. de Jong. We relate the spectral sequence $(E_r(f)_0\ d_r)$ to some new spectral sequences associated to the *transversal singularities* of f, these being the intersections of $f^{-1}(0)$ with transversals to each irreducible component of $f^{-1}(0)_{\text{sing}}$. We hope that these intricate local spectral sequences will play a fundamental role in understanding better even the isolated hypersurface singularities (see for instance the nice *characterization* (3.10′) of weighted homogeneous singularities). Concerning the numerical invariants in this case, we get interesting and effective formulas for the Euler characteristics $\chi(V)$ and $\chi(F)$ extending in highly nontrivial way the known formulas for the homogeneous case (we conjecture them to hold in general and check them under certain assumptions on the transversal singularities of f, see (3.19.ii)).

The last section is devoted to explicit computations with our spectral sequence. The first two of them are just simple illustrations of our technique, while the third offers a more subtle example, for which we know no other method to get even the Betti numbers for V. It is interesting to remark that if one wants to compute the Euler characteristic $\chi(V)$ in this case using Theorem (3.1) in Szafraniec [26], then one is led to compute bases of Milnor algebras (and signatures of bilinear forms defined on them) of a huge dimension ($\simeq 6^7$) and this is an impossible task even for a computer!

A more theoretical application (improving a result of Scherk [20]) is given in the end, the key point in the proof being again an explicit computation with the spectral sequence.

A basic *open problem* is to decide whether the spectral sequence $(E_r(f), d_r)$ or its local analog $(E_r(g, 0), d_r)$ degenerates always in a finite number of steps and, in the affirmative case, to determine a bound for this number in terms of other invariants of f or g.

1. Some spectral sequences

In this section we shall use many notations and results from Dolgachev [8] without explicit reference.

Let $\Delta\colon \Omega^k \to \Omega^{k-1}$ denote the contraction with the Euler vector field $\Sigma w_i x_i \partial/\partial x_i$. For $k \geqslant 1$ we put $\bar{\Omega}^k = \ker(\Delta\colon \Omega^k \to \Omega^{k-1}) = \operatorname{im}(\Delta\colon \Omega^{k+1} \to \Omega^k)$ and let $\Omega^k_{\mathbf{P}}$ denote the associated sheaf on $\mathbf{P}$. One has also the twisted sheaves $\Omega^k_{\mathbf{P}}(s)$, for any $s \in \mathbb{Z}$.

Let $i\colon U \to \mathbf{P}$ denote the inclusion and put $\Omega^k_U(s) = i^*\Omega^k_{\mathbf{P}}(s)$.

The Milnor fiber F is an affine smooth variety and according to Grothendieck [13] one has $H^\cdot(F) = H^\cdot(\Gamma(F, \Omega^\cdot_F))$. Let $p\colon F \to U$ denote the canonical projection and note that

$$p_*\Omega^\cdot_F = \bigoplus_{a=0}^{N-1} \Omega^\cdot_U(-a). \tag{1.1}$$

If we let $A^\cdot_a = \Gamma(U, \Omega^\cdot_U(-a))$ and $A^\cdot = \bigoplus_{a=0}^{N-1} A^\cdot_a$, then we clearly have

$$H^\cdot(F) = H^\cdot(A), \quad H^\cdot(F)_a = H^\cdot(A^\cdot_a). \tag{1.2}$$

There is a natural increasing filtration F_s on $A^\cdot_a$, related to the order of the pole a form in $A^\cdot_a$ has along V, namely

$$F_s A^j_a = 0 \text{ for } s < 0 \text{ and } F_s A^j_a = \{\omega/f^s; \omega \in \Omega^j_{sN-a}\} \text{ for } s \geqslant 0 \text{ similar to [12].} \tag{1.3}$$

But for obvious technical reasons it is more convenient to consider the decreasing filtration.

$$F^s A^j_a = F_{j-s} A^j_a. \tag{1.4}$$

The filtration F^s is compatible with d, exhaustive (i.e. $A_a^\cdot = \bigcup F^s A_a^\cdot$) and bounded above ($F^{n+1} A^\cdot = 0$). Here d denotes the differential of the complex $A_a^\cdot$ which is induced by the exterior differential d in $\Omega_F^\cdot$ via (1.1) and which is given explicitly by the formula

$$\mathrm{d}(\omega/f^s) = \mathrm{d}_f(\omega)\cdot f^{-s-1} \quad \text{where} \quad \mathrm{d}_f(\omega) = f\mathrm{d}\omega - (|\omega|/N)\mathrm{d}f \wedge \omega. \tag{1.5}$$

By the general theory of spectral sequences e.g. [16], p. 44 we get the next *geometric* spectral sequence.

(1.6) PROPOSITION. *There is an E_1-spectral sequence $(E_r(f)_a, d_r)$ with*

$$E_1^{s,t} = H^{s+t}(F^s A_a^\cdot / F^{s+1} A_a^\cdot)$$

and converging to the cohomology eigenspace $H^\cdot(F)_a$.

Moreover one can sum these spectral sequences for $a = 0, \ldots, N-1$ and get a spectral sequence $(E_r(f), d_r)$ converging to $H^\cdot(F)$. And $(E_r(f)_0, d_r)$ and $(E_r(f), d_r)$ are in fact spectral sequences of algebras converging to their limits as algebras. Note that $H^\cdot(F)_0 \simeq H^\cdot(U)$, either using the fact that $U = F/G$, G acting on F via the geometric monodromy or the fact that $\Omega_U^\cdot$ is a resolution of $\mathbb{C}$ [24].

We pass now to the construction of some purely *algebraic* spectral sequences. Let (B_a, d', d'') be the double complex $B_a^{s,t} = \Omega_{tN-a}^{s+t+1}$, $d' = d$ and $d''(\omega) = -|\omega|/N\, df \wedge \omega$ for a homogeneous differential form ω. Note that the associated total complex $B_a^\cdot$, with $B_a^k = \bigoplus_{s+t=k} B_a^{s,t}, D = d' + d''$ is precisely the complex $(\Omega_a^{\cdot -1}, D_f)$.

Similarly $B^\cdot = \bigoplus B_a^\cdot = (\Omega^{\cdot -1}, D_f)$.

Consider the decreasing filtration F^p on $B_a^\cdot$ given by $F^p B_a^k = \bigoplus_{s \geqslant p} B_a^{s,k-s}$ and similarly on $B^\cdot$. Using the contraction operator Δ, we define the next complex morphisms, compatible with the filtrations:

$$\bar{\delta}\colon B_a^\cdot \to A_a^\cdot \quad \text{and} \quad \bar{\delta}\colon B^\cdot \to A^\cdot,$$

$$\bar{\delta}(\omega) = \Delta(\omega) f^{-t} \quad \text{for} \quad \omega \in B_a^{s,t}.$$

Note that $B^\cdot$ and $A^\cdot$ are in fact differential graded algebras, but $\bar{\delta}$ is not compatible with the products.

(1.7) PROPOSITION. *There is an E_1-spectral sequence $('E_r(f)_a, d_r)$ with*

$$E_1^{s,t} = H^{s+t}(F^s B_a^\cdot / F^{s+1} B_a^\cdot)$$

and converging to the cohomology $H^\cdot(B_a^\cdot)$. The operator $\bar{\delta}$ induces a morphism $\delta_r\colon ('E_r(f)_a, d_r) \to (E_r(f)_a . d_r)$ of spectral sequences.

Moreover one can sum these spectral sequences $'E_r(f)_a$ and get a spectral sequence $('E_r(f), d_r)$ converging to $H^{\cdot}(B^{\cdot})$ and a morphism $('E_r(f), d_r) \to (E_r(f), d_r)$. The proof of these facts is standard e.g. [16], p. 49. Let $\tilde{E}_r(f)_0$ (resp. $\tilde{E}_r(f)$) denote the *reduced* spectral sequence associated to $E_r(f)_0$ (resp. $E_r(f)$) which is obtained by replacing the term at the origin $E_1^{0,0} = E_\infty^{0,0} = \mathbb{C}$ by zero. For $a \neq 0$, we put $\tilde{E}_r(f)_a = E_r(f)_a$.

We clearly have natural morphisms $\tilde{\delta}_r : 'E_r(f)_a \to \tilde{E}_r(f)_a, \tilde{\delta}_r : 'E_r(f) \to \tilde{E}_f(f)$ induced by δ_r. We can state now a basic result.

(1.8) THEOREM. *The morphisms $\tilde{\delta}_r$ are isomorphisms for $r \geqslant 1$ and they induce isomorphisms $H^{\cdot}(B_a) \cong \tilde{H}^{\cdot}(F)_a$ and $H^{\cdot}(B) \cong \tilde{H}^{\cdot}(F)$.*

Proof. Since $F^{n+1}B^{\cdot} = F^{n+1}A^{\cdot} = 0$, the filtrations F are strongly convergent [16], p. 50 and hence it is enough to show that $\tilde{\delta}_1$ is an isomorphism. The vertical columns in $'E_1(f)$ correspond to certain homogeneous components in the *Koszul complex* $K^{\cdot}$.

$$K^{\cdot} : 0 \to \Omega^0 \xrightarrow{df} \Omega^1 \xrightarrow{df} \cdots \xrightarrow{df} \Omega^{n+1} \to 0 \tag{1.9}$$

of the partial derivatives $f_i = (\partial f)/(\partial x_i)$, $i = 0, \ldots, n$ in S. To describe the vertical columns in $\tilde{E}_1(f)$ is more subtle. Note that $fK^{\cdot}$ is a subcomplex in $K^{\cdot}$ and let $\bar{K}^{\cdot}$ denote the quotient complex $K^{\cdot}/fK^{\cdot}$. There is a map $\bar{\Delta} : \bar{K}^{\cdot} \to \bar{K}^{\cdot -1}$ induced by Δ which is a complex morphism and hence $\tilde{K}^{\cdot} = \ker \bar{\Delta}$ is a subcomplex in $\bar{K}^{\cdot}$.

Let $\tilde{\Delta}$ denote the composition $K^{\cdot} \to \bar{K}^{\cdot} \xrightarrow{\bar{\Delta}} \tilde{K}^{\cdot -1}$.

Then the vertical lines in $\tilde{E}_1(f)$ correspond to certain homogeneous components in the cohomology groups $H^{\cdot}(\tilde{K}^{\cdot})$. The morphism $\tilde{\delta}_1$ corresponds to $\tilde{\Delta}^* : H^{\cdot}(K^{\cdot}) \to H^{\cdot}(\tilde{K}^{\cdot -1})$ and a well-defined inverse for $\tilde{\Delta}^*$ is given by the map

$$\nabla : H^{\cdot}(\tilde{K}^{\cdot -1}) \to H^{\cdot}(K^{\cdot}), \nabla[\Delta(\omega)] = [df \wedge \Delta(\omega)/(Nf)]. \tag{1.10}$$

To check this, use that $df \wedge \omega = 0$ implies $0 = \Delta(df \wedge \omega) = Nf\omega - df \wedge \Delta(\omega)$.

(1.11) EXAMPLE. Assume that f has an *isolated singularity* at the origin. Then $f_0, \ldots, f_n$ form a regular sequence in S and we get $'E_1^{s,t}(f)_a = 0$ for $s + t \neq n$ and

$$'E_1^{n-t,t}(f)_a \simeq H^{n+1}(K^{\cdot})_{tN-a} \simeq Q(f)_{tN-a-w}$$

where $Q(f) = S/(f_0, \ldots, f_n), w = w_0 + \cdots + w_n$. Moreover, the Poincaré series for $Q(f)$ (see for instance [7], p. 109) implies that $Q(f)_k = 0$ for $k > (n+1)N - 2w$. Hence in this case all our spectral sequences are finite and degenerate at the E^1-term (the degeneracy of the component $a = 0$ being equivalent to Griffiths' Theorem 4.3 in [12]). Note that one can have $'E_1^{-1,n+1}(f)_a \neq 0$. In general, one has the next result about the size of the spectral sequence $'E_r(f)$.

(1.12) Proposition. ${}'E_r^{s,t}(f) = 0$ *for any* $r \geqslant 1$ *and* $s + t < n - m$, *where* $m = \dim f^{-1}(0)_{\text{sing}}$.

Proof. The result follows using the description of ${}'E_1^{s,t}(f)$ in terms of the Koszul complex and Greuel generalized version of the de Rham-Lemma, see [11], (1.7).

(1.13) Corollary. $\tilde{H}^k(F) = 0$ *for* $k < n - m$.

This result is implied also by [15], but (1.12) will be used below in (2.8) in a crucial way.

Now we show that our complexes can be used to describe very explicitly the *Poincaré residue isomorphism* $R\colon H^{k+1}(\mathbb{C}^{n+1}\backslash F) \to \tilde{H}^k(F)$ and the *Sebastiani-Thom isomorphism.*

When X is a smooth complex manifold and D is a smooth closed hypersurface in X there is a Gysin exact sequence

$$\cdots \to H^{k+1}(X) \xrightarrow{i^*} H^{k+1}(X\backslash D) \xrightarrow{R} \tilde{H}^k(D) \to \cdots \tag{1.14}$$

where i^* is induced by the inclusion $i\colon X\backslash D \to X$ and R is the Poincaré residue, see for instance [24], Section 8.

Let $C_f^{\cdot}$ denote the complex $\Omega^{\cdot}$ with the differential D_f introduced above (up to a shift $C_f^{\cdot} = B^{\cdot}$!) and note that

$$\begin{aligned} &\alpha\colon C_f^{\cdot} \to \Gamma(\mathbb{C}^{n+1}\backslash F, \Omega^{\cdot}_{\mathbb{C}^{n+1}\backslash F}) \\ &\alpha(\omega) = \omega - (\mathrm{d}f \wedge \Delta(\omega))/N(f-1) \end{aligned} \tag{1.15}$$

is a morphism of differential graded algebras (i.e. $\mathrm{d}\alpha(\omega) = \alpha D_f(\omega)$ and $\alpha(\omega_1 \wedge \omega_2) = \alpha(\omega_1) \wedge \alpha(\omega_2)$).

Using the definition of the Poincaré residue as in [12], p. 290 it follows that

$$R\alpha(\omega) = (-1)^k/N\delta(\omega). \tag{1.16}$$

Since R is an isomorphism by using (1.14) in the case $X = \mathbb{C}^{n+1}, D = F$ and δ is an isomorphism by Theorem A, it follows that $\alpha\colon H^{\cdot}(C_f) \to H^{\cdot}(\mathbb{C}^{n+1}\backslash F)$ is an isomorphism too.

To discuss the Sebastiani-Thom isomorphism (see for instance [17]), we introduce a new complex associated to f, namely $\bar{C}_f$ which is the complex $\Omega^{\cdot}$ with the differential $\bar{D}_f\omega = \mathrm{d}\omega - \mathrm{d}f \wedge \omega$.

Define $\theta\colon \bar{C}_f \to C_f$ to be the $\mathbb{C}$-linear map which on a homogeneous form ω with $k = |\omega|$ acts by the formula $\theta(\omega) = \lambda(k)\cdot\omega$, where $\lambda(k) = 1$ for $k \leqslant N$ and $\lambda(k) = (k - N)\cdots(k - tN)\cdot N^{-t}$ for $tN < k \leqslant (t+1)N$, $t \geqslant 1$.

Then it is obvious that θ induces a complex isomorphism between the corresponding reduced complexes. In particular we get isomorphisms $\theta\colon H^t(\bar{C}_f) \to H^t(C_f)$ for any $t \geqslant 1$.

Let $\mathbf{w}' = (w'_0, \ldots, w'_{n'})$ be a new set of weights and $f' \in \mathbb{C}[y_0, \ldots, y_{n'}]$ be a homogeneous polynomial of degree N with respect to these weights. Then it is easy to check that

$$\bar{C}_{f+f'} \simeq \bar{C}_f \otimes \bar{C}_{f'} \tag{1.17}$$

and that there is no such result for $C_{f+f'}$.

Using the isomorphisms θ and Theorem A we get the Sebastiani-Thom isomorphism

$$\tilde{H}^k(F'') = \bigoplus_{s+t=k-1} \tilde{H}^s(F) \otimes \tilde{H}^t(F') \tag{1.18}$$

where F', F'' denote the Milnor fibers of f' and $f+f'$ respectively.

Keeping trace of the homogeneous components in (1.17) we get

$$\tilde{H}^{\cdot}(F'')_0 = \bigoplus_c \tilde{H}^{\cdot}(F)_c \otimes \tilde{H}^{\cdot}(F')_{N-c}$$

with $c = 0, \ldots, N-1$ and $\tilde{H}^{\cdot}(F')_N = \tilde{H}^{\cdot}(F')_0$. When $f' = y_0^N$, Example (1.11) shows that $\tilde{H}^{\cdot}(F')_0 = 0$ and $\tilde{H}^{\cdot}(F')_c = \langle \delta(y_0^{N-c-1}\,dy_0)\rangle$, a one-dimensional vector space for $c = 1, \ldots, N-1$. It follows that $\dim \tilde{H}^k(F'')_0 = \dim \tilde{H}^{k-1}(F)_{\neq 0}$ where $\tilde{H}^s(F)_{\neq 0} = \oplus_{c=1,N-1} H^s(F)_c$

This equality of dimensions is related to the next geometric setting. Let $H\colon y_0 = 0$ denote the hyperplane at infinity in the compactification $\mathbf{P}(\mathbf{w}, 1)$ of $\mathbb{C}^{n+1}$, let $V'' \subset \mathbf{P}(\mathbf{w}, 1)$ be the hypersurface given by $f(x) - y_0^N = 0$ and set $U'' = \mathbf{P}(\mathbf{w}, 1) \setminus V''$. Since $H \cap U'' = U, U'' \setminus H = \mathbb{C}^{n+1} \setminus F$, the Gysin sequence (1.14) applied to $X = U''$, $D = H \cap U''$ gives

$$\to H^k(U'') \xrightarrow{i^*} H^k(\mathbb{C}^{n+1} \setminus F) \xrightarrow{R} \tilde{H}^{k-1}(U) \to$$

(As a matter of fact U'' may be singular and then to apply (1.14) one has to do as follows. Let $q\colon \mathbf{P}^{n+1} \to \mathbf{P}(\mathbf{w}, 1)$ be the covering map induced by

$$(x_0 : \cdots : x_n : y) \mapsto (x_0^{w_0} : \cdots : x_n^{w_n} : y)$$

and let $\tilde{G}$ be the corresponding group of covering transformations.

If we set $\tilde{U} = q^{-1}(U''), \tilde{H} = q^{-1}(H)$, then there is a Gysin sequence associated to $X = \tilde{U}, D = \tilde{H} \cap \tilde{U}$. And the $\tilde{G}$-invariant part of this exact sequence is precisely the exact sequence which we have written above).

Note that

$$\begin{aligned}\dim H^k(\mathbb{C}^{n+1}\backslash F) &= \dim \tilde{H}^{k-1}(F) = \dim \tilde{H}^{k-1}(F)_{\neq 0} + \dim \tilde{H}^{k-1}(F)_0 \\ &= \dim H^k(U'') + \dim \tilde{H}^{k-}(U).\end{aligned}$$

It follows that the first and the last map in the above exact sequence are trivial. Note also that the geometric monodromy h acts on $\mathbb{C}^{n+1}\backslash F$ and hence it makes sense to define $H^s(\mathbb{C}^{n+1}\backslash F)_{\neq 0}$ as above.

It will be clear from what follows that the image of i^* is precisely $H^k(\mathbb{C}^{n+1}\backslash F)_{\neq 0}$ and hence we can write the next diagram of isomorphisms:

$$\begin{array}{ccccc}
H^k(F'')_0 \simeq H^k(U'') & \xrightarrow{i^*} & H^k(\mathbb{C}^{n+1}\backslash F)_{\neq 0} & \xrightarrow{R} & H^{k-1}(F)_{\neq 0} \\
\uparrow{\scriptstyle \delta} & & \nwarrow{\scriptstyle \alpha} & & \uparrow{\scriptstyle \delta} \\
H^{k+1}(C_{f+f'})_0 & & & & H^k(C_f)_{\neq 0} \\
\uparrow{\scriptstyle \theta} & & & & \uparrow{\scriptstyle \theta} \\
H^{k+1}(\bar{C}_{f+f'})_0 & & \xleftarrow{\quad\psi\quad} & & H^k(\bar{C}_f)_{\neq 0}
\end{array} \tag{1.19}$$

Here ψ is defined in a natural way: if $\omega \in H^k(\bar{C}_f)_c$ (i.e. ω is a sum $\omega_1 + \cdots + \omega_p$ of homogeneous forms such that $|\omega_i| \equiv -c$ modulo N) then $\psi(\omega) = \omega \wedge y_0^{c-1}\, dy_0$

The formula (1.16) tells us that the triangle in the diagram (1.19) is commutative up to a constant. The big rectangle in the diagram is commutative in a similar way by the next result.

(1.20) LEMMA. $Ri^*\delta\theta\psi = -1/N\delta\theta$.

Proof. We have to show that both sides of this equality yield the same result when applied to an element $\omega = \omega_1 + \cdots + \omega_p \in H^k(\bar{C}_f)_c$ as above. Since these computations are rather tedious, we treat here only the case $p = 2$ and let the reader check that the general case is completely similar.

So let $\omega = \omega_1 + \omega_2$ with $q = tN - c = |\omega_1|$ and $q + N = |\omega_2|$ (when $|\omega_2| - |\omega_1| > N$ the forms ω_1 and ω_2 are themselves cycles in $H^k(\bar{C}_f)_c$ and the proof is easier!).

The condition $\bar{D}_f\omega = 0$ is equaivalent to

(i) $df \wedge \omega_2 = 0$
(ii) $df \wedge \omega_1 = d\omega_2$
(iii) $0 = d\omega_1$.

It is easy to see that

$$i^*\delta\theta\psi(\omega) = (-1)^k t![\omega_1/t(f-1)^t + \omega_2/(f-1)^{t+1}].$$

To compute the residue of this element we proceed as follows. First we apply Δ to the equality (i) and get

(iv) $\omega_2/(f-1) + \omega_2 = \mathrm{d}f \wedge \Delta(\omega_2)/(N(f-1))$.

Next we can divide this equality by $(f-1)^s$ and get

(v) $\omega_2/(f-1)^{s+1} + \omega_2/(f-1)^s = -\mathrm{d}(\Delta(\omega_2)/Ns(f-1)^s) + \mathrm{d}\Delta(\omega_2)/Ns(f-)^s$.

If we apply Δ to (ii), we get

$$\mathrm{d}\Delta(\omega_2) = (q+N)\omega_2 + \mathrm{d}f \wedge \Delta(\omega_1) - Nf\omega_1.$$

This should be put in (v), one should apply once more this trick getting a term containing $\mathrm{d}\Delta(\omega_1)$ and then replace this by $q\omega_1$ as follows by applying Δ to (iii).

Let $A_s = \omega_1/s(f-1)^s + \omega_2/(f-1)^{s+1}$ and note that A_s is a closed form on $\mathbb{C}^{n+1}\backslash F$ for any $s \geqslant 1$. The above computation implies that the associated cohomology classes satisfy $[A_s] = ((q - N(s-1))/Ns)[A_{s-1}]$ and hence

$$R[A_t] = (\lambda(q)/t!)R[A_1] = ((-1)^{k-1}\lambda(q)/Nt!)[q/N\delta(\omega_2) + \delta(\omega_1)].$$

This ends the proof of (1.20) in this case.

(1.21) REMARK. There is a nice geometric consequence of the existence of the diagram (1.19). One can think of the weighted projective space $\mathbf{P}(\mathbf{w}, 1)$ as a *compactification* of $\mathbb{C}^{n+1}\backslash F$ such that the complement $\mathbf{P}(\mathbf{w}, 1)\backslash(\mathbb{C}^{n+1}\backslash F)$ consists of two irreducible components, namely V'' and H. Using the isomorphism α, it follows that any cohomology class in $H^\cdot(\mathbb{C}^{n+1}\backslash F)$ can be represented by a closed differential form on $\mathbb{C}^{n+1}\backslash F$ having a pole of order 1 along V'' and a pole (possibly of a higher order) along H.

On the other hand, the isomorphism $i^*\delta\theta\psi$ shows that any class in $H^\cdot(\mathbb{C}^{n+1}\backslash F)_{\neq 0}$ can be represented by a closed differential form on $\mathbb{C}^{n+1}\backslash F$ having a pole on V'' and no poles at all along H. It can be shown similarly that any class in $H^\cdot(\mathbb{C}^{n+1}\backslash F)_0$ can be represented by a form having a pole of order 1 along H and a pole along V''. It would be nice to have a more geometric understanding of this phenomenon.

In conclusion, the natural isomorphism $H^k(F) = H^k(F)_0 \oplus H^k(F)_{\neq 0} = H^k(U) \oplus H^{k+1}(U'')$ shows that it is enough to concentrate on the cohomology groups $H^\cdot(U)$ and this is what we do in the next two sections.

2. The relation with the Hodge filtration

Let us consider the decreasing filtration F^s on $H^\cdot(U)$ defined by the filtration F^s on $A_0^\cdot$, namely

$$F^sH^\cdot(U) = \text{im}\{H^\cdot(F^sA_0^\cdot) = H^\cdot(U)\}. \tag{2.1}$$

On the other hand there is on $H^\cdot(U)$ the decreasing *Hodge filtration* F_H^s introduced by Deligne [5].

(2.2) THEOREM. *One has* $F^sH^\cdot(U) \supset F_H^{s+1}H^\cdot(U)$ *for any* s *and* $F^0H^\cdot(U) = F_H^1H^\cdot(U)F_H^0H^\cdot(U) = H^\cdot(U)$.

Proof. Let $p: \mathbf{P}^n \to \mathbf{P}$ be the projection presenting $\mathbf{P}$ as the quotient of $\mathbf{P}^n$ under the group $G(\mathbf{w})$, the product of cyclic groups of orders w_i.

Then $\tilde{f} = p^*(f) = f(x_0^{w_0}, \ldots, x_n^{w_n})$ is a homogeneous polynomial of degree N and let $\tilde{U}$ be the complement of the hypersurface $\tilde{f} = 0$ in $\mathbf{P}^n$.

Since $H^\cdot(U)$ can be identified to the fixed part in $H^\cdot(\tilde{U})$ under the group $G(\mathbf{w})$ and since the monomorphism $p^*: H^\cdot(U) \to H^\cdot(\tilde{U})$ is clearly compatible with the filtrations F^s and F_H^s, it is enough to prove (2.2) for $\tilde{U}$.

To simplify the notation, we assume that $\mathbf{w} = (1, \ldots, 1)$ from the beginning. Then U is smooth and it is easier to describe the construction of the Hodge filtration [24].

Let $p: X \to \mathbf{P}^n$ be a proper modification with X smooth, $D = p^{-1}(V)$ a divisor with normal crossings in X and $\bar{U} = X \backslash D$ isomorphic to U via p.

From this point on it is more suitable to work with holomorphic differential forms on our algebraic varieties. If $\Omega_U^\cdot$ is this holomorphic sheaves complex, ${}^a\Omega_U^\cdot$ the algebraic version of it and $i: U \to \mathbf{P}^n$ is the inclusion, then one has inclusions $i_*({}^a\Omega_U^\cdot) \subset \Omega_{\mathbf{P}^n}^\cdot(*V) \subset i_*\Omega_U^\cdot$, where $\Omega_{\mathbf{P}^n}^\cdot(*V)$ denotes the sheaves of meromorphic differential forms on $\mathbf{P}^n$ with polar singularities along V. By Grothendieck [13], the inclusion $i_*({}^a\Omega_U^\cdot) \subset \Omega_{\mathbf{P}^n}^\cdot(*V)$ induces isomorphisms at the hypercohomology groups. And the same is true for the inclusions $\Omega_X^\cdot(\log D) \subset \Omega_X^\cdot(*D) \subset j_*\Omega_{\bar{U}}^\cdot$ where $j: \bar{U} \to X$ is the inclusion, $\Omega_X^\cdot(*D)$ is defined similarly to $\Omega_{\mathbf{P}^n}^\cdot(*V)$ and $\Omega_X^\cdot(\log D)$ is the complex of holomorphic differential forms with logarithmic poles along D[24].

Recall that there is a trivial filtration $\sigma_\geqslant$ on any complex $K^\cdot$, by defining $\sigma_{\geqslant s}K^\cdot$ to be the subcomplex of $K^\cdot$ obtained by replacing the first s terms in $K^\cdot$ by 0. The Hodge filtration is given by

$$F_H^sH^j(U) = \text{im}\{\mathbf{H}^j(\sigma_{\geqslant s}\Omega_X^\cdot(\log D)) \to \mathbf{H}^j(\Omega_X^\cdot(\log D))\} \tag{2.3}$$

via the identifications

$$\mathbf{H}^\cdot(\Omega_X^\cdot(\log D)) = \mathbf{H}(j_*\Omega_{\bar{U}}^\cdot) = \mathbf{H}^\cdot(\Omega_{\bar{U}}^\cdot) = H^\cdot(\bar{U}) = H^\cdot(U).$$

The filtration F^s on the complex $A_0^{\cdot}$ is related to a filtration F^s on the complex $\Omega^{\cdot}_{\mathbf{P}^n}(*V)$ defined in the following way: $F^s\Omega^j_{\mathbf{P}^n}(*V)$ is the sheaf of meromorphic j-forms on $\mathbf{P}^n$ having poles of order at most $j-s$ along V for $j \geqslant s$ and $F^s\Omega^j_{\mathbf{P}^n}(*V) = 0$ for $j < s$.

Note that $F^s\Omega^j_{\mathbf{P}^n}(*V) \simeq \Omega^j_{\mathbf{P}^n}((j-s)N)$ for $j \geqslant s$. We get next a filtration on the complex $\Omega^{\cdot}_X(*D) \simeq p^*(\Omega^{\cdot}_{\mathbf{P}^n}(*V))$ by defining $F^s\Omega^{\cdot}_X(*D) = p^*(F^s\Omega^{\cdot}_{\mathbf{P}^n}(*V))$.

At stalks level, a germ $\omega \in \Omega^j_X(*D)_x$ belongs to $F^s\Omega^j_X(*D)_x$ if and only if $p^*(u)^{j-s}\cdot\omega \in \Omega^j_{X,x}$, where $u = 0$ is a local equation for V around the point y $p(x)$.If $v_1, \ldots, v_n$ are local coordinates on X around x such that $v_1 \ldots v_k = 0$ is a local equation for D. then $p^*(u)$ vanishes on D and hence $p^*(u) = v_1^{a_1} \ldots v_k^{a_k} w$ for some germ $w \in \mathcal{O}_{X,x}$ and integers $a_i \geqslant 1$.

Using the definitiions, it follows that $\Omega^j_X(\log D) \subset F^s\Omega^j_X(*D)$ for $j > s$ and $j > 0$. And $\Omega^0_X(\log D) = \Omega^0_X \subset F^s\Omega^0_X(*D)$ for $s \leqslant 0$. We can state this as follows.

(2.4) LEMMA. (i) $\sigma_{\geqslant s+1}\Omega^{\cdot}_X(\log D \subset F^s\Omega^{\cdot}_X(*D)$ *for* $s > 0$;

(ii) $\Omega^{\cdot}_X(\log D) \subset F^0\Omega^{\cdot}_X(*D)$.

We can hence write the next commutative diagram

$$\begin{array}{ccccc}
\mathbf{H}^{\cdot}(\sigma_{\geqslant s+1}\Omega^{\cdot}_X(\log D)) & \rightarrow & \mathbf{H}^{\cdot}(F^s\Omega^{\cdot}_X(*D)) & \overset{p^*}{\underset{\sim}{\leftarrow}} & \mathbf{H}^{\cdot}(F^s\Omega^{\cdot}_{\mathbf{P}^n}(*V)) \\
\downarrow & & & & \downarrow \beta \\
\mathbf{H}^{\cdot}(\Omega^{\cdot}_X(\log D)) & \underset{\sim}{\longrightarrow} & \mathbf{H}^{\cdot}(\Omega^{\cdot}_X(*D)) & \overset{p^*}{\underset{\sim}{\leftarrow}} & \mathbf{H}^{\cdot}(\Omega^{\cdot}_{\mathbf{P}^n}(*V)).
\end{array}$$

Now $\mathbf{H}^{\cdot}(\Omega^{\cdot}_{\mathbf{P}^n}(*V)) = \mathbf{H}^{\cdot}({}^a\Omega^{\cdot}_U) = H^{\cdot}(A_0^{\cdot}) = H^{\cdot}(U)$. To compute $\mathbf{H}^{\cdot}(F^s\Omega^{\cdot}_{\mathbf{P}^n}(*V))$ we use the E_2-spectral sequence $E_2^{p,q} = H^p(H^q(\mathbf{P}^n, K^{\cdot}))$ converging to $\mathbf{H}^{\cdot}(K^{\cdot})$, where $K = F^s\Omega^{\cdot}_{\mathbf{P}^n}(*V)$ and Bott's vanishing theorem [8].

It follows that $E_2^{p,0} = H^p(F^sA_0^{\cdot})$, $E_2^{s,s} = H^s(\mathbf{P}^n, \Omega^s_{\mathbf{P}^n})$ and $E_2^{p,q} = 0$ in the other cases. The spectral sequence degenerates at E_2 since one can represent the generator of $E_2^{s,s}$ by a $\bar{\partial}$-harmonic form γ and hence $\bar{d}\gamma = 0$. On the other hand $\beta(\gamma) = 0$, since γ belongs to the kernel of the map $H^{2s}(\mathbf{P}^n) \xrightarrow{i^*} H^{2s}(U)$. In fact this map is zero for $s > 0$. To see this, it is enough to show that $i^*(c) = 0$, where $c = c_1(\mathcal{O}(1))$ is the first Chern class of the line bundle $\mathcal{O}(1)$ (in cohomology with complex coefficients!). But $Ni^*(c) = 0$, since it corresponds to the Chern class of $\mathcal{O}(N)|_U$ and this line bundle has a section (induced by f) without any zeros.

It follows that $\mathrm{im}(\beta) = F^sH^{\cdot}(\mathrm{U})$ and this gives the first part in (2.2).

The similar diagram associated to the inclusion (2.4. ii) gives $F^0H^{\cdot}(U) = F^0_HH^{\cdot}(U) = H^{\cdot}(\mathrm{U})$.

To see that $F^0_H = F^1_H$ we relate the mixed Hodge structure on $H^{\cdot}(U)$ to the mixed Hodge structure on $H^{\cdot}(V)$ Consider the exact sequence in cohomology with compact supports of the pair $(\mathbf{P}^n, V)$

$$\cdots \rightarrow H^k_c(U) \rightarrow H^k(\mathbf{P}^n) \rightarrow H^k(V) \rightarrow H^{k+1}_c(U) \rightarrow \cdots \tag{2.5}$$

This is an exact sequence of MHS (mixed Hodge structures) and it gives an isomorphism of MHS $H_c^{k+1}(U) \simeq H_0^k(V)$, the primitive cohomology of V [10]. Poincaré duality gives a natural identification (U is a **Q**-homology manifold):

$$H^s(U) = \mathrm{Hom}(H_c^{2n-s}(U), H_c^{2n}(U)).$$

Since $H_c^{2n}(U) \simeq H^{2n}(\mathbf{P}^n) \simeq \mathbb{C}(-n)$, we get the following relations among mixed Hodge numbers

$$h^{p,q}(H^s(U)) = h^{n-p,n-q}(H_0^{2n-s-1}(V)).$$

This gives $h^{0,q}(H^s(U)) = 0$ for any q and s, which shows that $F_H^0 H^{\cdot}(U) = F_H^1 H^{\cdot}(U)$, ending the proof of (2.5).

(2.6) REMARK. In spite of the fact that $F^s H^{\cdot}(U) = F_H^{s+1} H^{\cdot}(\mathrm{U})$ for any s in many cases (e.g. when V is a quasi-smooth hypersurface or when V is a nodal curve in $\mathbf{P}^2$), this equality *does not hold* in general. A simple example is the next: take $V: x[xy(x+y)+z^3] = 0$ the union of a smooth cubic curve in $\mathbf{P}^2$ with an inflexional tangent. Then it is easy to show that in this case $\dim F^1 H^2(U) = 2 > \dim F_H^2 H^2(U) = 1$.

There is a similar inclusion $F^s H^{\cdot}(F) \supset F_H^{s+1} H^{\cdot}(F)$ among the analogous filtrations on the cohomology of the Milnor fiber F. The proof of this fact can be reduced to (2.2) as follows. The geometric monodromy h is analgebraic map and hence h^* preserves both filtrations F^s and F_H^s on $H^{\cdot}(F)$. If we define $F^s H^{\cdot}(F)_a = F^s H^{\cdot}(F) \cap H^{\cdot}(F)_a$ it follows that $F^s H^{\cdot}(F) = \oplus_a F^s H^{\cdot}(F)_a$. And one has a similar result for the Hodge filtration F_H^s. In particular, it is enough to prove

(i) $F^s H^{\cdot}(F)_0 \supset F_H^{s+1} H^{\cdot}(F)_0$, and
(ii) $F^s H^{\cdot}(F)_{\neq 0} \supset F_H^{s+1} H^{\cdot}(F)_{\neq 0}$

where $F^s H^{\cdot}(F)_{\neq 0} = F^s H^{\cdot}(F) \cap H^{\cdot}(F)_{\neq 0} = \oplus_{a \neq 0} F^s H^{\cdot}(F)_a$ and similarly for F_H.

Now (i) is clearly implied by (2.2), since the isomorphism $H^{\cdot}(U) \xrightarrow[\sim]{p^*} H^{\cdot}(F)_0 \subset H^{\cdot}(F)$ is clearly compatible with both filtrations.

To get (ii) from (2.2) we use the diagram (1.19) and the next two facts.

The Poincaré residue map R is a morphism of MHS of type $(-1, -1)$ and hence

$$Ri^*(F_H^{s+1} H^{\cdot}(U'')) = F_H^s H^{\cdot}(F)_{\neq 0}.$$

Using the definition of the filtrations F^s and (1.20) it follows that

$$Ri^*(F^s H^{\cdot}(U'')) = F^{s-1} H^{\cdot}(F)_{\neq 0}.$$

Note also that the filtration F^s on $H^{\cdot}(F)$ is very close to the filtrations considered by Scherk and Steenbrink in the isolated singularity case in [21].

(2.7) COROLLARY. (i) $E_\infty^{s,t}(f)_0 =$ *for* $s < 0$ *and* $E_\infty^{s,t}(f)_a = 0$ *for* $s < -1$ *and* $a = 1\dots, N-1$.

(ii) *Any element in* $H^k(U)$ *can be represented by a differential k-form with a pole along* V *of order at most* k.

We note that (ii) can be regarded as an extension of Griffith's Theorem 4.2 in [12]. On the side of numerical computations of Betti numbers we get the following important consequence. Recall that $m = \dim f^{-1}(0)_{\text{sing}}$.

(2.8) THEOREM. *Let* $b_j^0(V) = \dim H_0^j(V)$ *denote the primitive Betti numbers of* V. *Then*

(i) $b_j^0(V) = 0$ *for* $j < n-1$ *or* $j > n-1+m$;
(ii) *For* $k \in [0, m]$ *and* $r > 1$ *one has*

$$b_{n-1+k}^0(V) = b_{n-k}(U) \leqslant \sum_{s=0}^{n-k-1} \dim E_r^{s,n-k-s}(f)_0.$$

When $k = m$ *and* $r \geqslant n - m$ *the above inequality is an equality.*

Proof. Use (1.6), (1.7), (1.8), (1.12) and (2.7).

There is also an analog of (2.8) for $\dim H^j(F)_a$ but we leave the details for the reader.

3. The case of a one-dimensional singular locus

We assume in this section that f has a one-dimensional singular locus, namely

$$f^{-1}(0)_{\text{sing}} = \{z \in \mathbb{C}^{n+1}; df(x) = 0\} = \{0\} \cup \bigcup_{i=1,p} \mathbb{C}^* a_i$$

for some points $a_i \in \mathbb{C}^{n+1}$, one in each irreducible component of $f^{-1}(0)_{\text{sing}}$.

If H_i is a small transversal to the orbit $\mathbb{C}^* a_i$ at the point a_i, then the isolated hypersurface singularity $(Y_i, a_i) = (H_i \cap f^{-1}(0), a_i)$ is called the *transversal singularity* of f along the brach $\mathbb{C}^* a_i$ of the singular locus.

The weighted homogeneity of f easily implies that the isomorphism class ($\mathscr{K}$-equivalence) of the singularity (Y_i, a_i) does not depend on the choice of a_i (in the orbit $\mathbb{C}^* a_i$) or of H_i.

In this section we get a better understanding of the sequence $(E_r(f)_0, d_r)$ by relating it to some spectral sequences associated to the transversal singularities (Y_i, a_i) for $i = 1, \dots, p$.

First we describe the construction of these new (local) spectral sequences.

Let $g: (\mathbb{C}^n, 0) \to (\mathbb{C}, 0)$ be an analytic function germ and let $(Y, 0) = (g^{-1}(0), 0)$ be the hypersurface singularity defined by g. Let $\Omega^{\cdot}_{g,0}$ denote the localization of the stalk at the origin of the holomorphic de Rham complex $\Omega^{\cdot}_{\mathbb{C}^n}$ with respect to the multiplicative system $\{g^s; s \geqslant 0\}$.

Choose $\varepsilon > 0$ small enough such that Y has a conic structure in the closed ball $B_\varepsilon = \{y \in \mathbb{C}^n; |y| \leqslant \varepsilon\}$ [4]. Let $S_\varepsilon = \partial B_\varepsilon$ and $K = S_\varepsilon \cap Y$ be the link of the singularity $(Y, 0)$. Then Thm. 2 in [13] implies the following.

(3.1) PROPOSITION. $H^{\cdot}(S_\varepsilon \backslash K) \simeq H^{\cdot}(\Omega^{\cdot}_{g,0})$.

One can construct a filtration F^s on $\Omega^{\cdot}_{g,0}$ in analogy to (1.4), namely

$$F^s\Omega^j_{g,0} = \{\omega/g^{j-s}; \omega \in \Omega^j_{\mathbb{C}^n,0}\} \text{ for } j \geqslant s \text{ and } F^s\Omega^j_{g,0} = 0 \text{ for } j < s.$$

(3.2) PROPOSITION. *There is an E_1-spectral sequence of algebras $(E_r(g, 0), d_r)$ with*

$$E_1^{s,t} = H^{s+t}(Gr^s_F\Omega^{\cdot}_{g,0})$$

and converging to $H^{\cdot}(S_\varepsilon \backslash K)$ as an algebra.

Assume from now on that $(Y, 0)$ is an isolated singularity and let $L^{\cdot} = (\Omega^{\cdot}_{\mathbb{C}^n,0}\, dg)$ denote the Koszul complex of the partial derivatives of g. In our case these derivatives form a regular sequence and hence $H^j(L^{\cdot}) = 0$ for $j < n$ and $H^n(L^{\cdot}) = M(g)$, the *Milnor algebra* of the singularity $(Y, 0)$, see for instance [7], p. 90. Let $I^{\cdot}$ denote the quotient complex $L^{\cdot}/gL^{\cdot}$ If g: $M(g) \to M(g)$ denotes the multiplication by g, it follows that $H^j(I^{\cdot}) = 0$ for $j < n - 1, H^{n-1}(I^{\cdot}) = \ker(g)$ and $H^n(I^{\cdot}) = \operatorname{coker}(g) = T(g)$, the *Tjurina algebra* of $(Y, 0)$, see [7], p. 90.

There is the next analog of (1.8), computing $E_1(g, 0)$ in terms of $H^{\cdot}(I^{\cdot})$.

(3.3) LEMMA. *The nonzero terms in $E_1(g, 0)$ are the following.*

(i) $E_1^{s,0}(g, 0) = \Omega^s_{\mathbb{C}^n,0}$ *for* $s \in [0, n]$

(ii) $E_1^{s,1}(g, 0) = \Omega^s_Y$ *for* $s \in [0, n - 3]$, *there is an exact sequence* $0 \to \Omega^{n-2}_Y \xrightarrow{u} E_1^{n-2,1}(g, 0) \xrightarrow{v} \ker(g) \to 0$ *and* $E_1^{n-1,1}(g, 0) = \Omega^n_{\mathbb{C}^n,0}/g\Omega^n_{\mathbb{C}^n,0}$, *where*

$$\Omega^{\cdot}_Y = (\Omega^{\cdot}_{\mathbb{C}^n,0})/(g\Omega^{\cdot}_{\mathbb{C}^n,0} + dg \wedge \Omega^{\cdot}_{\mathbb{C}^n,0}).$$

(iii) $E_1^{n-t-1,t}(g, 0) = \ker(g), E_1^{n-t,t}(g, 0) = T(g)$ *for* $t \geqslant 2$.

Proof. To get the more subtle point (ii), one uses the well-defined maps

$$u: \Omega^s_Y \to E_1^{s,1}(g, 0),\ u(\alpha) = [(dg \wedge \alpha)/g]$$
$$v: E_1^{s,1}(g, 0) \to H^{s+2}(L^{\cdot}), v[\beta/g] = [(dg \wedge \beta)/g]$$

and note that $\operatorname{im}(v) \subset \ker(g)$ for $s = n - 2$.

(3.4) COROLLARY. *The only (possibly) nonzero terms in* $E_2(g,0)$ *are* $E_2^{0,0} = E_2^{0,1} = \mathbb{C}$ *and* $E_2^{n-1-t,t}, E_2^{n-t,t}$ *for* $t \geq 1$.

Proof. Use the exactness of the de Rham complexes [11]:

$$0 \to \mathbb{C} \to \Omega^0_{\mathbb{C}^n,0} \to \cdots \to \Omega^n_{\mathbb{C}^n,0} \to 0$$

$$0 \to \mathbb{C} \to \Omega^0_Y \to \cdots \to \Omega^{n-1}_Y.$$

We can also describe the differentials

$$d_1^t : \ker(g) = E_1^{n-1-t,t} \xrightarrow{d_1} E_1^{n-t,t} = T(g).$$

An $(n-1)$ form α induces an element in $\ker(g)$ if $dg \wedge \alpha = g\beta$ and then

$$d_1^t[\alpha] = [d\alpha - t\beta] \tag{3.5}$$

(3.6) EXAMPLE. Assume that $(Y,0)$ is a *weighted homogeneous singularity* of type $(w_1, \ldots, w_n; N)$, i.e. $(Y,0)$ is defined in suitable coordinates by a weighted homogenous polynomial g of degree N with respect to the weights $\mathbf{w}$.

Then $M(g) = T(g) = \ker(g)$ and they are all graded $\mathbb{C}$-algebras.

Let $\alpha = \sum_{i=1,n}(-1)^{i+1} w_i x_i \, dx_1 \wedge \cdots \wedge d\hat{x}_i \wedge \cdots \wedge dx_n$ and note that $dg \wedge \alpha = N \cdot g\omega_n$, with $\omega_n = dx_1 \wedge \cdots \wedge dx_n$. It follows that the class of α generates $\ker(g)$ For a monomial $x^a = x_1^{a_1} \cdots x_n^{a_n}$ of degree $|x^a| = a_1 w_1 + \cdots + a_n w_n$ one has by (3.5)

$$d_1^t(x^a\alpha) = [(w + |x^a| - tN)x^a\omega_n]$$

with $w = w_1 + \cdots + w_n$.

It follows that $\ker d_1^t \simeq \operatorname{coker} d_1^t \simeq M(g)_{tN-w}$. Hence the E_2-term $E_2(g,0)$ has finitely many nonzero entries and the spectral sequence $E_r(g,0)$ degenerates at E_2 (compare to (1.11)).

The next result gives a large class of singularities having the E_3-term of the spectral sequence $E_r(g,0)$ with finitely many nonzero entries. The reader should have no difficulty in checking that this class contains in particular the next more familiar classes of singularities:

(i) all the non weighted homogeneous $\mathbb{R}$-*unimodal singularities*, see for instance [0], p. 184 for a complete list;

(ii) all the *semi weighted homogenous singularities* (see [7], p. 115 for a definition) of the form $g = g_0 + g'$ with g_0 weighted homogeneous of type $(w_1, \ldots, w_n; N)$, g' weighted homogeneous of type $(w_1, \ldots, w_n; N')$ and such that

$$N' \geq (n+1)N/2 - w_1 - \cdots - w_n.$$

To state the result, note that there is a linear map $d_1^0\colon \ker(g) \to T(g)$ defined by taking $t = 0$ in the formula (3.5).

(3.7) PROPOSITION. *Assume that the singularity $Y\colon g = 0$ satisfies the condition:*

(i) $g^2 = 0$ *in* $M(g)$ *and* $d_1^0 \mid (g) = 0$ (resp. (ii) $\mu(g) - \tau(g) = 1$).

Then $E_2^{n-1-t,t} = \ker d_1^t = (g)$ *for* $t \gg 0$ (*resp.* $\dim \ker d_1^t = 1$ *for* $t \gg 0$ *and the lines* $\ker d_1^t$ *in* $\ker(g)$ *converge to the line* $\mathbb{C}\text{-}g$ *when* $t \to \infty$) *and the* E_3*-term* $E_3(g, 0)$ *has finitely many nonzero entries.*

Proof. (i) Let $K \subset M(g)$ be a vector subspace which is a complement of the ideal $\operatorname{Ker}(g) \subset M(g)$.

Then multiplication by g induces a vector space isomorphism $K \xrightarrow{\sim} gK = (g)$. For t large enough, it is clear using (3.5) that $\ker d_1^t = (g)$ and that the canonical projection $M(g) \to T(g)$ induces an isomorphism $K \simeq \operatorname{coker} d_1^t$.

Via these isomorphisms we may regard d_2^t as an endomorphism of K for $t \gg 0$.

Next $d_1^t(ag) = 0$ implies that we may write $ag^2\omega_n = dg \wedge \alpha$ and the $(n-1)$-form α satisfies $d\alpha = dg \wedge \beta + \lambda g\omega_n$ for some $(n-1)$-form β and function germ λ. But then we have

$$d(\alpha/g^t) = (\lambda - t \cdot a)\omega_n/(g^{t-1}) + d(\beta/(t-1)g^{t-1}) - (d\beta)/((t-1)g^{t-1}))$$

This shows that the endomorphism d_2^t has a matrix of the form $-t \cdot Id + A + B(t-1)^{-1}$ for A, B some constant matrices. It follows that for $t \gg 0$ this matrix is invertible and this clearly ends the proof. The proof in case (ii) is similar.

Now we come back to our *global setting* and assume *first* that we are in the homogenous case, i.e $w_0 = \cdots = w_n = 1$. Let Z denote the singular locus of V.

Consider the restriction morphism

$$\rho\colon \Omega^{\cdot}_{\mathbb{P}^n}(*V) \to \Omega^{\cdot}_{\mathbb{P}^n}(*V)|_Z \tag{3.8}$$

and the associated morphisms

$$Gr_F^s\rho\colon Gr_F^s(\Omega^{\cdot}_{\mathbb{P}^n}(*V)) \to Gr_F^s(\Omega^{\cdot}_{\mathbb{P}^n}(*V)|_Z).$$

A moment thought shows that $Gr_F^s\rho$ is a quasi-isomorphism for $s < 0$. A computation using an E_2-spectral sequence shows that

$$\mathbf{H}^{\cdot}(Gr_F^s(\Omega^{\cdot}_{\mathbb{P}^n}(*V))) = H^{\cdot}(Gr_F^s A_0^{\cdot}).$$

Assume from now on that Z is a finite set, namely $Z = \{a_1, \ldots, a_p\}$. Note that the singularity (V, a_i) is precisely the transversal singularity of f along the line $\mathbb{C}^* \cdot a_i$ as defined in the beginning of this section.

Choose the coordinates on $\mathbf{P}^n$ such that $H\colon x_0 = 0$ is transversal to V and $Z \subset \mathbf{P}^n \backslash H \simeq \mathbb{C}^n$. We denote again by a_i the corresponding points in $\mathbb{C}^n$ and let $g(y) = f(1, y)$.

Then $\Omega^{\cdot}_{\mathbf{P}^n}(*V)|_Z = \oplus_{j=1,p}\Omega^{\cdot}_{g,a_j}$, this identification being compatible with the F filtrations. Thus we get

$$\mathbf{H}^{\cdot}(Gr^s_F(\Omega^{\cdot}_{\mathbf{P}^n}(*V)|_Z)) = \bigoplus_{j=1,p} H^{\cdot}(Gr^s_F(\Omega^{\cdot}_{g,a_j})).$$

We can restate these considerations in the next form.

(3.9) THEOREM. *The restriction map ρ induces a morphism $\rho_r\colon E_r(f)_0 \to \oplus_{j=1,p}E_r(g, a_j)$ of spectral sequences such that at the E_1-level $\rho_1^{s,t}$ is an isomorphism for $s < 0$.*

(3.10) COROLLARY. *For a projective hypersurface $V\colon f = 0$ with isolated singularities the next statements are equivalent*

(i) *all the singularities of V are weighted homogeneous;*
(ii) *$E_2^{s,t}(f)_0 = 0$ for $s < 0$;*
(iii) *$E_2^{s,t}(f)_0 \neq 0$ for finitely many pairs (s, t).*

Proof. Using (3.6) and (3.9) we get (i) $\Rightarrow$ (ii). The implication (ii) $\Rightarrow$ (iii) is obvious. To prove (iii) $\Rightarrow$ (i) we compute the Euler Poincaré characteristic $\chi(U)$ in two ways. First we use the fact that $U = \mathbf{P}^n \backslash V$ and the well-known formula for $\chi(V)$ given in (3.12) below and get

$$\chi(U) = \chi(U_0) + (-1)^{n-1} \sum_{i=1,p} \mu(V, a_i)$$

where U_0 is the complement of a smooth hypersurface V_0 in $\mathbf{P}^n$.

Next using (1.8) and a standard property of spectral sequences we get

$$\chi(U) = 1 + \sum (-1)^{s+t} \dim E_2^{s,t}(f)_0$$

where the sum is finite by our assumption. Choose $m > n$ such that $E_2^{s,t}(f)_0 = 0$ for $t > m$. Then

$$\begin{aligned}\chi(U) - 1 &= (-1)^{n-1} \sum_{t=1,m} (\dim E_2^{n-1-t,t}(f)_0 - \dim E_2^{n-t,t}(f)_0)\\ &= (-1)^{n-1} \sum_{t=1,m} (\dim E_1^{n-1-t,t}(f)_0 - \dim E_1^{n-t,t}(f)_0)\\ &= (-1)^{n-1} [\dim E_1^{n-1-m,m}(f)_0 + \varphi]\end{aligned}$$

with

$$\varphi = \sum_{t=1,m} \dim E_1^{n-t,t-1}(f)_0 - \dim E_1^{n-t,t}(f)_0).$$

By (3.3.iii) and (3.9) it follows that

$$\dim E_1^{n-1-m,m}(f)_0 = \sum_{i=1,p} \tau(V, a_i)$$

where $\tau(V, a_i) = \dim T(g, a_i) = \dim \ker(g, a_i)$ are the corresponding Tjurina numbers. On the other hand, using the connection of $E_1(f)$ with the Koszul complex, it is easy to see that the sum φ does not depend on f. Since one can compute $\chi(U_0)$ in the same way, it follows that

$$\chi(U) = \chi(U_0) + (-1)^{n-1} \sum_{i=1,p} \tau(V, a_i).$$

Comparing the two formulas for $\chi(U)$ we get $\mu(V, a_i) = \tau(V, a_i)$ for any $i = 1, \ldots, p$ and hence by *K. Saito's Theorem* (see for instance [7], p. 119 for a discussion) all the singularities (V, a_i) are weighted homogeneous.

Since for any isolated hypersurface singularity $(Y, 0)$ there is a projective hypersurface V having just one singular point a_1 and such that $(V, a_1) \simeq (Y, 0)$, see for instance [2], we get the next result using (3.6), (3.9) and (3.10).

(3.10′) COROLLARY. *For an isolated hypersurface singularity* $(Y, 0)$ *defined by* $g = 0$ *in* $(\mathbb{C}^n, 0)$, *the next statements are equivalent:*

(i′) $(Y, 0)$ *is a weighted homogeneous singularity;*
(ii′) *the spectral sequence* $E_r(g, 0)$ *degenerates at* E_2;
(iii′) $E_2^{s,t}(g, 0) \neq 0$ *for finitely many pairs* (s, t).

We *conjecture* in analogy with (3.10′) that the statements in (3.10) are equivalent to the next stronger version of (ii):

(iv) *the spectral sequence* $E_r(f)_0$ *degenerates at* E_2.

(3.11) REMARK. Let f be a homogenous polynomial such that V has an isolated singularity of the type considered in (3.7). Then $E_r(f)_0$ surely does not degenerate at E_2. Note that $f\colon (\mathbb{C}^{n+1}, 0) \to (\mathbb{C}, 0)$ is *concentrated* in the terminology of [25], p. 206 and our spectral sequence $E_r(f)_0$ is a subobject in the *huge* spectral sequence considered in [25], p. 209. Hence in this case that spectral sequence does not degenerate at E_2 and this gives a negative answer to the question at the top of p. 209 in [25].

By Theorem (2.8) the interesting Betti numbers for V in the isolated singularities case are just $b_{n-1}(V), b_n(V)$ and we can get $b_n(V)$ from $E_{n-1}(f)_0$.

But one has a simple formula for the *Euler-Poincaré characteristic* in this case [6]:

$$\chi(V) = \chi(V_0) + (-1)^n \sum_{i=1,p} \mu(V, a_i) \tag{3.12}$$

where V_0 denotes a smooth hypersurface in $\mathbf{P}^n$ of degree N and $\mu(V, a_i) = \dim M(g, a_i)$ are the corresponding Milnor numbers.

In this way we get $b_{n-1}(V)$ knowing $b_n(V)$. We remark that there is a formula for $\chi(F)$ similarto (3.12) and which appears in the special case $n = 2$ as Theorem 6. A in [9].

(3.13) PROPOSITION. $\chi(F) = 1 + (-1)^n[(N-1)^{n+1} - N \sum_{i=1,p} \mu(V, a_i)]$.

Proof. If $\bar{F}$ denotes the closure of F in $\mathbf{P}^{n+1}$, one has $\chi(F) = \chi(\bar{F}) \backslash \chi(V)$. One then use (3.12) and the remark that the singularities of $\bar{F}$ are just the N-fold suspensions of the singularities of V and hence

$$\mu(\bar{F}, (a_i:0)) = (N-1)\mu(V, a_i).$$

(3.14) REMARK. An important invariant of the singularity f is the *zeta function* $Z(h)$ of the monodromy operator h. Explicitly one has

$$Z(h) = \prod_{i \geqslant 0} \det(1 - th^* \mid H^i(F))^{(-1)^{i+1}} = \exp \sum_{k \geqslant 1} \wedge (h^k) t^k / k$$

where $\wedge(h^k)$ denotes the Lefschetz number of the map h^k. Using the second expression above for $Z(h)$ it follows that for any homogeneous polynomial f one has

$$Z(h) = (1 - t^N)^{-\chi(F)/N}.$$

When V has only isolated singularities, this formula may be used to compute $\dim H^n(F)_a$ for $a = 1, \ldots, N-1$ assuming that we know $\dim H^{n-1}(F)_a$ via computations with the spectral sequence $E_r(f)$ as in the remark after (2.8).

Next we describe briefly the *additional facts* necessary in order to treat the case when f has *arbitrary weights* $\mathbf{w} = (w_0, \ldots, w_n)$.

First we have to include a group action in the local setting. Let $G \subset U(n)$ be a finite group and consider the induced action on $\mathbb{C}^n$. Then the ball B_ε and the sphere S_ε are G-invariant subsets. Assume that Y:$g = 0$ is a reduced hypesurface singularity which is also G-invariant (i.e. $y \in Y, \gamma \in G \Rightarrow \gamma(y) \in Y$ for a representative Y of $(Y, 0)$ in B_ε).

There is an associated action of G on $\Omega^{\cdot}_{\mathbb{C}^n,0}$ given by $\gamma \cdot \omega = (\gamma^{-1})^* \omega$ And there is character $\chi_Y \colon G \to \mathbb{C}^*$ such that $\gamma \cdot g = \chi_Y(\gamma) g$ for any $\gamma \in G$. In this situation we call $(Y, 0)$ a G-singularity. Note that this setting is larger than in Wall [27] where one takes $\chi_Y = 1$, but coincides (in the case of G cyclic) to the hyperquotient singularity notion of M. Reid [19].

Let $(\Omega^{\cdot G}_{g,0}, d)$ be the subcomplex in $(\Omega^{\cdot}_{g,0}, d)$ consisting of the fixed elements under the obvious action of G. If $K^{\cdot}$ is any complex of $\mathbb{C}$-vector spaces with G-actions compatible with the differentials, then there is a natural isomorphism $H^{\cdot}(K^{\cdot G}) = H^{\cdot}(K^{\cdot})^G$ which says that taking cohomology commutes with taking the fixed parts under G. Moreover in Proposition (3.1) both cohomology groups have natural G-actions and the isomorphism considered there is compatible with these actions. It follows that

$$H^{\cdot}(\Omega^{\cdot G}_{g,0}) = H^{\cdot}(\Omega^{\cdot}_{g,0})^G = H^{\cdot}(S_\varepsilon \backslash K)^G = H^{\cdot}((S_\varepsilon \backslash K)/G). \tag{3.15}$$

Next, using again the above commutativity, we get an E_1-spectral sequence $(E_r(g,0)^G, d_r)$ consisting of the fixed parts of the spectral sequence described in (3.2) and converging to $H^{\cdot}((S_\varepsilon \backslash K)/G)$.

Assume now that $(Y, 0)$ is an isolated singularity and note that G acts on the complex $L^{\cdot}$ considered above. Since the G-action commutes with the differentials in $L^{\cdot}$ up-to multiplicative constants, it follows that there is an induced action on the cohomology $H^{\cdot}(L^{\cdot})$. And one has exactly as in Wall [27] an isomorphism of G-vector spaces

$$H^n(L^{\cdot}) = M(g) \otimes \mathbb{C}\omega_{n+1}$$

with $\omega_{n+1} = \mathrm{d}x_0 \wedge \cdots \wedge \mathrm{d}x_n$. Let χ_0 be the character of the action of G on $\mathbb{C}\omega_{n+1}$. If W is any G-vector space and $\chi \colon G \to \mathbb{C}^*$ is a character we set

$$W^\chi = \{w \in W; \gamma \cdot w = \chi(\gamma) w \text{ for all } \gamma \in G\}.$$

With this notation, note that

$$\omega/g^t \in \Omega^{\cdot G}_{g,0} \text{ if and only if } \omega \in \Omega^{\cdot \chi_Y^t}_{\mathbb{C}^n,0}.$$

Combining these remarks we get the next analog of (3.3.iii):

$$E_1^{n-t-1,t}(g,0)^G \simeq \ker(g)^{\chi_Y^t \chi_0^{-1}}$$

$$E_1^{n-t,t}(g,0)^G \simeq T(g)^{\chi_Y^t \chi_0^{-1}} \tag{3.16}$$

for all $t \geqslant 2$, where $\ker(g)$ and $T(g)$ have the obviously induced G-actions.

We consider now the global setting. Let $a \in \mathbb{C}^{n+1} \backslash \{0\}$ be a point in the singular locus $f^{-1}(0)_{\text{sing}}$. Let G_a be the *isotropy subgroup* of a with respect to the $\mathbb{C}^*$-action on $\mathbb{C}^{n+1}$ given by

$$t \cdot x = (t^{-w_0}x_0, \dots, t^{-w_n}x_n).$$

Then G_a is the finite cyclic group of the unity roots of order

$$k_a = \text{g.c.d.}\{w_j;\ \text{the component } a_j \text{ of a is nonzero}\}.$$

Take H to be a transversal to the orbit $\mathbb{C}^* . a$ at the point a which is G_a-invariant. For instance, we may assume that $a_0 \neq 0$ and then take $H\colon x_0 - a_0 = 0$. We identify the germs $(\mathbb{C}^n, 0)$ and (H, a) via the isomorphism φ given by $(y_1, \dots, y_n) \mapsto (a_0, y_1, \dots, y_n)$. Then the transversal singularity $(Y, a) = (H \cap f^{-1}(0), a)$ is in an obvious way a G_a-*singularity* and moreover

$$\chi_Y = N, \quad \chi_0 = w_0 + \cdots + w_n = w$$

under the identification of the (multiplicative) group of the characters of G_a with the (additive) group $\mathbb{Z}/k_a\mathbb{Z}$ (the character $t \mapsto t^m$ corresponds to the class of m modulo $k_a\mathbb{Z}$, denoted again by m!).

Note that the germ $(\mathbf{P}, a)$ (resp. (V, a)) can be identified to $(H/G_a, a)$ (resp. $(Y/G_a, a)$) and hence the latter is a *hyperquotient singularity* in the sense of Reid [19]. It follows that $\Omega^{\bullet}_{\mathbf{P},a} \simeq \Omega^{\bullet G_a}_{\mathbb{C}^n,0}$ and $\Omega^{\bullet}_{\mathbf{P}}(*V)_a \simeq \Omega^{\bullet G_a}_{g_a,0}$ where $g_a(y) = f(a_0, y_1, \dots, y_n)$ is a local equation for (Y, a), compare with [24], Section 5.

Let $Z \subset V$ be the finite set corresponding to the singular locus $f^{-1}(0)_{\text{sing}}$. Then we have, (with exactly the same proof) the next analog of Theorem (3.9):

(3.17) THEOREM. *The restriction map* $\rho\colon \Omega^{\bullet}_{\mathbf{P}}(*V) \to \Omega^{\bullet}_{\mathbf{P}}(*V)|_Z$ *induces a morphism* $\rho_r\colon E_r(f)_0 \to \oplus_{a \in Z} E_r(g_a, a)^{G_a}$ *of spectral sequences such that at the* E_1*-level* $\rho_1^{s,t}$ *is an isomorphism for* $s < 0$.

As an application we derive now *new formulas* for the Euler characteristics $\chi(V)$ and $\chi(F)$ similar to (3.12), (3.13). Our result should be compared to the more explicit formulas of Siersma [22] (obtained in the very special case when $f^{-1}(0)_{\text{sing}}$ is a complete intersection and all the transversal singularities are of type A_1) and, on the other hand, to the very general formulas of Yomdin [28] (which involve some numerical invariants defined topologically and hence difficult to compute in general concrete cases).

Consider the Poincaré series

$$P(t) = ((1 - t^{N-w_0}) \cdots (1 - t^{N-w_n}))/((1 - t^{w_0}) \cdots (1 - t^{w_n})) = \sum_{k \geqslant 0} c_k(\mathbf{w}, N)t^k$$

associated to the weighted homogenity type $(\mathbf{w}, N)$. Define next the *virtual Euler characteristics of order m* of V and F by the formulas:

$$\chi_m(V(\mathbf{w}, N)) = n + (-1)^{n-1} \sum_{s=1,m} c_{sN-w}(\mathbf{w}, N)$$

$$\chi_m(F(\mathbf{w}, N)) = 1 + (-1)^n \sum_{s=1,mN-w} c_s(\mathbf{w}, N) \tag{3.18}$$

where $w = w_0 + \cdots + w_n$.

Note that if there is a weighted homogeneous polynomial of type $(\mathbf{w}, N)$ having an isolated singularity at the origin and if V' (resp. F') denotes the corresponding hypesurface in $\mathbf{P}$ (resp. Milnor fiber) then

$$\chi_m(V(\mathbf{w}, N)) = \chi(V') \quad \text{for} \quad m \geqslant n$$

(resp. $\chi_m(F(\mathbf{w}, N)) = \chi(F')$ for $m \geqslant n + 1$). To see this you may find useful to read first the proof of (3.19. ii) below!

(3.19) PROPOSITION. (i) *Assume that a polynomial f' as above exists. Then*

$$\chi(V) = \chi(V') + (-1)^n \sum_{a \in Z} \dim M(g_a)^{-w}$$

$$\chi(F) = \chi(F') + (-1)^{n+1} \sum_{a \in Z} \sum_{j=1,N} \dim M(g_a)^{-N-w+j}.$$

(ii) *Assume that any transversal singularity $g_a = 0$ for $a \in Z$ is either weighted homogeneous or satisfies the assumptions in* (3.7). *Then*

$$\chi(V) = \chi_m(V(\mathbf{w}, N)) + (-1)^n \sum_{a \in Z} \dim M(g_a)^{mN-w}$$

$$\chi(F) = \chi_m(F(\mathbf{w}, N)) + (-1)^{n+1} \sum_{a \in Z} \sum_{j=1,N} \dim M(g_a)^{(m-1)N-w+j}$$

for all m large enough. When all the singularities g_a are weighted homogeneous, it is enough to take $m \geqslant n + 1$.

Proof. On a *formal level*, note that the formulas in (i) are a special case of the formulas in (ii), obtained by taking m divisible by all $k_a = |G_a|, a \in Z$. The proof of (i) is purely topological and independent of our previous results. Let $a, H, \ldots$, be as above. We may take f' close enough to f such that for all $a \in Z$ the intersection $F_a = B_\varepsilon \cap (f' \circ \varphi)^{-1}(0)$ can be identified to the Milnor fiber of the singularity (Y, a). Note also that F_a is G_a-invariant. Let $B'(a)$ be the image of the small ball B_ε

under the natural projection $(\mathbb{C}^n, 0) \to (\mathbf{P}, a)$. Then there is a homeomorphism $V \backslash \mathring{B}' \simeq V' \backslash \mathring{B}'$ where $B' = \bigcup_{a \in Z} B'(a)$. Moreover $B'(a) \cap V$ is contractible, while $B'(a) \cap V'$ can be identified to F_a/G_a and hence has middle Betti number

$$b_{n-1}(F_a/G_a) = \dim H^{n-1}(F_a)^G = \dim M(g_a)^{-w}$$

as in [27]. Then a Mayer-Vietories argument gives the result for $\chi(V)$. The result for $\chi(F)$ then follows from that for $\chi(V)$ as in the proof of (3.13).

To prove (ii) we use basically the same argument as in the proof of (3.10) (if necessary starting the computations with the E_3-term of the spectral sequence $E_r(f)_0$) together with (3.17) and (3.16). First we express the sum φ from the proof of (3.10) in terms of the Poincaré series $P(t)$. Following Siersma [22], we define a *new grading* on $\Omega^{\bullet}$ by setting for a homogeneous p-form $\omega \in \Omega^p$:

$$\deg \omega = (n - p + 1)N - w + |\omega|$$

where $|\omega|$ denote the degree of ω as defined in our introduction. Then multiplication by $\mathrm{d}f$ becomes a map of degree 0 and one has

$$P(t) = \sum_{k=0,n+1} (-1)^{n+1-k} P(\Omega^k)(t)$$

where $P(\Omega^k)$ is the Poincaré series of Ω^k with respect to this new grading [22]. Then it is obvious that

$$\varphi = -\sum_{s=1,m} c_{sN-w}(\mathbf{w}, N).$$

To treat the case of transversal singularities covered by (3.7) one has to use the next isomorphisms of vector spaces, which are clear by the proof of (3.7):

$$\begin{aligned} E_1^{n-1-m,m}(g,0)^G + E_2^{n-2-m,m+1}(g,0)^G &= \ker(g)^{mN-w} + (g)^{(m+1)N-w} \\ = \ker(g)^{mN-w} + K^{mN-w} &= (\ker(g) + K)^{mN-w} = M(g)^{mN-w}. \end{aligned}$$

The case of singularities in (3.7.ii) can be treated similarly.

(3.20) EXAMPLE. The polynomial $f = x_0^{265} + x_0 x_1^{11} + x_0 x_2^8 + x_2 x_3^4$ has degree $N = 265$ with respect to the weights $\mathbf{w} = (1, 24, 33, 58)$. The singular set Z consists of one point, namely $a = (0, -1, 1, 0)$ with transversal singularity g_a of type A_3. The corresponding isotropy group G_a is $\mathbb{Z}/3\mathbb{Z}$ and acts on $M(g_a)$ such that $\dim M(g_a)^i = 1$ for any i. It is konwn that the Poincaré series $P(t)$ is a polynomial in this case, in spite of the fact that there is no isolated singularity f'

of this homogeneity type $(\mathbf{w}, N)$, see [0], p. 201. It follows that

$$\chi_m(F(\mathbf{w}, N)) = 1 - P(1) = -66515 \text{ for } m \gg 0 \text{ and}$$
$$\chi(F) = \chi_m(F(\mathbf{w}, N)) + 265 = -66250.$$

Using a computer to determine the coefficients of $P(t)$, one gets $\chi(V) = 254$.

(3.21) REMARKS. (i) We conjecture that the formulas (3.19.ii) hold for *any* transversal singularities.

(ii) If all the transversal singularities (Y, a) for $a \in Z$ have links which are $\mathbb{Q}$-homology spheres, then the hypersurface V is a $\mathbb{Q}$-homology manifold and hence satisfies the Poincaré duality over $\mathbb{Q}$. In this case $b_n(V) = b_n(\mathbf{P}^{n-1})$ and the remaining interesting Betti number $b_{n-1}(V)$ can be determined from $\chi(V)$ once this Euler characteristic is known.

For *concrete computations* it is useful to use the following general remark. Assume that $f_1, \ldots, f_n$ is a regular sequence in S(this can be always achieved by a linear change of coordinates in the homogeneous case!). Then the Koszul complex $K^\bullet$ (1.9) is quasi-isomorphic to the complex

$$0 \to Q_1(f) \xrightarrow{f_0} Q_1(f) \to 0 \tag{3.22}$$

where $Q_1(f) = S/(f_1, \ldots, f_n)$ and f_0 denotes multiplication by f_0. An indication of the dimensions of $H^{n+1}(K^\bullet)_k \simeq Q(f)_{k-n-1}$ and $H^n(K^\bullet)_k \simeq \ker(f_0)_{k-n}$ can be obtained from the exact sequence

$$0 \to \ker(f_0)_k \to Q_1(f)_k \xrightarrow{f_0} Q_1(f)_{k+N-1} \to Q(f)_{k+N-1} \to 0 \tag{3.23}$$

since the Poincaré series of $Q_1(f)$ is known.

4. Explicit computations

(4.1) EXAMPLE (Computation of $H^1(U)$).

Let $f = f_1^{a_1} \ldots f_k^{a_k}$ be the decomposition of f in distinct irreducible factors. Then it is known that $b_1(U) = b_{2n-2}^0(V) = k - 1$ and it is easy to check that the closed forms

$$\omega_i = (\mathrm{d}f_i)/(f_i) - (N_i/N(\mathrm{d}f)/(f)$$

where $N_i = \deg(f_i), i = 1, \ldots, k$ generate $H^1(U)$ with only one relation: $\Sigma a_i \omega_i = 0$ Compare to (2.7.ii).

(4.2) EXAMPLE (with isolated singularities for V).

Let $f = xyz(x + y + z)$, $n = 2$. Then V consists of 4 lines in general position in $\mathbf{P}^2$ and its topology is simple to describe. However, the dimensions of the eigenspaces $H^{\cdot}(F)_a$ are more subtle invariants.

First we compute explicit bases for the homogeneous components of $Q(f)$:

$$Q(f)_0 = \langle 1 \rangle, Q(f)_1 = \langle x, y, z \rangle, Q(f)_2 = \langle x^2, y^2, z^2, xy, yz, zx \rangle$$

$$Q(f)_3 = \langle x^3, y^3, z^3, x^2y, y^2z, z^2x, xyz \rangle \quad \text{and}$$

$$Q(f)_k = \langle x^k, y^k, z^k, x^{k-1}y, y^{k-1}z, z^{k-1}x \rangle \quad \text{for } k \geqslant 4.$$

Then we look for the elements in $H^2(K^{\cdot})$ and define:

$$\omega_{xy} = x(x + 2y + z)\,\mathrm{d}y \wedge \mathrm{d}z + y(2x + y + z)\,\mathrm{d}x \wedge \mathrm{d}z$$

and ω_{yz}, ω_{zx} by cyclic symmetry.

Then $\mathrm{d}f \wedge \omega_{xy} = \mathrm{d}f \wedge \omega_{yz} = \mathrm{d}f \wedge \omega_{zx} = 0$ and these three forms give a basis for $H^2(K^{\cdot})_4$.

The six forms $x\omega_{xy}, y\omega_{xy}, y\omega_{yz}, z\omega_{yz}, z\omega_{zx}, x\omega_{zx}$ generate $H^2(K^{\cdot})_5$ with one relation among them (their sum is trivial).

And the six forms $x^k\omega_{xy}, y^k\omega_{xy}, \ldots$, form a basis for $H^2(K^{\cdot})_{k+4}$ for any $k \geqslant 2$.

It is now easy to compute $d_k^1 : H^2(K^{\cdot})_k \to H^3(K^{\cdot})_k$ and the nontrivial kernels and cokernels are listed below together with $E_2^{0,0}(f)_0$:

$$E_2^{0,0}(f)_0 = E_2^{0,2}(f)_2 = E_2^{0,2}(f)_3 = E_2^{1,1}(f)_1 = \mathbb{C},$$

$$E_2^{0,1}(f)_0 = E_2^{1,1}(f)_0 = \mathbb{C}^3.$$

The computations also show that the spectral sequence degenerates at E_2 and hence we get the complete results. One can restate them by saying that the monodromy operator h^* acts trivially on $H^0(F) = \mathbb{C}, H^1(F) = \mathbb{C}^3$ and its action on $H^2(F) = \mathbb{C}^6$ has characteristic polynomial $(t - 1)^3(t + 1)(t^2 + 1)$.

(4.3) EXAMPLE (with nonisolated singularities for V).

An irreducible cubic surface in $\mathbf{P}^3$ with nonisolated singularities is projectively equivalent to one of the next normal forms [3]

(i) a cone on the nodal cubic curve;
(ii) a cone on the cuspidal cubic curve;
(iii) $S' : x^2z + y^2t = 0$;
(iv) $S : x^2z + y^3 + xyt = 0$.

The topology of the surfaces (i)–(iii) can be described easier e.g. using [18], so that we concentrate on the last case: $f=x^2z+y^3+xyt$. The homogeneous components of $Q(f)$ are given by

$$Q(f)_0 = \langle 1\rangle \quad \text{and} \quad Q(f)_k = \langle z,t\rangle_k + \langle z^{k-1}x, z^{k-1}y, z^{k-2}y^2\rangle \quad \text{for } k \geqslant 1,$$

where $\langle z,t\rangle_k$ denotes the vector space of all homogeneous polynomials in z,t of degree k. Hence $\dim Q(f)_k = k+4$ for $k \geqslant 2$. Consider now the differential forms:

$$\begin{aligned}
\omega_1 &= x\,dx \wedge dy \wedge dz + y\,dx \wedge dy \wedge dt,\\
\omega_2 &= x\,dx \wedge dz \wedge dt + t\,dx \wedge dy \wedge dt - 3y\,dx \wedge dy \wedge dz,\\
\omega_3 &= t\,dx \wedge dy \wedge dz - 2z\,dx \wedge dy \wedge dt + x\,dy \wedge dz \wedge dt,
\end{aligned}$$

Then some tedious computations show that:

$$\begin{aligned}
H^3(K^\cdot)_4 &= \langle \omega_1, \omega_2, \omega_3\rangle,\\
H^3(K^\cdot)_5 &= \langle z,t\rangle_1\omega_2 + \langle z,t\rangle_1\omega_3 + \langle x\omega_2, y\omega_2, y\omega_3\rangle,\\
H^3(K^\cdot)_{k+4} &= \langle z,t\rangle_k\omega_2 + \langle z,t\rangle_k\omega_3 + \langle z^{k-1}x, z^{k-1}y, z^{k-2}y^2\rangle\omega_3,
\end{aligned}$$

for $k \geqslant 2$. This last vector space has dimension $2k+5$. And similarly one gets

$$H^2(K^\cdot)_{k+4} = \langle z,\, t\rangle_k\omega \quad \text{with}$$

$$\omega = (6yz - t^2)\,dx \wedge dy - xt\,dx \wedge dz - yt\,dx \wedge dt - 3xy\,dy \wedge dz - 3y^2\,dy \wedge dt.$$

After these complicated formulas it comes as a surprise that the spectral sequence $E_r(f)$ degenerates at E_2 and the only nonzero terms are $E_\infty^{0,0}(f)_0 = E_\infty^{0,2}(f)_1 = E_\infty^{0,2}(f)_2 = \mathbb{C}$.

It follows that $H^\cdot(S) \simeq H^\cdot(\mathbf{P}^2)$ and hence S has the same *rational homotopy type* as $\mathbf{P}^2$, according to Berceanu [1], who has proved that a projective complete intersection (with arbitrary singularities) is an intrinsically formal space.

Concerning the Milnor fiber one has $H^0(F) = \mathbb{C}$ with trivial action of h^*, $H^2(F) = \mathbb{C}^2$ with the characteristic polynomial of h^* equal to t^2+t+1 and $H^1(F) = H^3(F) = 0$.

Our next result is an improvement of Corollary (3.11) in Scherk [20] (to see the connexion between these two results have a look at the exact sequences (1.3) in [20]!).

Let $(Y,0)$ be an isolated hypersurface singularity given by $g=0$ in $\mathbb{C}^n$. We define the *μ-constant determinacy order* of $(Y,0)$ (denote by $\mu-\det(Y,0)$) to be the smallest integer $s>0$ such that the family $g^t = g+th$ $(t\in[0,1])$ is μ-constant for any $h\in(y_1,\dots,y_n)^s$ with small enough coefficients. Note that μ-$\det(Y,0)$ can be easily computed for large classes of singularities (e.g. weighted homogeneous or Newton nondegenerate singularities) and is always less or equal to the strongly $\mathscr{K}$-determinancy order $O(g)$ see [7], p. 75. In Scherk's notation, one has.

$$S = \min\{j; (y_1, \dots, y_n)^{j+1} \subset (g, g_1, \dots, g_n)\} \leqslant O(g) - 2.$$

(4.4) PROPOSITION. *Let* $V \subset \mathbf{P}^n$ *be a hypersurface having just one singular point* a *and such that* $N = \deg(V) > \mu\text{-det}(V, a)$.

Then $b_{n-1}(V) = b_{n-1}(V_0) - \mu(V, a)$ and $b_n(V) = b_n(V_0)$, *where* V_0 *is a smooth hypersurface in* $\mathbf{P}^n$ *with* $\deg(V_0) = N$.

Proof. Choose the coordinates on $\mathbf{P}^n$ such that $a = (1{:}0{:}\cdots{:}0)$ and $H\colon x_0 = 0$ is transversal to V. If $f = 0$ is an equation for V, then we set $g(y) = f(1, y_1, \dots, y_n) = g_2(y) + \cdots + g_N(y)$, with g_k a homogeneous polynomial of degree k. Using the assumptions, we can find a continuous family

$$g^t(y) = g_2^t(y) + \cdots + g_N^t(y) \quad \text{for } t \in [0, 1]$$

with the properties:

(i) $g^0 = g, g_k^t = g_k$ for $k < N - 1$;
(ii) For any $t > 0$, the hypersurfaces in $\mathbf{P}^{n-1}$

$$W_i^t\colon g_i^t = 0 \quad \text{for } i = N - 1, N$$

are smooth and intersect transversally;
(iii) g^t is a μ-constant family;
(iv) The projective hypersurfaces V^t with the affine equations $g^t = 0$ have no singularities except a.

According to [6], the cohomology of V^t is determined by a *lattice morphism*

$$\varphi^t\colon L_1^t \hookrightarrow L^t \to \bar{L}^t = L^t/\text{Rad}\, L^t$$

where L_1^t (resp. L^t) is the Milnor lattice of the singularity $g^t = 0$ (resp. $g_N^t = 0$). When t varies, these Milnor lattices are constant and hence the morphism φ^t has to be constant too.

Hence $H^{\boldsymbol{\cdot}}(V) = H^{\boldsymbol{\cdot}}(V^1)$ and so we can assume from the beginning that g_{N-1}, g_N satisfy the condition (ii).

Let $\varphi\colon L_1 \overset{i}{\hookrightarrow} L \to \bar{L}$ be the lattice morphism in this case. We have to show that $i(L^1) \cap \text{Rad}\, L = 0$, where i is the embedding of Milnor lattices arising from the small deformation $g^r(y) = g(r \cdot y) \cdot r^{-N}$ $(r \gg 0)$ of the singularity $g_N = 0$, see [6], proof of (1.2).

But we may think of g^r as being a even smaller deformation (of order r^{-2}) of the germ $g' = g_{N-1} \cdot r^{-1} + g_N$, which is a small deformation of g_N. If L' denotes the Milnor lattice of the singularity $g' = 0$, then the inclusion i above factorizes as $L_1 \hookrightarrow L' \overset{j}{\hookrightarrow} L$ and hence it is enough to show that

(v) $j(L') \cap \operatorname{Rad} L = 0$.

Now j is related to the cohomology of the hypersurface $V' \subset \mathbf{P}^n$ with the affine equation $g_{N-1} + g_N = 0$. Note that V' has just one singular point too, namely a. By a μ-constant argument as above, we can assume that

$$g_k(y) = y_1^k \cdots + y_n^k \quad \text{for } k = N-1, N.$$

Next (v) is equivalent to $H_0^n(V') = 0$ and we show this using the spectral sequence $E_r(f')$ for $f' = x_0 g_{N-1}(x_1, \dots, x_n) + g_N(x_1, \dots, x_n)$. It is enough to show that d_1 is injective. And this follows easily using the fact that a base for $H^n(K)$ is given by the forms $x_n^{a_0} \dots x_n^{a_n} \cdot \omega$ with $a_i < N-2$ for $i = 1, \dots, n$ and $\omega = \omega_1 \wedge \cdots \wedge \omega_n$, where the 1-forms

$$\omega_k = x_k \, dx_0 + [(N-1)x_0 + Nx_k] \, dx_k$$

are the obvious solution of the equation

$$df = \sum_{k=1,n} x_k^{N-2} \omega_k.$$

Compare to [22], [25], but note that here the transversal type is not A_1 for $N > 3$.

Note added in proof

The proof of Theorem (2.2) above contains an error on p. 11 lines 7 and 8. It is possible to repair this in some special cases, e.g. when all the singularities of V are isolated and weighted homogeneous. A more general result implying Theorem (2.2) has been proved by the author and P. Deligne (to whom I am very grateful for pointing out the above mentioned error!). For details, see our preprint "Hodge and order of the pole filtrations for singular hypersurfaces".

References

0. V.I. Arnold, S.M. Gusein-Zade, A.N. Varchenko, Singularities of Differentiable Maps, vol. I, *Monographs in Math.* 82, Boston – Basel – Stuttgart, Birkhauser 1985.
1. B. Berceanu, Formal algebraic varieties, unpublished manuscript.
2. E. Brieskorn, Die Monodromie der Isolierten Singularitaten von Hyperflachen. *Manuscr. Math.* 2 (1970) 103–161.

3. J.W. Bruce and C.T.C.Wall, On the classification of cubic surfaces, *J. London Math. Soc.* (2) 19, (1979) 245–256.
4. D. Burghelea and A. Verona, Local homological properties of analytic sets. *Manuscr. Math.* 7, (1972) 55–66.
5. P. Deligne, Theorie de Hodge II, III. *Publ. Math. IHES* 40, 5–58 (1971) and 44, (1974) 5 – 77.
6. A. Dimca, On the homology and cohomology of complete intersections with isolated singularities. *Compositio Math.* 58, (1986) 321–339.
7. A. Dimca, Topics on Real and Complex Singularities, Braunschweig-Wiesbaden: *Vieweg* 1987.
8. I. Dolgachev, Weighted projective varieties. In: Carrell, J.B. (ed) Group Actions and Vector Fields, Proceedings 1981. (*Lect. Notes Math.*, vol. 956, pp. 34–71) Berlin Heidelberg New York: Springer 1982.
9. H. Esnault, Fibre de Milnor d'un cone sur une courbe plane singuliere, *Invent. Math.* 68, (1982) 477–496.
10. A. Fujiki, Duality of mixed Hodge structures of algebraic varieties, *Publ. RIMS* Kyoto Univ. 16, (1980), 635–667.
11. G.-M. Greuel, Der Gauss-Manin-Zusammenhang isolierter Singularitaten von vollstandigen Durchschnitten. *Math. Ann.* 214, (1975) 235–266.
12. P. Griffiths, On the periods of certain rational integrals I, II, *Ann. Math.* 90, (1969) 460–541.
13. A. Grothendieck, On the de Rhan cohomology of algebraic varieties, *Publ. Math. IHRES* 29, (1966) 351–358.
14. H. Hamm, Ein Beispiel zur Berechnung der Picard-Lefschetz-Monodromie fur nichtisolierte Hyperflachensingularitaten *Mat. Ann.* 214, (1975) 221–234.
15. M. Kato, Y. Matsumoto, On the connectivity of the Milnor fiber of a holomorphic function at a critical point. In: Manifolds, Proceedings Tokyo 1973, pp. 131–136, Univ. of Tokyo Press 1975.
16. J. McClearly, User's Guide to Spectral Sequences, Publish or Perish: 1985.
17. M. Oka, On the homotopy types of hypersurfaces defined by weighted homogeneous polynomials. *Topology* 12, (1973) 19–32.
18. M. Oka, On the cohomology structure of projective varieties. In: Manifolds, Proceedings Tokyo 1973, pp. 137–143, Univ. of Tokyo Press 1975.
19. M. Reid, Young person's guide to canonical singularities, Proc. AMS summer Institute Bowdoin 1985, *Proc. Symp. Pure Math.* 46, AMS 1987.
20. J. Scherk, On the monodromy theorem for isolated hypersurface singularities, *Invent. math.* 58, (1980) 289–301.
21. J. Scherk, J.H.M. Steenbrink, On the mixed Hodge structure on the cohomology of the Milnor fiber. *Math. Ann.* 271, (1985) 641–665.
22. D. Siersma, Quasihomogeneous singularities with transversal type A_1, Contemporary Mathematics 90, *Amer. Math. Soc.* (1989) 261–294.
23. J.H.M. Steenbrink, Intersection form for quasihomogeneous singularities. *Compositio Math.* 34, (1977) 211–223.
24. J.H.M. Steenbrink, Mixed Hodge Structures and Singularities (book to appear).
25. D. van Straten, On the Betti numbers of the Milnor fiber of a certain class of hypersurface singularities. In: Greuel, G.-M., Trautmann, G. (eds) Singularities, Representations of Algebras and Vector Bundles, Proceedings, Lambrecht 1985 (Lect. Notes Math. Vol. 1273, pp. 203–220) Berlin Heidelberg New York: Springer 1987.
26. Z. Szafraniec, On the Euler characteristic of complex algebraic varieties, *Math. Ann.* 280, (1988) 177–183.
27. C.T.C. Wall, A note on symmetry of singularities, *Bull. London Math.* Soc. 12, 169–175 (1980).
28. I.N. Yomdin, Complex varieties with 1-dimensional singular locus, *Siberian Math. J.* 15, (1974) 1061–1082.

Compositio Mathematica **76**: 49–67, 1990.

Mixed Hodge structures on the intersection cohomology of links

ALAN H. DURFEE[1]* & MORIHIKO SAITO[2]**

[1]*Mount Holyoke College, South Hadley, MA USA-01075;* [2]*RIMS Kyoto University, Kyoto 606 Japan*

Received 4 November 1988; accepted in revised form 16 March 1990

Keywords: Mixed Hodge structures, links of singularities, intersection homology, mixed Hodge modules, semipurity, topology of algebraic varieties.

Abstract. The theory of mixed Hodge modules is applied to obtain results about the mixed Hodge structure on the intersection cohomology of a link of a subvariety in a complex algebraic variety. The main result, whose proof uses the purity of the intersection complex in terms of mixed Hodge modules, is a generalization of the semipurity theorem obtained by Gabber in the l-adic case. An application is made to the local topology of complex varieties.

Introduction

Let X be a complex algebraic variety, assumed irreducible and of dimension n, and let Z be a closed subvariety. This paper studies the mixed Hodge structure on the intersection cohomology of the link of Z in X, derives a semipurity result, and deduces some topological consequences. The mixed Hodge structure is obtained using the theory of mixed Hodge modules developed by the second author.

Although the concept of 'link' of Z in X is intuitively obvious, its precise meaning is unclear. In this paper, we will define it as the nearby level set of a suitable nonnegative real valued distance function which vanishes exactly on Z. If Z is compact, a reasonable concept of link results if the distance function is assumed to be real analytic. Another stronger, concept results if N is a neighborhood of Z in X and if there is a proper continuous retraction map r from ∂N to Z such that the closure of N is the total space of the mapping cylinder of r; the distance function is then the projection of the mapping cylinder to $[0, 1]$. For most of this paper, however, we will use a third, weaker type of link and distance function which combines topological and homological properties. In fact, the homological notions of link alone are enough for most of our results. These homological notions are functors which can be canonically defined in the derived category and fit well with the theory of mixed Hodge modules which we

* Partially supported by NSF grant DMS-8701328, the Max-Planck-Institut für Mathematik and the Università di Pisa.
** Partially supported by the Max-Planck-Institut für Mathematik and the Institut des Hautes Études Scientifiques.

will be using. The material on the various definitions of link is in the beginning of Section 2.

Whichever definition is used, a link L is an oriented topological pseudomanifold of (real) dimension $2n - 1$ with odd-dimensional strata. In particular, its Goresky-MacPherson middle-perversity intersection homology is defined. If L is a rational homology manifold, then intersection homology is ordinary homology. Furthermore, $N\backslash Z$ is a rational homology manifold if and only if L is.

This paper applies the theory of mixed Hodge modules to put a mixed Hodge structure on the intersection cohomology with rational coefficients of a link L. Various elementary properties of these groups are derived. The main result is as follows:

THEOREM 4.1. *If* $\dim X = n$ *and* $\dim Z \leqslant d$, *then* $IH_c^k(L)$ *has weights* $\leqslant k$ *for* $k < n - d$.

Duality then immediately shows that $IH^k(L)$ has weights $> k$ for $k \geqslant n + d$. The notation here is as follows: The intersection cohomology group $IH_c^k(L)$ is (topologically) the homology group of geometric $(2n - 1 - k)$-dimensional intersection chains with compact support, and $IH^k(L)$ is the similar group with closed support.

For varieties over finite fields and Z a point this result was proved by Gabber [Ga]. The result of Gabber is equivalent to the local purity of the intersection complex by definition and self duality; it implies the purity of intersection cohomology by Deligne's stability theorem for pure complexes under direct images for proper morphisms [De2]. This local purity also follows from the existence of the weight filtration on mixed perverse sheaves, since intersection complexes are simple [BDD 4.3.1].

For varieties over the complex numbers, Z a point and $X\backslash Z$ smooth, the above result for mixed Hodge structures on ordinary cohomology was deduced by several people [St1, El] using the characteristic 0 decomposition theorem. Later Steenbrink and Navarro found a more elementary proof using Hodge theory [St2, Na]. We show that Theorem 4.1 follows naturally from the second author's theory of mixed Hodge modules combined with the theory of gluing t-structures from [BBD]. The proof is in the spirit of the second proof of local purity in the l-adic case. (See the end of 1.4.)

This theorem is then used to show that certain products in the intersection homology of a link must vanish (Theorems 5.1 and 5.2). This result is a generalization of [DH], which treated the case where Z is a point and $X\backslash Z$ is smooth. For example, Theorem 5.2 implies that the five-torus $S^1 \times \cdots \times S^1$ is not a link of a compact curve Z in a three-fold X, and Theorem 5.1 implies that certain pseudomanifolds L cannot be links of points in a complex variety. The only previous result is this area is, we believe, one of Sullivan: If L is a link of a compact subvariety, then the Euler characteristic of L vanishes [Su]. Sullivan's proof is entirely topological and holds for any compact oriented pseudomanifold L with only odd-dimensional strata. The above examples are independent of this result, and hence provide new restrictions on the topology of complex algebraic varieties.

Although the results and proof of this paper are given in terms of mixed Hodge modules, they can actually be read in three settings:

The derived category of sheaves on X: On a first reading, this paper can be understood in the derived category $D(X)$ of sheaves of vector spaces over a field on X, together with the derived functors $f_*, f_!, \ldots$ (Following recent convention, we omit the R or L when referring to derived functors.) We use no more properties than those summarized in [GM2 Sect. 1]. Various properties of the intersection homology of links are stated and proved in this language. Of course no conclusions can be drawn about weights (Sect. 4) or the resulting corollaries on the topological structure of varieties (Sect. 5).

The category of mixed Hodge modules on X: The additional material we need about mixed Hodge modules is basically the same as the formalities of [BBD]. This material is summarized in the first section of this paper.

The category of mixed l-adic sheaves on a variety in characteristic p: Lastly, this paper can be read in the setting of [BBD], with the conclusions of Sect. 4 about weights for varieties in characteristic p. Of course, the concept of link as topological space makes no sense here, so the isomorphisms of 2.11 should be taken as the definition of cohomology, homology and intersection cohomology of 'link' in this case. However the applications of Sect. 5 to the topology of varieties can still be obtained by the methods of reduction modulo p as in [BBD Sect. 6].

Throughout the paper references are given for each of these three settings.

The first author wishes to thank the Max-Planck-Institut für Mathematik, Bonn, the Università di Pisa, and the Katholieke Universiteit, Nijmegen for their cordial hospitality during his sabbatical year when a first draft of this paper was written. The second author would like to thank the Max-Planck-Institut für Mathematik and the Institut des Hautes Études Scientifiques.

1. Background material

General references for the following material on intersection homology and derived functors are [GM2, Bo, GM3 §1, Iv]. All groups will be assumed to have the rational numbers as coefficients, unless otherwise indicated. Following the convention of [BBD], we will use the same symbol for a functor and its right or left derived functor.

1.1. Suppose that W is a topological pseudomanifold of dimension m with strata of even codimension. For example, W can be a complex algebraic variety or a link of a subvariety. The following groups (all with rational coefficients) can be associated to W:

$H^k(W)$, resp. $H_c^k(W)$: The kth cohomology group, resp. cohomology group with compact supports.

$H_k(W)$, resp. $H_k^{BM}(W)$: The kth homology group, resp. Borel-Moore homology group (homology with closed supports).

$IH_k(W)$, resp. $IH_k^{BM}(W)$: The kth intersection homology group with middle perversity k-dimensional chains with compact support, resp. closed support [GM2, GM3 1.2; Bo I]. Since W has even codimensional strata, the middle perversity is well defined.

$IH^k(W)$, resp. $IH_c^k(W)$: The kth intersection cohomology group, defined (topologically) as $IH_{m-k}^{BM}(W)$, resp. cohomology with compact supports, defined as $IH_{m-k}(W)$.

1.2. We also have the following:

$a_W: W \to pt$

$\mathbf{Q}_W = (a_W)^*\mathbf{Q}$ = the constant sheaf on W

$\mathbf{D}_W = (a_W)^!\mathbf{Q}$ = the dualizing sheaf on W

$\mathbf{IC}_W^{\mathrm{top}}$ = the intersection complex on W [GM2 2.1]

These complexes have the following properties:

$$H^k(W) = H^k(W, \mathbf{Q}_W) \text{ and } H_c^k(W) = H_c^k(W, \mathbf{Q}_W)$$

$$H_k(W) = H_c^{-k}(W, \mathbf{D}_W) \text{ and } H_k^{BM}(W) = H^{-k}(W, \mathbf{D}_W)$$

$$IH_k(W) = IH_c^{m-k}(W) = H_c^{-k}(W, \mathbf{IC}_W^{\mathrm{top}}) \text{ and}$$

$$IH_k^{BM}(W) = IH^{m-k}(W) = H^{-k}(W, \mathbf{IC}_W^{\mathrm{top}})$$

1.3. Let X be a complex algebraic variety (a reduced separated scheme of finite type over $\mathbf{C}$), assumed irreducible and of complex dimension n. A general reference for the following material is [BBD].

Let

$$\mathbf{IC}_X = \mathbf{IC}_X^{\mathrm{top}}[-n]$$

as objects of $D_c^b(X)$ as in [BBD]. If $j: U \to X$ is the inclusion of a smooth dense Zariski-open set, then

$$\mathbf{IC}_X = j_{!*}\mathbf{Q}_U[n].$$

We also have

$D_c^b(X) =$ The derived category whose objects are bounded complexes of sheaves of **Q**-modules with constructible cohomology.

$\operatorname{Perv}(X) =$ The full subcategory of perverse sheaves over **Q**.

1.4. General references for the theory of mixed Hodge modules are [Sa1, Sa2]. Many of their formal properties are similar to those of mixed perverse sheaves [BBD 5.1].

We have

$\mathrm{MHM}(X) =$ the abelian category of mixed Hodge modules on X

rat: $\mathrm{MHM}(X) \to \operatorname{Perv}(X)$, an additive, exact, faithful functor which assigns the underlying perverse sheaf over **Q**.

Functors f_*, $f_!$, f^*, $f^!$, $\mathbf{D}$, $\otimes$, $\boxtimes$ on $D^b\mathrm{MHM}(X)$ compatible with the corresponding derived functors on $D_c^b(X)$ and with the corresponding perverse functors on $\operatorname{Perv}(X)$ via

$$\begin{array}{ccccc} D^b\mathrm{MHM}(X) & \xrightarrow{\text{rat}} & D^b\operatorname{Perv}(X) & \xrightarrow{\text{real}} & D_c^b(X) \\ \downarrow{\scriptstyle H^0} & & \downarrow{\scriptstyle H^0} & \swarrow{\scriptstyle {}^p\underline{\mathbf{H}}^0} & \\ \mathrm{MHM}(X) & \xrightarrow{\text{rat}} & \operatorname{Perv}(X) & & \end{array}$$

where 'real' is an equivalence of categories [Be, BBD 3.1.10]. See also [BBD 1.3.6, 1.3.17(i), 3.1.14].

Adjoint relations for f^*, f_* and $f_!$, $f^!$, the natural morphism $f_! \to f_*$ and the usual relations $\mathbf{D}^2 = \mathrm{id}$, $\mathbf{D}f_* = f_!\mathbf{D}$ and $\mathbf{D}f^* = f^!\mathbf{D}$.

The fact that the category $\mathrm{MHM}(pt)$ is the category of polarizable mixed Hodge structures (over **Q**).

Let $\mathbf{Q}^H \in \mathrm{MHM}(pt)$ be the mixed Hodge structure of weight 0 and rational structure $\operatorname{rat}(\mathbf{Q}^H) = \mathbf{Q}$. In this language the cohomology of X has a mixed Hodge structure since we can write

$$H^{\cdot}(X) = H^{\cdot}((a_X)_*(a_X)^*\mathbf{Q}^H)$$

$$H_c^{\cdot}(X) = H^{\cdot}((a_X)_!(a_X)^*\mathbf{Q}^H)$$

and so forth. These mixed Hodge structures coincide with those of Deligne

[Del] at least if X is globally embeddable into a smooth variety (for example, if X is quasiprojective). We let

$$\mathbf{Q}_X^H = (a_X)^*\mathbf{Q}^H$$
$$\mathbf{D}_X^H = (a_X)^!\mathbf{Q}^H$$
$$\mathbf{IC}_X^H = j_{!*}\mathbf{Q}_U^H[n]$$

note that

$$\mathbf{DIC}_X^H = \mathbf{IC}_X^H(n)$$

where (n) denotes the Tate twist, which can be defined by $\boxtimes\mathbf{Q}^H(n)$.

As in the category of mixed complexes of l-adic sheaves, there is a weight filtration in MHM(X) and the notions of 'weight $\leqslant k$', etc., in D^bMHM(X) such that [BBD 5.1.8, 5.1.14, 5.3.2; Sal 4.5]:

In MHM(pt) these are the usual notions of mixed Hodge theory.

$f_!$, f^* preserve weight $\leqslant k$.

f_*, $f^!$ preserve weight $\geqslant k$.

$j_{!*}$ preserves weight $= k$.

In particular, this implies the (local) purity of the intersection complex $\mathbf{IC}_X^H$, since $\mathbf{Q}_U^H[n]$ is pure and $\mathbf{IC}_X^H = j_{!*}(\mathbf{Q}_U^H[n])$ for j as in 1.3. This is the same argument as in the l-adic case, which uses stability by direct images and subquotients. Note that local purity can also be proved using only the existence of the weight filtration, since the weight filtration of $\mathbf{IC}_X^H$ must be trivial by the simplicity of $\mathbf{IC}_X$ and the faithfulness and exactness of the forgetful functor rat.

In the l-adic case, the existence of the weight filtration is proved [BBD 5.3.5] after showing the purity of intersection complexes with twisted coefficients [BBD 5.3.2]; in fact, this existence is not used in the definition of 'weight $\leqslant k$', etc., nor in the proof of its stability by the functors as above. However, in the mixed Hodge module case, the existence of the weight filtration is more or less assumed from the beginning, since the Hodge filtration and the rational structure are not together strong enough to determine the weight filtration uniquely. The latter fact is of course even true for mixed Hodge structures.

2. Links

We define a link L of a subvariety Z in a variety X to be the level set of a suitable distance function d whose zero locus is Z. We put three kinds of conditions on this distance function, and consequently get three different kinds of links: a 'weak topological link', which will be used for the rest of this paper, an 'analytic link', which can be defined if Z is compact, and a 'topological link', which is the strongest kind of link. These links are stratified topological pseudomanifolds [GM2 1.1]. We expect that an analytic link is a topological link. We also expect that an analytic link is unique in some sense as stratified topological pseudomanifold.

Since we cannot expect these links to be in general unique, we also introduce functors which describe the cohomology, homology and intersection cohomology of a link intrinsically. These are analogous with the vanishing cycle functors [DE3]. The 'local link cohomology functor' assigns a complex of sheaves on Z to a complex of sheaves on $X\backslash Z$. We will only apply this functor to the constant sheaf, the dualizing complex and the intersection complex on $X\backslash Z$. The 'global link cohomology functors' are obtained by composing the local link cohomology functor with the direct image (with and without compact supports) of the map of Z to a point. These link cohomology functors are intrinsically defined. Furthermore, they are naturally related to mixed theories (Hodge or l-adic). The connection between the topological types of link and the cohomological types of link is that a topological link determines a functor which is canonically isomorphic to the local link cohomology functor, and that a weak topological link correspondingly determines a functor canonically isomorphic to the global link cohomology functor.

2.1. Let X be an irreducible complex algebraic variety of dimension n, and Z a closed subvariety. Let

$$i: Z \hookrightarrow X,$$

$$j: U = X\backslash Z \hookrightarrow X.$$

In each of the following three definitions, there will be an open neighborhood N of Z in X and a *distance function*

$$d: N \to [0, 1)$$

with $Z = d^{-1}(0)$. Let N^* be defined as $N\backslash Z$ with stratification induced by a complex analytic Whitney stratification of X.

The link L will be of (real) dimension $2n-1$, have odd-dimensional strata, and have an orientation induced from that of X. We will have

$$L = d^{-1}(\varepsilon)$$

for $0 < \varepsilon$ suitably small. Let

$$k\colon L \hookrightarrow U$$

be the inclusion whose image is $d^{-1}(\varepsilon)$. Also, N^* will be a rational homology manifold if and only if L is.

2.2. A stratified topological pseudomanifold L is a *weak topological link* if there exist an open neighborhood N of Z in X and a continuous distance function $d\colon N \to [0, 1)$ such that $Z = d^{-1}(0)$ and such that the following conditions are satisfied:

(i) There is a stratified homeomorphism $\alpha\colon N^* \simeq L \times (0, 1)$ such that d is identified with the second projection.
(ii) For any F in $D_c^b(N^*)$ such that $\underline{\mathbf{H}}^i F$ is locally constant on each stratum of N^*, the natural morphisms $(\underline{\mathbf{H}}^{\cdot} d_* j_* F)_0 \to H^{\cdot}(Z, i^* j_* F)$ are isomorphisms.

Note that condition (ii) is always satisfied if d is proper.

2.3. Suppose Z is compact. A stratified topological pseudomanifold L is an *analytic link* of Z in X, if there exists a nonnegative real analytic distance function $d\colon N \to [0, 1)$ where N is an open neighborhood of Z in X, such that $Z = d^{-1}(0)$ and L is isomorphic to $d^{-1}(\varepsilon)$ as stratified space (up to a refinement of stratification) for $0 < \varepsilon \ll 1$. (The stratification of $d^{-1}(\varepsilon)$ is induced by a complex Whitney stratification of X compatible with Z).

We may assume that d is proper over its image $[0, \delta)$ by taking two relatively compact open neighborhoods U_1, U_2 of Z in N such that $\bar{U}_1 \subset U_2$, and restricting d to $U_1 \backslash d^{-1}d(\overline{U_2 \backslash U_1})$.

When X is quasiprojective, analytic links can be shown to exist by arguments similar to [Du].

2.4. PROPOSITION. *An analytic link is a weak topological link.*

Proof. Condition (ii) is satisfied by the above remark. By the curve selection lemma we may assume that the restriction of d to the strata disjoint with Z is smooth. Thus, by the Thom isotopy theorem, the restriction of the Whitney stratification to $d^{-1}(\varepsilon)$ is well-defined and $N \backslash Z$ is isomorphic to $d^{-1}(\varepsilon) \times (0, \delta)$ as stratified space (See [Du]). Thus condition (i) is checked. □

2.5. A stratified topological pseudomanifold L is a *topological link* of Z in X, if there exist a continuous proper surjective map (the *retraction* of L onto Z)

$$r\colon L \to Z$$

and a homeomorphism

$$\alpha\colon N \to \tilde{N}$$

of an open neighborhood N of Z in X with an open neighborhood $\tilde{N}$ of Z in the open mapping cylinder $\mathrm{Cyl}^\circ(r)$ of r which is the identity on Z and induces a stratification preserving homeomorphism $N\backslash Z \simeq \tilde{N}\backslash Z$. Here $\mathrm{Cyl}^\circ(r)$ is defined as $Z \amalg L \times [0, 1)/r(x) \sim (x, 0)$, and the stratifications of $N\backslash Z$ and $\tilde{N}\backslash Z$ in $L \times (0, 1)$ are induced by a complex analytic Whitney stratification of X (defined on a neighborhood of Z) compatible with Z and the stratification of L, respectively.

This definition is inspired by [De3]. In this case we define the distance function d from Z to be the composition of α with the natural projection $\mathrm{Cyl}^\circ(r) \to [0, 1) \subset \mathbf{R}$. We define the retraction $\tilde{r}$ of N onto Z to be the composition of α with the natural morphism $\mathrm{Cyl}^\circ(r) \to Z$.

In the definition of topological link, we may assume that $\tilde{N} = \mathrm{Cyl}^\circ(r)$ and that ∂N is identified with L by replacing d with $(d/g)\circ r$ and N with the subset $\alpha^{-1}[Z \amalg \{(x, a) \in L \times (0, 1)\colon a < g(r(x))\}]$, where g is a continuous function on Z whose value is in $(0, 1]$ and goes to zero appropriately at infinity. Then $r = \tilde{r}k$ where k is as in 2.1.

If $X\backslash Z$ is smooth, the existence of topological links can be shown by reducing to the case where X is smooth and Z is a divisor with normal crossings. The existence of topological links in general will be treated elsewhere.

2.6. PROPOSITION. *A topological link is a weak topological link.*

Proof. We may assume $\tilde{N} = \mathrm{Cyl}^\circ(r)$ as above. Then the condition (i) is satisfied. For (ii) we have

$$(\underline{\mathbf{H}}^\cdot d_* j_* F)_0 = \varinjlim H^\cdot(d^{-1}(0, \delta), F)$$

and

$$H^\cdot(Z, i^* j_* F) = \varinjlim H^\cdot(V\backslash Z, F)$$

where V runs over open neighborhoods of Z in X. Since every V contains an open neighborhood V_g corresponding by α to $Z \amalg \{(x, a) \in L \times (0, 1)\colon a < g(r(x))\}$ for some continuous function g, and $H^\cdot(V_g\backslash Z, F)$ is independent of g by the hypothesis on F, we get the assertion. □

When Z is compact we expect that an analytic link is a topological link. The argument is similar to the proof of the Thom isotopy theorem, but also the limit of integral curves must be controlled using a good partition of unity. This problem will be treated elsewhere.

2.7. Assume that the stratifications of the objects in $D_c^b(U)$ and $D_c^b(Z)$ are algebraic. In the notation of 2.1, the *local link cohomology functor* Λ_Z of Z in X is defined for $F \in D_c^b(U)$ by

$$\Lambda_Z F = i^* j_* F$$

as an object of $D_c^b(Z)$. We also let

$$\Lambda'_Z F = i^! j_! F.$$

Note that the canonical isomorphism of 3.1 implies that

$$\Lambda'_Z F[+1] = \Lambda_Z F$$

and that there also is the duality isomorphism

$$\mathbf{D}\Lambda_Z = \Lambda'_Z \mathbf{D}$$

We will take F to be $\mathbf{Q}_U$, $\mathbf{D}_U$ and $\mathbf{IC}_U$ in the applications of this paper.

2.8. With the above notation, the *global link cohomology functors* of Z in X are defined by

$$(a_Z)_* \Lambda_Z F \quad \text{and} \quad (a_Z)_! \Lambda_Z F$$

for F as above, where $(a_Z)_*\colon Z \to pt$. Note that

$$\mathbf{D}(a_Z)_* \Lambda_Z F = (a_Z)_! \Lambda'_Z \mathbf{D} F.$$

2.9. PROPOSITION. *If L is a topological link of Z in X and if F is in $D_c^b(U)$ and $\underline{\mathbf{H}}^i F$ is locally constant on each stratum of the stratification of X in 2.5, then in $D^b(Z)$ there is a natural functorial isomorphism*

$$\Lambda_Z F = r_* k^* F.$$

Proof. Let $\tilde{r}'$ be the restriction of $\tilde{r}$ to N^*. We have natural morphisms of $\tilde{r}'_* F$ to $i^* j_* F$ and $r_* k^* F$, and since r is proper it is easy to check that they are quasi-isomorphisms. □

2.10. PROPOSITION. *If L is a weak topological link of Z in X, if F is in $D_c^b(U)$ and if $\underline{\mathbf{H}}^i F$ is locally constant on each stratum of the stratification of X in 2.2(i), then there are natural functorial isomorphisms in $D^b(pt)$*

$$(a_Z)_* \Lambda_Z F = (a_L)_* k^* F \quad \text{and} \quad (a_Z)_! \Lambda_Z F = (a_L)_! k^* F$$

Proof. The natural morphism $k^! \mathbf{Q}_U \otimes k^* F \to k^! F$ gives a natural isomorphism $k^* F[-1] = k^! F$. So by duality it is enough to show the first isomorphism. There is a commutative diagram

$$\begin{array}{ccccccc}
Z & \xrightarrow{i} & N & \xleftarrow{j} & N^* & \xleftarrow{k} & L \\
\downarrow{\scriptstyle a_Z} & & \downarrow{\scriptstyle d} & & \downarrow{\scriptstyle d'} & & \downarrow{\scriptstyle a_L} \\
0 & \xrightarrow{i'} & [0,1) & \xleftarrow{j'} & (0,1) & \xleftarrow{k'} & \varepsilon
\end{array}$$

By 2.2(ii) and the commutativity of the diagram we have

$$(a_Z)_*\Lambda_Z F = i'^* d_* j_* F = i'^* j'_* d'_* F.$$

By 2.2(i), $\underline{\mathbf{H}}^{\cdot} d'_* F$ is constant on $(0, 1)$, and the natural morphisms

$$i'^* j'_* d'_* F \leftarrow \mathbf{R}\Gamma([0, 1), j'_* d'_* F) = \mathbf{R}\Gamma((0, 1), d'_* F) \rightarrow (a_L)_* K^* F$$

are quasi-isomorphisms. □

If L is a topological link, the isomorphisms of 2.10 coincide with the direct images of the isomorphisms of 2.9.

2.11. PROPOSITION. *If L is a weak topological link, then*

(i) $H^k(L) = H^k((a_Z)_*\Lambda_Z \mathbf{Q}_U)$
(ii) $H_k(L) = H^{-k}((a_Z)_!\Lambda'_Z \mathbf{D}_U)$
(iii) $IH^{n+k}(L) = H^k((a_Z)_*\Lambda_Z \mathbf{IC}_U) = H^{k+1}((a_Z)_*\Lambda'_Z \mathbf{IC}_U)$

and the same for $H^k_c(L)$, $H^{BM}_k(L)$ *and* $IH^{n+k}_c(L)$ *with* $(a_Z)_*$ *and* $(a_Z)_!$ *exchanged.*

Proof. This follows immediately from 2.10. For the isomorphism (i), let $F = \mathbf{Q}_U$. For (ii), let $F = \mathbf{D}_U$ and use $k^!\mathbf{D}_U = \mathbf{D}_L$ [Bo V10.11]. For (iii), let $F = \mathbf{IC}^{\text{top}}_U$ and use $k^!\mathbf{IC}^{\text{top}}_U = \mathbf{IC}^{\text{top}}_L$ [GM2 5.4.1] and $\mathbf{IC}^{\text{top}}_U = \mathbf{IC}_U[n]$. The proofs in the other cases are similar. □

3. Mixed Hodge structures

In this section we derive various elementary properties for the mixed Hodge structure on the cohomology, homology and intersection homology of a link. Throughout the section, the notation is that of 2.1.

By the general theory of mixed Hodge modules and making the obvious modifications (replacing $\mathbf{Q}_U$ by $\mathbf{Q}^H_U$ and so forth), the right hand side of the six isomorphisms in 2.11 have mixed Hodge structures. We define the mixed Hodge structures on the cohomology, homology and intersection cohomology groups associated to a weak topological link by these isomorphisms. Various maps and operations on the right-hand groups are compatible with the corresponding operations on a weak topological link by [GM II]. Note that the results of this section and the following sections are actually proved for the groups on right-hand of 2.11. We define the mixed Hodge structures on the intersection homology groups by setting

$$IH_k(L) = IH_c^{2n-1-k}(L)(n),$$
$$IH_k^{BM}(L) = IH^{2n-1-k}(L)(n)$$

If U is smooth, these mixed Hodge structures have already been found by the methods of Deligne [El, St1, Na]. The mixed Hodge structures here agree with these, because $K := C(\Omega^{\cdot}_{\tilde{X}} \to \varepsilon_*\Omega^{\cdot}_{\tilde{D}.} \oplus \Omega_{\tilde{X}}(\log D))$ together with natural Hodge and weight filtrations is isomorphic to the filtered de Rham complex of the underlying filtered complex of $\mathscr{D}$-Modules of $C(\tilde{j}_!\mathbf{Q}_U^H \to \tilde{j}_*\mathbf{Q}_U^H)$ in a bifiltered derived category, where $\pi: (\tilde{X}, D) \to (X, Z)$ is an embedded resolution with $\tilde{j}: U = X\backslash Z \to \tilde{X}$, and $\varepsilon: \tilde{D}_{\bullet} \to \tilde{X}$ the semisimplicial space defined by the intersections of the irreducible components of D. In fact (K, F) is a filtered differential complex with weight filtration W as in [Sa3 2.5.8], and it can be checked that the filtered $\mathscr{D}$-Modules associated to $K_1 := C(\Omega^{\cdot}_{\tilde{X}} \to \varepsilon_*\Omega^{\cdot}_{\tilde{D}.})[-1]$ and $K_2 := \Omega_{\tilde{X}}(\log D)$ (together with the weight filtration) are canonically isomorphic to the underlying filtered $\mathscr{D}$-modules of $\tilde{j}_!\mathbf{Q}_U^H$ and $\tilde{j}_*\mathbf{Q}_U^H$ respectively by the calculations as in [Sa2 §3]. Moreover the morphism $K_1 \to K_2$ induced by the natural morphism $\Omega^{\cdot}_{\tilde{X}} \to \Omega_{\tilde{X}}(\log D)$ corresponds to the natural morphism $\tilde{j}_!\mathbf{Q}_U^H \to \tilde{j}_*\mathbf{Q}_U^H$ by the inverse of the de Rham functor [Sa3 §2], which is the identity on U.

3.1. LEMMA. *In the notation of* 2.1:

(i) *For* $F \in D^b\mathrm{MHM}(U)$ *there are isomorphisms in* $D^b\mathrm{MHM}(X)$

$$i_*i^*j_*F = \mathrm{cone}(j_!F \to j_*F) = i_!i^!j_!F[+1]$$

(ii) *For* $G \in D^b\mathrm{MHM}(X)$ *there are isomorphisms in* $D^b\mathrm{MHM}(Z)$

$$i^*j_*j^*G = \mathrm{cone}(i^!G \to i^*G) = i^!j_!j^!G[+1]$$

Proof. (i) For $G \in D^b\mathrm{MHM}(X)$ there is a distinguished triangle [Sa1 4.4.1, BBD 1.4.34, GM2 1.11]

$$j_!j^!G \to G \to i_*i^*G.$$

Letting $G = j_*F$ gives

$$j_!F \to j_*F \to i_*i^*j_*F$$

which gives the first isomorphism of (i). The dual of this triangle with $\mathbf{D}F$ replacing F is

$$i_! i^! j_! F \to j_! F \to j_* F$$

which shows the second isomorphism of (i).

(ii) The dual of the first triangle in the proof of (i) is

$$i_* i^! G \to G \to j_* j^* G.$$

Taking i^* gives

$$i^! G \to i^* G \to i^* j_* j^* G$$

which shows the first isomorphism of (ii).

The second follows by duality as above. □

3.2. PROPOSITION. *If L is a weak topological link of Z in X, the natural morphisms*

$$H^k(L) \to IH^k(L) = IH^{BM}_{2n-1-k}(L)(-n) \to H^{BM}_{2n-1-k}(L)(-n)$$

(and similarly with compact supports) are morphisms of mixed Hodge structures. These morphisms are isomorphisms if $X \backslash Z$ is a rational homology manifold in a neighborhood of Z.

Proof. There is a morphism in $D^b\mathrm{MHM}(U)$ [Sa 4.5.8; GM2 5.1]

$$\mathbf{Q}^H_U[n] \to \mathbf{IC}^H_U.$$

The dual of this is

$$\mathbf{IC}^H_U(n) \to \mathbf{D}^H_U[-n]$$

which combines with the first morphism to give

$$\mathbf{Q}^H_U[n] \to \mathbf{IC}^H_U \to \mathbf{D}^H_U(-n)[-n].$$

Taking $i^* j_*$ and $H^{k-n}(Z, -)$ gives

$$H^k(Z, i^* j_* \mathbf{Q}^H_U) \to H^{k-n}(Z, i^* j_* \mathbf{IC}^H_U) \to H^{k-2n}(Z, i^* j_* \mathbf{D}^H_U)(-n)$$

so we get the first assertion. The statement with compact supports is similar. The last assertion is well-known, and follows immediately from the definition of intersection complex [GM2]. □

3.3. PROPOSITION. *If* L *is a weak topological link of* Z *in* X, *the duality isomorphism* [Bo p. 149]

$$IH^k(L) \simeq [IH_c^{2n-1-k}(L)(n)]^*$$

is an isomorphism of mixed Hodge structures.

Proof. In $D^b\mathrm{MHM}(pt)$

$$\mathbf{D}(a_X)_* i_! i^! j_! \mathbf{IC}_U^H = (a_X)_! i_* i^* j_* \mathbf{IC}_U^H(n) = (a_X)_! i_! i^! j_! \mathbf{IC}_U^H(n)[+1]$$

where the second isomorphism is by 3.1(i). Taking cohomology, we get the proposition. □

3.4. PROPOSITION. *If* L *is a weak topological link of* Z *in* X, *the natural intersection pairings* [GM2 5.2]

$$H^p(L) \otimes H^q(L) \to H^{p+q}(L)$$

$$H^p(L) \otimes IH^q(L) \to IH^{p+q}(L)$$

$$H^p(L) \otimes H_q^{BM}(L) \to H_{q-p}^{BM}(L)$$

$$IH^p(L) \otimes IH^q(L) \to H_{2n-1-p-q}^{BM}(L)(-n)$$

(and similarly with compact supports) are morphisms of mixed Hodge structures.

Proof. Let $\Delta\colon X \to X \times X$ be the diagonal embedding. We have the natural morphisms

$$\mathbf{Q}_X^H \otimes F := \Delta^*(\mathbf{Q}_X^H \boxtimes F) \simeq F$$

for $F = \mathbf{Q}_X^H$, $\mathbf{IC}_X^H$ and $\mathbf{D}_X^H$, and

$$\mathbf{IC}_X^H \otimes \mathbf{IC}_X^H := \Delta^*(\mathbf{IC}_X^H \boxtimes \mathbf{IC}_X^H) \to \mathbf{D}_X^H(-n) = a_X^! \mathbf{Q}^H(-n)$$

In fact the first is clear by $\mathbf{Q}_X^H \boxtimes = pr_2^*$ and the functoriality of pull-backs. The second is equivalent to

$$(a_X)_! \mathbf{IC}_X^H \otimes \mathbf{IC}_X^H \to \mathbf{Q}^H(-n)$$

by the adjoint relation. For a nonsingular Zariski-open dense subset V of X we have

$$H^i_c(V, \mathbf{IC}_V \otimes \mathbf{IC}_V) \xrightarrow{\sim} H^i_c(X, \mathbf{IC}_X \otimes \mathbf{IC}_X) \quad \text{for } i \geqslant 0$$

using a stratification and the support condition of $\mathbf{IC}_X$. Moreover they are equal to 0 for $i > 0$ and $\mathbf{Q}(-n)$ for $i = 0$, because $\mathbf{IC}_V \otimes \mathbf{IC}_V = \mathbf{Q}_V[2n]$ and V is connected since X is irreducible. So the morphism is obtained by

$$\mathrm{Hom}((a_X)_! \mathbf{IC}^H_X \otimes \mathbf{IC}^H_X, \mathbf{Q}^H(-n)) \xrightarrow{\sim} \mathrm{Hom}(H^0_c(X, \mathbf{IC}^H_X \otimes \mathbf{IC}^H_X), \mathbf{Q}^H(-n)) \xrightarrow{\sim} \mathbf{Q}.$$

By the adjoint relation for j^*j_*, we have the morphisms

$$j_*\mathbf{IC}^H_U \otimes j_*\mathbf{IC}^H_U \to j_*\mathbf{D}^H_U(-n)$$

and

$$i^*j_*\mathbf{IC}^H_U \otimes i^*j_*\mathbf{IC}^H_U \to i^*j_*\mathbf{D}^H_U(-n)$$

because $\Delta^* \circ (i \times i)^* = i^* \circ \Delta^*$ (same for the others). Then the assertion follows from

$$(a_{X \times X})_*(F \boxtimes G) = (a_X)_* F \otimes (a_X)_* G$$

and the same for $(a_X)_!$, and the canonical morphism

$$F \boxtimes G \to \Delta_* \Delta^*(F \boxtimes G) = \Delta_*(F \otimes G).$$

Note $\Delta_! = \Delta_*$ because Δ is a closed embedding. □

In the l-adic case this result follows by an argument similar to [GM 5.2]. In general we have a canonical isomorphism

$$\mathrm{Hom}(F \otimes G, \mathbf{D}^H_X) \simeq \mathrm{Hom}(F, \mathbf{D}G)$$

(i.e. $\mathrm{Hom}(\mathbf{Q}^H_X, \underline{\mathrm{Hom}}(F, G)) \simeq \mathrm{Hom}(F, G)$ by duality) for $F, G \in D^b\mathrm{MHM}(X)$, and we can use it to obtain the morphism $\mathbf{IC}^H_X \otimes \mathbf{IC}^H_X \to \mathbf{D}^H_X(-n)$.

If S is the family of supports in U defined by $S = \{C \subset U : C \text{ is closed in } X\}$, then $H^k_S(U, F) = H^k(X, j_! F)$; see [Go]. If X is compact, then $H^k_S(U, F) = H^k_c(U, F)$.

3.5. PROPOSITION. *If L is a weak topological link of Z in X, the following are exact sequences of mixed Hodge structures:*

(i) $\cdots \to H^k_S(U) \to H^k(U) \to H^k(L) \to \cdots$
(ii) $\cdots \to IH^k_S(U) \to IH^k(U) \to IH^k(L) \to \cdots$
(iii) $\cdots \to H^k_Z(X) \to H^k(Z) \to H^k(L) \to \cdots$

Proof. By 3.1(i) for $F \in D^b\mathrm{MHM}(U)$ there is a distinguished triangle

$$j_!F \to j_*F \to i_*i^*j_*F.$$

The long exact sequence of hypercohomology on X with $F = \mathbf{Q}^H_U$ gives (i), and with $F = \mathbf{IC}^H_U$ and 3.1(i) gives (ii).

By 3.1(ii) there is for $G \in D^b\mathrm{MHM}(X)$ a distinguished triangle

$$\to i^!G \to i^*G \to i^*j_*j^*G.$$

The long exact sequence of hypercohomology on Z with $G = \mathbf{Q}^H_X$ gives (iii). □

4. Semipurity

This section contains the main result of this paper.

4.1. THEOREM. *Let L be a weak topological link of Z in X. If* $\dim X = n$ *and* $\dim Z \leqslant d$, *then the intersection cohomology* $IH^k_c(L)$ *has weights* $\leqslant k$ *for* $k < n - d$.

For the case Z a point, see also [Sa2 1.18].

Proof. We use the notation of 3.4. By [BBD 1.4.23(ii)]

$$j_{!*}\mathbf{IC}_U = \tau^Z_{\leqslant -1} j_*\mathbf{IC}_U$$

where $\tau^Z_{\leqslant -1}$ is from [BBD 1.4.13]. Now $\mathbf{IC}_X = j_{!*}\mathbf{IC}_U$, so the above together with the distinguished triangle of [BBD 1.4.13] with $X = j_*\mathbf{IC}_U$ gives a triangle

$$\to \mathbf{IC}_X \to j_*\mathbf{IC}_U \to i_*\, {}^p\tau_{>-1} i^*j_*\mathbf{IC}_U \to$$

Taking i^* gives

$$\to i^*\mathbf{IC}_X \to i^*j_*\mathbf{IC}_U \to {}^p\tau_{>-1} i^*j_*\mathbf{IC}_U \to$$

so $i^*\mathbf{IC}_X \to i^*j_*\mathbf{IC}_U$ is identified with ${}^p\tau_{\leqslant -1} i^*j_*\mathbf{IC}_U \to i^*j_*\mathbf{IC}_U$ [BBD 1.3.3].

Since rat: $\mathrm{MHM}(X) \to \mathrm{Perv}(X)$ is faithful and $\mathrm{rat} \circ \tau = {}^p\tau \circ \mathrm{rat}$, we have

$$\to i^*\mathbf{IC}_X^H \to i^*j_*\mathbf{IC}_U^H \to \tau_{>-1}i^*j_*\mathbf{IC}_U^H \to$$

Thus there is a long exact sequence

$$\to H_c^{k-n}(Z, i^*\mathbf{IC}_X^H) \to H_c^{k-n}(Z, i^*j_*\mathbf{IC}_U^H) \to H_c^{k-n}(Z, \tau_{>-1}i^*j_*\mathbf{IC}_U^H) \to$$

The middle term above is $IH_c^k(L)$. The left term has weight $\leqslant k$, since $\mathbf{IC}_X^H$ has weight n, which implies $i^*\mathbf{IC}_X$ has weight $\leqslant n$. The right term is zero for $k - n < -d = -\dim Z$, since $\underline{\mathbf{H}}^i({}^p\tau_{>-1}i^*j_*\mathbf{IC}_U) = 0$ for $i < -d$ as a consequence of the support condition for a perverse sheaf. □

4.2. COROLLARY. *The group $IH^k(L)$ has weights $>k$ for $k \geqslant n + d$.*

This follows immediately from 4.1 and 3.3.

5. Applications

In this section the results of the previous sections are applied to the topology of a complex variety. We use again the notation of 2.1. Let L be a weak topological link of Z in X. Let

$$v\colon IH^k(L) = IH^{BM}_{2n-1-k}(L)(-n) \to H^{BM}_{2n-1-k}(L)(-n)$$

be the morphism from 3.2. The following theorem is concerned with the composite

$$IH_c^p(L) \times IH_c^q(L) \overset{\cup}{\to} H_{2n-1-p-q}(L)(-n) \overset{u}{\to} H^{BM}_{2n-1-p-q}(L)(-n)$$

where the first morphism is the cup product from 3.4, and the second morphism is the obvious one.

5.1. THEOREM. *Let L be a weak topological link of Z in X, and assume that $\dim X = n$ and $\dim Z \leqslant d$. If $\alpha \in IH_c^p(L)$, $\beta \in IH_c^q(L)$, and $u(\alpha \cup \beta)$ is the image of v, with $p, q < n - d$ and $p + q \geqslant n + d$, then $u(\alpha \cup \beta) = 0$.*

Proof. Weight $\alpha \leqslant p$ and weight $\beta \leqslant q$ by 4.1, so weight $u(\alpha \cup \beta) \leqslant p + q$ by 3.4. Since weight $(\operatorname{im} v) > p + q$ by 4.2 and 3.2, this implies $u(\alpha \cup \beta) = 0$. □

Let

$$w\colon H_c^k(L) \to H^k(L)$$

be the natural map.

5.2. THEOREM. *Let X, Z and L be as in 5.1. Suppose $X \setminus Z$ is a rational homology manifold in a neighborhood of Z. If $k_1, \ldots, k_m < n - d$ and $k = k_1 + \cdots + k_m \geqslant n + d$, then the composition*

$$H_c^{k_1}(L) \times \cdots \times H_c^{k_m}(L) \xrightarrow{\cup} H_c^k(L) \xrightarrow{w} H^k(L)$$

is the zero morphism.

The proof is similar to the proof of 5.1.

Note that since $H_c^{2n-1}(L)$ has weight $2n$ by 3.3, the above theorem is true without the composition with w when $k = 2n - 1$. Also note that if Z is compact, both u and w are isomorphisms.

5.3. EXAMPLES. Let

$$T^k = \underbrace{S^1 \times \cdots \times S^1}_{k}$$

(i) If $n \geqslant 2$ and $d = 0$, then $L = (T^{2n-1}$ with two odd-dimensional submanifolds identified) cannot be a weak topological link of Z in X: In Theorem 5.1 take $p = q = n - 1$. The map $IH^{n-1}(L) \otimes IH^{n-1}(L) \to H_1(L)(-n)$ is nonzero and factors through $IH_1(L)(-n) \to H_1(L)(-n)$ as is seen by using $IH(L) \simeq H(T^{2n-1})$ [GM1 4.2] and the diagram

$$\begin{array}{ccccc} H^{n-1}(T^{2n-1}) \otimes H^{n-1}(T^{2n-1}) & \to & H_1(T^{2n-1})(-n) & \xleftarrow{\simeq} & IH_1(T^{2n-1})(-n) \\ \downarrow \simeq & & \downarrow & & \downarrow \simeq \\ IH^{n-1}(L) \otimes IH^{n-1}(L) & \to & H_1(L)(-n) & \leftarrow & IH_1(L)(-n) \end{array}$$

(ii) If Z is compact and $n - d > 1$, then T^{2n-1} cannot be a weak topological link of Z in X: In Theorem 5.2 take $k_i = 1$.

Note that the above examples L are compact oriented topological pseudomanifolds with odd-dimensional strata, so that they are not excluded from being links by [Su].

Bibliography

[Be] A. Beilinson: On the derived category of perverse sheaves, in: *K-Theory, arithmetic and geometry*, Lecture Notes in Math. 1289, Springer-Verlag, Berlin, 1987, 27–41.

[BBD] A. Beilinson, J. Bernstein, P. Deligne: *Faisceaux pervers*, Asterisque 100 (1982).

[Bo] A. Borel et al., *Intersection cohomology*. Birkhäuser Boston 1984.

[De1] P. Deligne: Théorie de Hodge III. *Publ. Math.* IHES 44 (1974) 5–78.

[De2] P. Deligne: La conjecture de Weil II. *Publ. Math.* IHES 52 (1980) 137–252.

[De3] P. Deligne: *Le formalisme des cycles évanescents.* In SGA7 XIII and XIV, Lecture Notes in Math. 340, Springer-Verlag, Berlin, 1973, pp. 82–115 and 116–164.

[Du] A. Durfee: Neighborhoods of algebraic sets. *Trans. Amer. Math. Soc.* 276 (1983) 517–530.

[DH] A. Durfee and R. Hain: Mixed Hodge structures on the homotopy of links. *Math. Ann.* 280 (1988) 69–83.

[E1] F. Elzein: Mixed Hodge structures. *Proc. Symp. Pure Math.* 40 (1983) 345–352.

[Ga] O. Gabber: *Purité de la cohomologie de MacPherson-Goresky*, rédigé par P. Deligne. (IHES preprint Feb. 1981).

[GM1] M. Goresky and R. MacPherson: Intersection homology theory. *Topology* 19 (1980) 135–162.

[GM2] M. Goresky and R. MacPherson: Intersection homology II. *Inv. Math.* 72 (1983) 77–129.

[GM3] M. Goresky and R. MacPherson: Morse theory and intersection homology theory. *Astérisque 101-2* (1982) 135–192.

[Go] R. Godement: *Théorie des faisceaux.* Hermann, Paris 1958.

[IV] B. Iverson, *Cohomology of sheaves.* Springer, Berlin, 1986.

[Na] V. Navarro-Aznar: Sur la théorie de Hodge des variétés algébriques à singularités isolées. *Asterisque 130* (1985) 272–307.

[Sa1] M. Saito: Mixed Hodge modules. Publ. RIMS, Kyoto Univ. 26 (1990), 221–333.

[Sa2] M. Saito: Introduction to mixed Hodge modules. Preprint RIMS-605 (1987), to appear in Astérisque.

[Sa3] M. Saito: Modules de Hodge polarisables. *Publ. RIMS*, Kyoto Univ. 24 (1988), 849–995.

[St1] J. Steenbrink: Mixed Hodge structures associated with isolated singularities. *Proc. Symp. Pure Math.* 40 (1983) 513–536.

[St2] J. Steenbrink: *Notes on mixed Hodge theory and singularities.* To appear, Astérisque.

[Su] D. Sullivan: *Combinatorial invariants of analytic spaces*, In: Proceedings of Liverpool Singularities Symposium I, Lecture Notes in Math. 192, Springer-Verlag, Berlin, 1971, 165–168.

Compositio Mathematica **76**: 69–85, 1990.

Effective bounds for semipositive sheaves and for the height of points on curves over complex function fields*

HÉLÈNE ESNAULT[1] & ECKART VIEHWEG[2]

[1]*Max-Planck-Institut für Mathematik, Gottfried-Claren-Str. 26, D-5300 Bonn 3, FRG;* [2]*FB6, Mathematik, Universität-GH-Essen, Universitätsstr 3, D-4300 Essen 1, FRG*

Received September 1988; accepted 20 July 1989

In this note we prove an effective version of the positivity theorems for certain direct image sheaves for fibre spaces over curves and apply it to obtain bounds for the height of points on curves of genus $g \geqslant 2$ over complex function fields. Similar positivity theorems over higher dimensional basis and their applications to moduli spaces [13] were presented by the second author at the conference on algebraic geometry, Humboldt Universität zu Berlin, 1988.

Let X be a complex projective surface, Y be a curve and $f: X \to Y$ be a surjective, non isotrivial morphism with connected general fibre F. In 1963 Y. Manin [6] showed that the number of $\mathbb{C}(Y)$ rational points of F is finite if the genus g of F is larger than 1. A $\mathbb{C}(Y)$ rational point $p \in F$ gives rise to a section $\sigma: Y \to X$ of f. If one assumes that the fibres of f do not contain exceptional curves the height of p with respect to ω_F is $h(p) = h(\sigma(Y)) = \deg(\sigma^* \omega_{X/Y})$.

It is well known that Manin's theorem "the Mordell conjecture over function fields" can be proved by bounding $h(\sigma(Y))$ from above for semistable morphisms f. The main result of this note is:

THEOREM 1. *Assume that $f: X \to Y$ is relatively minimal. Let q be the genus of Y, $g \geqslant 2$ the genus of F and s the number of singular fibres of f. Then for all sections σ of f one has*

$$h(\sigma(Y)) < 2 \cdot (2g-1)^2 \cdot (2q-2+2s).$$

If moreover f is semistable, then

$$h(\sigma(Y)) < 2 \cdot (2g-1)^2 \cdot (2q-2+s).$$

In fact, if f is not semistable, a closer look to the semistable reduction of f gives a slightly better bound (see Corollary 4.10).

*Supported by "Heisenberg Programm", DFG.

Effective bounds for the height were first given by A. Parshin [8] and, in arbitrary characteristics, by L. Szpiro [9]. In [15] A. Parshin announced that, using H. Grauert's proof [4], it is possible to bound $h(\sigma(Y))$ by a polynomial of degree 13 in g. As S. Lang and Y. Miyaoka pointed out, one can use the Miyaoka-Yau inequality to get a bound, linear in g.

The proof of Theorem 1 presented in this paper is given in two steps: First we show, that the sheaf $f_*\omega^2_{X/Y}(\sigma(Y))$ can not have an invertible quotient sheaf of low degree, if $h(\sigma(Y))$ is large. Then, in Section 4, we use the Kodaira-Spencer map to show that this sheaf always has a quotient of degree $2q - 2 + 2s$. The existence of global one forms is only used in this second step. Of course, it would be quite interesting to get along without using Ω^1_X at all. May be, combining methods from P. Vojta's proof of Manin's theorem in [14] and from this paper, this could be done. In fact, the methods used in the first part are overlapping with those used in [1] to prove Dyson's lemma in several variables. Hence the relation between [14] and this paper might be quite close.

The experts will see immediately that the second step in our proof is not too different from the arguments used by Y. Manin [6], H. Grauert [4] and L. Szpiro [9]. The "Parshin-construction" used in [8] and [9], however, is replaced by the effective bounds on the "positivity of certain direct image sheaves". This part (see 2.4 for the exact statement) is presented in the first two sections of this article. Without having any other application, we took Theorem 1 as a pretext allowing us to work out for fibre spaces over curves an effective version of the results of [11]. The reader not familiar with the notations used there should have a look to S. Mori's survey article [7].

In Section 3 we just evaluate the constants obtained for general fibre spaces in the special case of families of curves and we verify the assumptions made in 2.4 in this special case.

The motivation to write this note grew out of discussions with A. Parshin during his and our stay at the Max-Planck-Institute for Mathematics in Bonn. The details were worked out during our stay at the TATA-Institute in Bombay.

1. The lower degree of direct images of sheaves

Let Y be a nonsingular compact curve defined over $\mathbb{C}$ and $\mathscr{F}$ be a coherent locally free sheaf on Y.

DEFINITION 1.1. (a) The *lower degree* of $\mathscr{F}$ is defined as

$$\mathrm{ld}(\mathscr{F}) = \min\{\deg(\mathscr{N});\ \mathscr{N} \text{ invertible quotient sheaf of } \mathscr{F}\}.$$

(b) The *stable lower degree* of $\mathcal{F}$ is

$$\operatorname{sld}(\mathcal{F}) = \inf\left\{\frac{\operatorname{ld}(\tau^*\mathcal{F})}{\deg(\tau)};\ \tau\colon Y' \to Y \text{ a finite map of non singular curves}\right\}.$$

If $\mathcal{F} = 0$ we put $\operatorname{ld}(\mathcal{F}) = \operatorname{sld}(\mathcal{F}) = \infty$.

(c) $\mathcal{F}$ is called *semi-positive* if $\operatorname{sld}(\mathcal{F}) \geqslant 0$ (Fujita, [3]).

1.2. Some properties

(a) If $\mathcal{L}$ is an invertible sheaf of degree d, then

$$\operatorname{ld}(\mathcal{F} \otimes \mathcal{L}) = \operatorname{ld}(\mathcal{F}) + d \quad \text{and}$$
$$\operatorname{sld}(\mathcal{F} \otimes \mathcal{L}) = \operatorname{sld}(\mathcal{F}) + d.$$

(b) If $\rho\colon Y'' \to Y$ is a non singular covering then

$$\operatorname{sld}(\rho^*\mathcal{F}) = \deg(\rho)\cdot\operatorname{sld}(\mathcal{F})$$

(c) $\mathcal{F}$ is ample if and only if $\operatorname{sld}(\mathcal{F}) > 0$.

(d) The following three conditions are equivalent:

(i) $\operatorname{sld}(\mathcal{F}) \geqslant 0$.

(ii) $\mathcal{F}$ is weakly positive over Y (see [11]).

(iii) If $\mathcal{H}$ is an ample invertible sheaf on Y then for all $\gamma > 0$ the sheaf $S^\gamma(\mathcal{F}) \otimes \mathcal{H}$ is ample.

(e) For all $\gamma > 0$ $\operatorname{ld}(\otimes^\gamma(\mathcal{F})) \leqslant \operatorname{ld}(S^\gamma\mathcal{F}) \leqslant \gamma\cdot\operatorname{ld}(\mathcal{F})$ and

$$\operatorname{sld}(\otimes^\gamma(\mathcal{F})) \leqslant \operatorname{sld}(S^\gamma\mathcal{F}) \leqslant \gamma\cdot\operatorname{sld}(\mathcal{F}).$$

Proof. (a) and (e) follow directly from the definition. (c) If $\mathcal{F}$ is ample then $S^\gamma(\mathcal{F}) \otimes \mathcal{H}^{-1}$ will be ample for some $\gamma \gg 0$ and $\operatorname{sld}(\mathcal{F}) \geqslant 1/\gamma\cdot\deg(\mathcal{H})$. If $\operatorname{sld}(\mathcal{F}) \geqslant \varepsilon > 0$ then $\mathcal{O}_{\mathbb{P}}(1)$ on $\mathbb{P} = \mathbb{P}(\mathcal{F})$ satisfies the Seshadri criterium for ampleness (see for example: R. Hartshorne, Ample subvarieties of algebraic varieties, Lecture Notes in Math. *156*, springer 1970, or [3] §2). The proof of (d) is similar. In [11] Section 1 and [13] Section 3 the reader can find some generalizations for higher dimensional Y.

In (b) it is obvious that $\operatorname{sld}(\rho^*\mathcal{F}) \geqslant \deg(\rho)\cdot\operatorname{sld}(\mathcal{F})$. On the other hand, if $\tau\colon Y' \to Y$ is another nonsingular covering we can find $\tau'\colon Y''' \to Y$ dominating both, τ and ρ. One as

$$\operatorname{sld}(\rho^*\mathcal{F}) \leqslant \frac{\operatorname{ld}(\tau'^*\mathcal{F})}{\deg(Y''' \to Y'')} \leqslant \frac{\operatorname{ld}(\tau^*\mathcal{F})\cdot\deg(Y''' \to Y')}{\deg(Y''' \to Y'')} = \frac{\operatorname{ld}(\tau^*\mathcal{F})\cdot\deg(\rho)}{\deg(\tau)}.$$

(1.3) In our estimations of the stable lower degree we will frequently use vanishing theorems for integral parts of $\mathbb{Q}$ divisors. Let $g: Z \to X$ be a morphism of complex projective manifolds. For a normal crossing divisor $D = \Sigma \nu_i D_i$ and $e \in \mathbb{R}$ we write $[e \cdot D] = \Sigma [e \cdot \nu_i] \cdot D_i$ where $[e \cdot \nu_i]$ is the integral part of $e \cdot \nu_i$. Recall that an invertible sheaf $\mathscr{L}$ on Z is called numerically effective if $\deg(\mathscr{L}|_C) \geqslant 0$ for all curves C in Z. We will say that $\mathscr{L}$ is *g-numerically effective* if $\deg(\mathscr{L}|_C) \geqslant 0$ for all curves C in Z with $\dim(g(C)) = 0$. $\kappa(\mathscr{L})$ denotes the Iitaka dimension of $\mathscr{L}$ (see [7] for example) and $\omega_{Z/X} = \omega_Z \otimes g^* \omega_X^{-1}$ the difference of the canonical sheaves.

LEMMA 1.4. (a) *Assume that there exists an effective normal crossing divisor D on Z such that $\mathscr{L}^N(-D)$ is g-numerically effective. If for a general fibre G of g $\kappa(\mathscr{L}(-[D/N]) \otimes \mathcal{O}_G) = \dim G$ then for $i > 0$, $R^i g_*(\omega_{Z/X} \otimes \mathscr{L}(-[D/N])) = 0$.*

(b) *Assume that g is birational.*

If Γ is an effective divisor on X, $D = g^*\Gamma$ a normal crossing divisor and $N > 0$, then $R^i g_* \omega_{Z/X} \otimes \mathcal{O}_Z(-[D/N]) = 0$ for $i > 0$. If moreover Γ is a normal crossing divisor then $g_* \omega_{Z/X} \otimes \mathcal{O}_Z(-[D/N]) = \mathcal{O}_X(-[\Gamma/N])$.

Proof. (a) The assumptions imply that $\mathscr{L}^N(-D) \otimes g^* \mathscr{H}^N$ will be numerically effective and $\kappa(\mathscr{L}(-[D/N]) \otimes g^* \mathscr{H}) = \dim Z$ for all "very very" ample invertible sheaves $\mathscr{H}$ on X. As in [10], 2.3, the vanishing theorem due to Kawamata [5] and the second author ([10] and [2], 2.13) implies (a) by using the Leray spectral sequence. (b) Is shown in [10], 2.3.

(1.5) Let in the sequel Y be a nonsingular compact curve, X a projective manifold of dimension n and $f: X \to Y$ a surjective morphism. The general fibre of f will be denoted by F. Let $\mathscr{L}$ be an invertible sheaf on X.

All estimates of $\mathrm{sld}(f_*(\mathscr{L} \otimes \omega_{X/Y}))$ will follow from the following corollary of Fujita's positivity theorem [3] (see [11] 5.1):

LEMMA 1.6. *Let D be an effective normal crossing divisor on X and $N > 0$ such that $\mathscr{L}^N = \mathcal{O}_X(D)$. Then $f_*(\omega_{X/Y} \otimes \mathscr{L}(-[D/N]))$ is semi-positive.*

Proof. The case $\mathscr{L} = \mathcal{O}_X$ is Fujita's original theorem. It may be easily obtained by Hodge theory on cyclic covers ([12], 6 and 8).

The general case follows from this one, applied to the cyclic cover given by $\mathscr{L}^N = \mathcal{O}_X(D)$ (as in [2], 2.7 or [11], 2.2).

Recall the following notation: If $U \subset X$ is open, we call $\mathscr{L}$ *very ample with respect to U* if $H^0(X, \mathscr{L}) \otimes_{\mathbb{C}} \mathcal{O}_X \to \mathscr{L}$ is surjective over U and the natural map $U \to \mathbb{P}(H^0(X, \mathscr{L}))$ is an embedding. Correspondingly we call $\mathscr{L}$ *ample with respect to U* if for some $a > 0$ $\mathscr{L}^a$ is very ample with respect to U (see [13], 1.16).

We will call $\mathscr{L}$ *numerically effective with respect to U* if there exists a birational morphism $g: X' \to X$, isomorphic over U, and an invertible numerically effective sheaf $\mathscr{L}'$ and an inclusion $\mathscr{L}' \to g^* \mathscr{L}$, isomorphic over U.

Obviously, if $\mathcal{L}$ is ample with respect to U it is numerically effective with respect to U. Moreover, if $\mathcal{H}$ is ample with respect to U and $\mathcal{L}$ numerically effective with respect to U then $\mathcal{H} \otimes \mathcal{L}^a$ is ample with respect to U, for all $a \geqslant 0$.

COROLLARY 1.7. *Let D be a normal crossing divisor on X and $N > 0$, such that $\mathcal{L}^N(-D)$ is numerically effective with respect to a neighbourhood of F and $\kappa(\mathcal{L}^N(-D)) = n$. Then $f_*(\omega_{X/Y} \otimes \mathcal{L}(-[D/N]))$ is semi-positive.*

Proof. By 1.4(b) the statement is compatible with blowing ups. As $\kappa(\mathcal{L}^N(-D)) = n$, we may assume (for example as in [2], 2.12) that there exists an effective divisor Γ on X such tha $D + \Gamma$ has normal crossings and such that $\mathcal{L}^N(-D - \Gamma)$ is ample. Then for $M \geqslant 0$ $\mathcal{L}^{N \cdot M}(-M \cdot D - \Gamma)$ will be ample with respect to a neighbourhood of F. Replacing N, D and Γ by some common multiple we can find a divisor H, smooth on F such that (blowing up a little bit more) $D' = H + M \cdot D + \Gamma$ has normal crossings and $\mathcal{L}^{N \cdot M} = \mathcal{O}_X(D')$. For M big enough

$$\left[\frac{D'}{M \cdot N}\right]\bigg|_F = \left[\frac{H + M \cdot D + \Gamma}{M \cdot N}\right]\bigg|_F = \left[\frac{D}{N}\right]\bigg|_F$$

and therefore 1.7 follows from 1.6.

DEFINITION 1.8. Let Z be a manifold $\mathcal{M}$ be an invertible sheaf and Γ be an effective divisor.

(a) Let $\tau: Z' \to Z$ be a blowing up such that $\Gamma' = \tau^*\Gamma$ is a normal crossing divisor. We define

$$e(\Gamma) = \min\left\{e \in \mathbb{N} - \{0\}; \tau_* \omega_{Z'/Z}\left(-\left[\frac{\Gamma'}{e}\right]\right) = \mathcal{O}_Z\right\}.$$

(b) $e(\mathcal{M}) = \max\{e(\Gamma); \Gamma \text{ zero divisor of } s \in H^0(Z, \mathcal{M})\}$.

By 1.4(b) the definition of $e(\Gamma)$ is independent of the blowing up choosen. In Section 2 we will give upper bounds for $e(\Gamma)$.

1.8.1. Especially one obtains $e(\mathcal{M}) < \infty$.

For $e \geqslant e(\Gamma)$ and τ as above one has $\tau_* \omega_{Z'/Z}(-[\Gamma'/e]) = \mathcal{O}_Z$. If $\Gamma = \Sigma \nu_i \Gamma_i$ is a normal crossing divisor, then $e(\Gamma) = \max\{\nu_i\} + 1$.

COROLLARY 1.9. *Assume that $\mathcal{L}$ is numerically effective with respect to some neighbourhood of F and $\kappa(\mathcal{L}) = n$. Assume moreover that for some invertible sheaf $\mathcal{H}$ on Y of degree h and some $N > 0$ one has an inclusion $\varphi: f^*\mathcal{H} \to \mathcal{L}^N$. Then*

$$\operatorname{sld}(f_*(\mathcal{L} \otimes \omega_{X/Y})) \geqslant h \cdot \min\left\{\frac{1}{N}, \frac{1}{e(\mathcal{L}^N|_F)}\right\}.$$

Remark. Especially, if $h > 0$ the sheaf $f_*(\mathscr{L} \otimes \omega_{X/Y})$ will be ample by 1.2(c) and 1.8.1. A similar result for higher dimensional Y can be found in [11] 5.4.

Proof. Let $M = \max\{N, e(\mathscr{L}^N|_F)\}$ (1.8.1). We can choose a cover $\sigma\colon Y' \to Y$ of degree M such that the fibres of f over the ramification locus are non singular. Then $X' = X \times_Y Y'$ is non singular and

$$pr_{2*}(pr_1^*\mathscr{L} \otimes \omega_{X'/Y'}) = \sigma^* f_*(\mathscr{L} \otimes \omega_{X/Y}).$$

By 1.2(b) it s enough to show 1.9 for $X' \to Y'$. Hence, by abuse of notations we may assume that M divides h. Let $\mathscr{N} \in \operatorname{Pic}^\circ(Y)$. One has for example by Seshadri's criterion $\kappa(\mathscr{L} \otimes f^*\mathscr{N}) = \kappa(\mathscr{L})$. Using 1.2(a) we may replace $\mathscr{L}$ by $\mathscr{L} \otimes f^*\mathscr{N}$ and $\mathscr{H}$ by $\mathscr{H} \otimes \mathscr{N}^N$. Therefore we may assume that $\mathscr{H} = \mathscr{O}_Y(h \cdot p)$ for some point $p \in Y$. Let Γ be the zero divisor of φ and let $\tau\colon X' \to X$ be a blowing up such that $\tau^*(\Gamma + h \cdot f^{-1}(p))$ becomes a normal crossing divisor. Let $f' = f \circ \tau, \Gamma' = \tau^*\Gamma$ and $\mathscr{L}' = \tau^*\mathscr{L}$. Since $M \geqslant e(\Gamma|_F)$ the inclusion

$$f'_*\left(\mathscr{L}'\left(-\left[\frac{\Gamma'}{M}\right]\right) \otimes \omega_{X'/Y}\right) \to f_*(\mathscr{L} \otimes \omega_{X/Y})$$

is surjective at the general point of Y. This implies

$$\operatorname{sld}(f_*(\mathscr{L} \otimes \omega_{X/Y})) \geqslant \operatorname{sld}\left(f'_*\left(\mathscr{L}'\left(-\left[\frac{\Gamma'}{M}\right]\right) \otimes \omega_{X'/Y}\right)\right).$$

One has $\mathscr{L}'^M(-\Gamma' - h \cdot f'^{-1}(p)) = \mathscr{L}'^{M-N}$ and by 1.6 (if $M = N$) or 1.7 (if $M > N$) the sheaf

$$f'_*\left(\mathscr{L}'\left(-\left[\frac{\Gamma'}{M}\right] - \frac{h}{M} \cdot f'^{-1}(p)\right) \otimes \omega_{X'/Y'}\right)$$

$$= f'_*\left(\mathscr{L}'\left(-\left[\frac{\Gamma'}{M}\right]\right) \otimes \omega_{X'/Y}\right) \otimes \mathscr{O}_Y\left(-\frac{h}{M} \cdot p\right)$$

is semipositive. From 1.2, a we obtain $\operatorname{sld}(f_*(\mathscr{L} \otimes \omega_{X/Y})) \geqslant h/M$.

2. Bounds for $e(\mathscr{L})$ and the main theorem

(2.1) Consider a complex projective manifold V and an effective divisor Γ on V. We write $\mathscr{M} = \mathscr{O}_V(\Gamma)$ and choose a blowing up $\tau\colon V' \to V$ such that $\tau^*\Gamma = \Gamma'$ has normal crossings. Let us write $\mathscr{C}(e) = \operatorname{coker}(\tau_*\omega_{V'}(-[\Gamma'/e]) \to \omega_V)$. By 1.4(b) $\mathscr{C}(e)$ is independent of the blowing up choosen.

LEMMA 2.2 *Let H be a smooth prime divisor of V which is not a component of Γ. Then* $\operatorname{Supp}(\mathscr{C}(e)) \cap H = \emptyset$ *for* $e \geqslant e(\Gamma|_H)$.

Proof. We may assume by 1.4(b) that Γ' intersects the proper transform H' of H transversally. Then $[\Gamma'/e]|_{H'} = [\Gamma'|_{H'}/e]$. One has a commutative diagram

$$\begin{array}{ccccc} \tau_*\omega_{V'}\left(-\left[\dfrac{\Gamma'}{e}\right] + H'\right) & \xrightarrow{\alpha} & \tau_*\omega_{H'}\left(-\left[\dfrac{\Gamma'|_{H'}}{e}\right]\right) & \xrightarrow{\beta_H} & \omega_H \\ \downarrow & & & & \| \\ \tau_*\omega_{V'}\left(-\left[\dfrac{\Gamma'}{e}\right]\right)\otimes\mathcal{O}(H) & \xrightarrow{\beta} & \omega_V(H) & \longrightarrow & \omega_H \end{array}$$

By the vanishing Theorem 1.4(b), α is surjective. If $e \geqslant e(\Gamma|_H)$, β_H is surjective. Then β has to be surjective in a neighbourhood of H.

PROPOSITION 2.3. *Let Z_i, $i = 1, \ldots, r$, be projective manifolds, $\mathscr{H}_i$ be a very ample invertible sheaf on Z_i and $m, d > 0$ such that $c_1(\mathscr{H}_i)^{\dim(Z_i)} \leqslant d/m$. Let V be the r-fold product $Z_1 \times \cdots \times Z_r$ and $\mathscr{M} = \otimes_{i=1}^r pr_i^*\mathscr{H}_i^m$. Then $e(\mathscr{M}) \leqslant d + 1$.*

Proof. Consider first the case $r = 1$. Set $Z = Z_1$, $\mathscr{H} = \mathscr{H}_1$. We prove 2.3 by induction on $\dim Z$. If Z is a curve Γ is an effective divisor on it of degree $\leqslant d$. Therefore $[\Gamma/d + 1] = 0$. Assume $\dim Z \geqslant 2$. Choose, Γ, τ as in 2.1 and H a smooth hyperplane section with $\mathscr{H} = \mathcal{O}(H)$. Then $c_1(\mathscr{H}|_H)^{\dim Z - 1} \leqslant d/m$. If H is not a component of Γ, then by induction and 2.2 $\operatorname{Supp} \mathscr{C}(d + 1)$ does not meet H. As we may find such a H containing any given point, we obtain $\mathscr{C}(d + 1) = 0$.

We proceed by induction on r. We assume that 2.3 holds for $T = Z_1 \times \cdots \times Z_{r-1}$ and $\mathscr{L} = \otimes_{i=1}^{r-1} pr_i^*\mathscr{H}_i^m$.

If Z_r is a point, then 2.3 holds by induction. Assume that Z_r is a curve. Choose Γ and τ as in 2.1. Take a point $p \in Z_r$, and define $D = T \times p \simeq T$. Let ν be the maximal integer such that $\nu \cdot D \leqslant \Gamma$. As $\deg \mathscr{H}_r^m \leqslant d$, one has $0 \leqslant \nu \leqslant d$. We may assume that the proper transform D' of D in V' meets $\Delta' = \Gamma' - \nu \cdot \tau^* D$ transversally. From the inequality

$$\Delta' \geqslant \Gamma' - (d + 1)\cdot(\tau^* D - D') - \nu \cdot D' = \Delta' - (d + 1 - \nu)\cdot(\tau^* D - D')$$

one obtains $-[\Delta'/d + 1] + D' \leqslant -[\Gamma'/d + 1] + \tau^* D$. The multiplicity of D' is one on both sides of the inequality. One has thereby a commutative diagram

$$\begin{array}{ccccc}
\tau_*\omega_{V'}\left(-\left[\dfrac{\Delta'}{d+1}\right]+D'\right) & \xrightarrow{\alpha} & \tau_*\omega_{D'}\left(-\left[\dfrac{\Delta'|_{D'}}{d+1}\right]\right) & \xrightarrow{\beta_D} & \omega_D \\
\downarrow & & & & \| \\
\tau_*\omega_{V'}\left(-\left[\dfrac{\Gamma'}{d+1}\right]\right)\otimes\mathcal{O}(D) & \xrightarrow{\beta} & \omega_V(D) & \longrightarrow & \omega_D
\end{array}$$

By the vanishing Theorem 1.4(b) α is surjective. As $\mathcal{O}_D(\Gamma - \nu\cdot D) = \mathcal{O}_D(\Gamma) \simeq \mathcal{L}$, β_D is an isomorphism by induction. Therefore Supp $\mathcal{C}(d+1)$ does not meet D. Moving p, we obtain 2.3 for dim $Z_r = 1$. Assume dim $Z_r \geqslant 2$. Choose F a general hyperplane section in Z_r with $\mathcal{O}(F) = \mathcal{H}_r$. As $c_1(\mathcal{H}_{r|F})^{\dim F} \leqslant d/m$, we have $e(\mathcal{M}|_H) \leqslant d+1$ by induction for $H = T \times F$, and 2.2 implies that for all Γ with $\mathcal{M} = \mathcal{O}_V(\Gamma)$, supp $\mathcal{C}\,(d+1)$ does not meet H. As we may find such a F such that $H = T \times F$ is not a component of Γ and contains any given point, we obtain 2.3.

The main result of this note is the following theorem, which for $\mathcal{L} = \omega_{X/Y}^{k-1}$ is an effective version of a special case of [11], 6.2.

THEOREM 2.4. *Let Y be a nonsingular compact curve; X be a projective manifold of dimension n and $f: X \to Y$ be a surjective morphism. Let $\mathcal{L}$ be an invertible sheaf on X with $\kappa(\mathcal{L}) = \dim X$. Assume that for some $N > 0$ the sheaf $\mathcal{L}^N|_F$ is very ample on the general fibre F of f and that $\mathcal{L}$ is numerically effective with respect to some neighbourhood of F. Write $d = c_1(\mathcal{L}|_F)^{n-1}$. Then for all $m > 0$*

$$\operatorname{sld}(f_*(\mathcal{L}\otimes\omega_{X/Y}))\cdot(m\cdot N^{n-1}\cdot d+1) \geqslant \frac{\deg(f_*(\mathcal{L}^{m\cdot N}))}{\operatorname{rank}(f_*(\mathcal{L}^{m\cdot N}))}.$$

Proof. Let us start with the case $m = 1$:

Let $r = \operatorname{rank}(f_*\mathcal{L}^N)$, X^r the r-fold product $X \times_Y X \cdots \times_Y X$ and $f^r: X^r \to Y$ the induced map. If $\mathcal{N}$ is any locally free sheaf on X we obtain by flat base change $f^r_*(\otimes^r_{i=1} pr_i^*\mathcal{N}) = \otimes^r f_*\mathcal{N}$. f is a flat Gorenstein morphism and $\omega_{X/Y}$ the same as the dualizing sheaf of f. Therefore $\omega_{X^r/Y} = \otimes^r_{i=1} pr_i^*\omega_{X/Y}$ (see [11], 3.5, for similar constructions). Let $\sigma: X^{(r)} \to X^r$ be a desingularization, isomorphic on the general fibre $F \times \cdots \times F$, and $f^{(r)} = f^r \circ \sigma$. For $\mathcal{M} = \sigma^*(\otimes^r_{i=1} pr_i^*\mathcal{L})$ we have inclusions $\otimes^r{}_{i=1} pr_i^*\mathcal{L}^N \to \sigma_*\mathcal{M}^N$ and

$$\sigma_*(\mathcal{M}\otimes\omega_{X^{(r)}/Y}) = \left(\bigotimes_{i=1}^r pr_i^*\mathcal{L}\right)\otimes \mathcal{H}om_{\mathcal{O}_{X^r}}(\sigma_*\mathcal{O}_{X^{(r)}}, \omega_{X^r/Y}) \to$$

$$\to \left(\bigotimes_{i=1}^r pr_i^*\mathcal{L}\right)\otimes\omega_{X^r/Y} = \bigotimes_{i=1}^r pr_i^*(\mathcal{L}\otimes\omega_{X/Y}).$$

The induced inclusions

$$\bigotimes^{r} f_*\mathscr{L}^N \to f_*^{(r)}\mathscr{M}^N \text{ and } f_*^{(r)}(\mathscr{M}\otimes\omega_{X^{(r)}/Y}) \to \bigotimes^{r} f_*(\mathscr{L}\otimes\omega_{X/Y})$$

are both isomorphisms at the general point of Y. Especially one has

$$\operatorname{sld}\bigotimes^{r} f_*(\mathscr{L}\otimes\omega_{X/Y}) \geqslant \operatorname{sld} f_*^{(r)}(\mathscr{M}\otimes\omega_{X^{(r)}/Y}).$$

One has $\kappa(\mathscr{M}) = \dim X^{(r)}$ and $\mathscr{M}^N$ is very ample on the general fibre $F\times\cdots\times F$ of $f^{(r)}$. We have a natural inclusion

$$\det(f_*\mathscr{L}^N) \to \bigotimes^{r} f_*\mathscr{L}^N \to f_*^{(r)}\mathscr{M}^N$$

$$\text{and}\quad f^{(r)*}\det(f_*\mathscr{L}^N) \to f^{(r)*}f_*^{(r)}\mathscr{M}^N \to \mathscr{M}^N.$$

By 1.9 (applied to $f_*^{(r)}(\mathscr{M}\otimes\omega_{X^{(r)}/Y})$ and $\mathscr{H} = \det(f_*\mathscr{L}^N)$ one has

$$\operatorname{sld}(f_*^{(r)}(\mathscr{M}\otimes\omega_{X^{(r)}/Y})) \geqslant \deg(f_*\mathscr{L}^N)\cdot\min\left\{\frac{1}{N}, \frac{1}{e(\mathscr{M}^N|_F)}\right\}.$$

Up to now we did not use that $\mathscr{L}^N$ is very ample and the last inequality holds for all exponents. Especially replacing N by $m\cdot N$ we find that (using 1.2(e))

$$\operatorname{rank}(f_*(\mathscr{L}^{m\cdot N}))\cdot\operatorname{sld}(f_*(\mathscr{L}\otimes\omega_{X/Y})) \geqslant \deg(f_*(\mathscr{L}^{m\cdot N}))\cdot\min\left\{\frac{1}{m\cdot N}, \frac{1}{e(\mathscr{L}^{m\cdot N}|_F)}\right\}.$$

On the other hand we have shown in 2.3 that

$$e(\mathscr{L}^{m\cdot N}) \leqslant m\cdot c_1(\mathscr{L}^N|_F)^{n-1} + 1 = m\cdot N^{n-1}\cdot d + 1.$$

COROLLARY 2.5. *Under the assumptions of* 2.4 *let* $\mathscr{L}$ *be even numerically effective on* X. *Then*

$$\operatorname{sld}(f_*(\mathscr{L}\otimes\omega_{X/Y})) \geqslant \frac{c_1(\mathscr{L})^n}{n\cdot N^{n-2}\cdot d^2}.$$

Proof. If $\mathscr{L}$ is numerically effective, the dimension of the higher cohomology groups of $\mathscr{L}^{m\cdot N}$ is bounded from above by a polynomial of degree $n-1$ in m.

Since the Leray spectral sequence gives an inclusion

$$H^1(Y, f_* \mathscr{L}^{m \cdot N})) \to H^1(X, \mathscr{L}^{m \cdot N}),$$

the same holds true for $h^1(Y, f_* \mathscr{L}^{m \cdot N})$.

The Riemann-Roch-Theorem for vector-bundles on Y and for invertible sheaves on X implies that

$$\frac{1}{n!} \cdot c_1(\mathscr{L})^n \cdot N^n \cdot m^n = h^0(X, \mathscr{L}^{N \cdot m}) + O(m^{n-1}) = h^0(Y, f_* \mathscr{L}^{N \cdot m}) + O(m^{n-1})$$
$$= O(m^{n-1}) + \deg(f_* \mathscr{L}^{N \cdot m}) - \operatorname{rank}(f_* \mathscr{L}^{N \cdot m}) \cdot (q - 1).$$

Then, using in the same way the Riemann-Roch on F and taking the limit over m we get 2.5 from 2.4.

REMARK 2.6. Especially for those, mostly interested in the case that f is a family of curves, it might look quite complicated that the proof of 2.5 and 2.4 forced us to consider higher dimensional fibre spaces. In fact, if one is just interested in 2.5 this is not necessary and we sketch in the sequel a proof which is avoiding the products in 2.3 and 2.4: If p is a point on Y the Riemann-Roch theorem shows that $h^0(Y, f_* \mathscr{L}^{N \cdot m} \otimes \mathcal{O}_Y(-h \cdot p))$ is larger than or equal to $\deg(f_* \mathscr{L}^{N \cdot m}) - \operatorname{rank}(f_* \mathscr{L}^{N \cdot m}) \cdot (h + q - 1)$. Therefore, whenever we have

$$h < \deg(f_* \mathscr{L}^{N \cdot m}) \cdot \operatorname{rank}(f_* \mathscr{L}^{N \cdot m})^{-1} - q + 1,$$

we will find an inclusion of $\mathcal{O}_Y(+h \cdot p)$ in $f_* \mathscr{L}^{N \cdot m}$.

Applying 1.9 and 2.3 (for $r = 1$) we obtain the same inequality as in 2.4, except that we have to add a $-q$ on the right hand side. Since in the proof of 2.5 we were taking the limit over m anyway, this is enough to obtain 2.5.

3. Examples and applications

The first application of 2.4 is not really needed in the proof of theorem one and it is just added for historical reasons.

THEOREM 3.1. *Let $f: X \to Y$ be a surjective morphism with general fibre F, where X is a projective manifold of dimension n and Y a non singular curve, and let $\nu > 1$. Assume that for $N > 0$ ω_F^N is very ample and that f is non isotrivial. Then, for all multiples m of $\nu - 1$*

$$\operatorname{sld}(f_* \omega_{X/Y}^{\nu}) \cdot (m \cdot (N \cdot c_1(\omega_F))^{n-1} + 1) \geqslant \frac{\deg(f_* \omega_{X/Y}^{N \cdot m})}{\operatorname{rank}(f_* \omega_{X/Y}^{N \cdot m})}.$$

Moreover for $\nu > 1, f_*\omega^{\nu}_{X/Y}$ *is ample, (if it is not trivial).*

Proof. Results due to J. Kollár and the second author show that $\kappa(\omega_{X/Y}) = \dim(X)$ (see [7] or [13], §1(c) for example). In fact, one first shows that $\deg(f_*\omega^{N\cdot m}_{X/Y}) > 0$ for $m \gg 0$ and then one uses methods similar to those used in the proof of 2.4 to show that $f_*\omega^{\mu}_{X/Y}$ is ample for $\mu \gg 0$. Then $\omega_{X/Y}$ will be ample with respect to a neighbourhood of F and the inequality follows then from 2.4 for $\mathscr{L} = \omega^{\nu-1}_{X/Y}$. By 1.4(a) $f_*\omega^{\nu}_{X/Y}$ will be ample whenever it is not trivial.

EXAMPLE 3.2. Assume that X is a surface, and moreover that f is a non isotrivial family of curves of genus $g \geqslant 2$.

(a) For $N > 1$ one has rank $f_*\omega^N_{X/Y} = (2N-1)\cdot(g-1)$. Applying 3.1 for $\nu = 2$, one obtains

$$\operatorname{sld}(f_*\omega^2_{X/Y})\cdot(N\cdot(2g-2)+1)\cdot(2N-1)\cdot(g-1) \geqslant \deg f_*\omega^N_{X/Y}.$$

(b) Let $\tilde{f}\colon \tilde{X} \to Y$ be the relative minimal model of f. By definition the fibres of $\tilde{f}$ do not contain any (-1)-curves and $\omega_{\tilde{X}/Y}$ is $\tilde{f}$-numerically effective. On the other hand, if B is a curve on X which dominates Y, then $\tilde{f}^*\tilde{f}_*\omega_{\tilde{X}/Y} \to \omega_{\tilde{X}/Y} \to \omega_{\tilde{X}/Y}|_B$ is non trivial. Since the sheaf on the left hand side is the pullback of a semipositive sheaf, $\deg(\omega_{\tilde{X}/Y}|_B) \geqslant 0$. Therefore $\omega_{\tilde{X}/Y}$ is numerically effective. From 2.5 we obtain that $\operatorname{sld}(f_*\omega^2_{X/Y})\cdot(2g-2)^2 \geqslant \frac{1}{2}c_1(\omega_{\tilde{X}/Y})^2$.

(c) One has $c_1(\omega_{\tilde{X}/Y})^2 > 0$ as $\kappa(\omega_{\tilde{X}/Y}) = 2$ and since $\omega_{\tilde{X}/Y}$ is numerically effective (see [10], §3). If one does not want to use this non trivial fact, one can get along with $c_1(\omega_{\tilde{X}/Y})^2 \geqslant 0$ if one replaces all strict inequalities in the sequel by "$\geq$". The weak inequality follows directly from (b).

(3.3) From now on $f\colon X \to Y$ will denote a non isotrivial family of curves of genus $g \geqslant 2$ and $\sigma\colon Y \to X$ a section. Let $C = \sigma(Y)$ and let $\tilde{f}\colon \tilde{X} \to Y$ be the relative minimal model. The image $\tilde{C}$ of C in $\tilde{X}$ intersects the fibres of $\tilde{f}$ in smooth points. Therefore we may assume that all fibres of f are normal crossing divisors, and that C does not meet any exceptional divisor contained in the fibres. Of course, $h(\tilde{C}) = c_1(\omega_{\tilde{X}/Y})\cdot\tilde{C}$ is the same as $c_1(\omega_{X/Y})\cdot C$ under this assumption.

LEMMA 3.4. *For* $N > 1$ *one has*

$$\deg(f_*\omega_{X/Y}(C)^N) = \deg(f_*\omega^N_{X/Y}) + \tfrac{1}{2}N(N-1)\cdot h(C).$$
$$\operatorname{rank}(f_*\omega_{X/Y}(C)^N) = N\cdot(2g-1)-(g-1).$$

Proof. We may assume here that f is relatively minimal, i.e. $\tilde{X} = X$. Then $\omega_{X/Y}$ as well as $\omega_{X/Y}(C)$ are f-numerically effective (see 1.3). Then by 1.4(a) we have for $0 \leqslant \mu < N$ exact sequences

$$0 \to f_*\omega^N_{X/Y}(\mu\cdot C) \to f_*\omega^N_{X/Y}((\mu+1)C) \to \omega^N_{X/Y}((\mu+1)C)|_C \to 0.$$

Since $\omega_{X/Y}(C)|_C = \mathcal{O}_C$, the sheaf on the right hand side is

$$\mathcal{O}_C(-(N-\mu-1)\cdot C) = \omega_{X/Y}^{N-\mu-1}|_C.$$

Adding up we obtain

$$\deg(f_*\omega_{X/Y}(C)^N) - \deg(f_*\omega_{X/Y}^N) = \sum_{\mu=0}^{N-1} (N-\mu-1)\cdot h(C).$$

The second equality is trivial.

COROLLARY 3.5. *Under the assumptions made in* 3.3 *we have for* $N \geqslant 2$

$$\begin{aligned}\mathrm{sld}(f_*\omega_{X/Y}^2(C))\cdot(N\cdot(2g-1)+1)\cdot(N\cdot(2g-1)-(g-1))\\ \geqslant \deg(f_*\omega_{X/Y}^N) + \frac{N(N-1)}{2}\cdot h(C)\end{aligned}$$

and

$$\mathrm{sld}(f_*\omega_{X/Y}^2(C))\cdot(2g-1)^2 \geqslant \tfrac{1}{2}(c_1(\omega_{\tilde{X}/Y})^2 + h(C)) > \tfrac{1}{2}h(C).$$

Proof. Since $\omega_{\tilde{X}/Y}$ is numerically effective, the same holds for $\omega_{\tilde{X}/Y}(\tilde{C})$. Moreover, if $h(C) \neq 0$, $c_1(\omega_{\tilde{X}/Y}(\tilde{C}))^2 = c_1(\omega_{\tilde{X}/Y})^2 + h(C) > 0$ and hence $\kappa(\omega_{\tilde{X}/Y}(\tilde{C})) = 2$. The first inequality follows from 3.4 and 2.4 applied to $\mathcal{L} = \omega_{\tilde{X}/Y}(C)$ and the second one from 2.5 applied to $\mathcal{L} = \omega_{\tilde{X}/Y}(\tilde{C})$.

REMARK. Since the arguments used in 3.4 also show that $\omega_{X/Y|C}$ is a quotient of $f_*\omega_{\tilde{X}/Y}^2(C)$ we can state as well $h(C) \geqslant \mathrm{sld}(f_*\omega_{\tilde{X}/Y}^2(C))$ and

COROLLARY 3.6. *Using the notations and assumptions made in* 3.3

$$h(C)\cdot(2\cdot(2g-1)^2 - 1) \geqslant c_1(\omega_{\tilde{X}/Y})^2.$$

4. Effective bounds for the height

We want to finish the proof of Theorem 1.

(4.1) Let $f\colon X \to Y$ be a family of curves. Let $S = \{y \in Y; f^{-1}(y) \text{ singular}\}$ and $D = f^*(S)$. We assume that D is a normal crossing divisor (i.e. an effective divisor, locally in the analytic topology with nonsingular components meeting transversally). Recall that f is called semistable when D is a reduced divisor.

Let $\Omega_X^1\langle D\rangle = \Omega_X^1\langle D_{\mathrm{red}}\rangle$ be the sheaf of differential forms with logarithmic poles along D. The natural inclusion $f^*\Omega_Y^1\langle S\rangle \to \Omega_X^1\langle D\rangle$ splits locally. In fact, f is locally given by $t = x^\alpha \cdot y^\beta$, where x and y are parameters on X and

t a parameter on Y. Then $\mathrm{d}t/t=\alpha\cdot \mathrm{d}x/x+\beta\cdot \mathrm{d}y/y$ is part of a local bases of $\Omega^1_X\langle D\rangle$. The quotient sheaf, denoted by $\Omega^1_{X/Y}\langle D\rangle$, is therefore invertible. Comparing the highest wedge products one finds

$$\Omega^1_{X/Y}\langle D\rangle = \omega_X(D_{\mathrm{red}})\otimes f^*\omega_Y(S)^{-1} = \omega_{X/Y}(D_{\mathrm{red}} - D).$$

As $C=\sigma(Y)$ meets D only in points which are smooth on D, the cokernel $\Omega^1_{X/Y}\langle D+C\rangle$ of $f^*\Omega^1_Y\langle S\rangle \to \Omega^1_X\langle D+C\rangle$ will be invertible as well and one has $\Omega^1_{X/Y}\langle D+C\rangle = \Omega^1_{X/Y}\langle D\rangle \otimes \mathcal{O}_X(C)$.

LEMMA 4.2. *If f is non isotrivial, there is a nonzero map*

$$\gamma\colon f_*(\omega_{X/Y}\otimes\Omega^1_{X/Y}\langle D\rangle)\to\Omega^1_Y\langle S\rangle.$$

Moreover, if C is a section γ factors through

$$\gamma'\colon f_*(\omega_{X/Y}\otimes\Omega^1_{X/Y}\langle D+C\rangle)\to\Omega^1_Y\langle S\rangle.$$

Proof of Theorem 1. If f is semistable then $\Omega^1_{X/Y}\langle D\rangle=\omega_{X/Y}$. 4.2 implies that $\mathrm{sld}(f_*(\omega^2_{X/Y}(C))\leqslant 2q-2+s$. In general one has an inclusion

$$f_*(\omega^2_{X/Y}(C))\to f_*(\omega_{X/Y}\otimes\Omega^1_{X/Y}\langle D+C\rangle)\otimes\mathcal{O}_Y(S)$$

and using 4.2 and 1.2(a) one obtains $\mathrm{sld}(f_*\omega^2_{X/Y}(C))\leqslant 2\cdot q-2+2\cdot s$. In both cases 3.5 gives $(2\cdot g-1)^{-2}\cdot h(C)\leqslant 2\cdot\mathrm{sld}(f_*\omega^2_{X/Y}(C))$.

The construction of γ': (4.3) We have a commutative diagram of exact sequences:

$$\begin{array}{ccccccc}
 & & 0 & & 0 & & \\
 & & \downarrow & & \downarrow & & \\
0\to f^*\Omega^1_Y\langle S\rangle & \to & \Omega^1_X\langle D\rangle & \longrightarrow & \Omega^1_{X/Y}\langle D\rangle & \to & 0\\
\downarrow{=} & & \downarrow & & \downarrow & & \\
0\to f^*\Omega^1_Y\langle S\rangle & \to & \Omega^1_X\langle D+C\rangle & \to & \Omega^1_{X/Y}\langle D+C\rangle & \to & 0\\
 & & \downarrow & & \downarrow & & \\
 & & \mathcal{O}_C & \xrightarrow{=} & \mathcal{O}_C & & \\
 & & \downarrow & & \downarrow & & \\
 & & 0 & & 0 & &
\end{array}$$

(4.4) Applying Rf_* to the diagram 4.3 we obtain

$$\begin{array}{ccc} f_*\Omega^1_{X/Y}\langle D\rangle & \xrightarrow{\ \delta\ } & R^1f_*\mathcal{O}_X \otimes \Omega^1_Y\langle S\rangle \\ \downarrow & & \downarrow{=} \\ f_*\Omega^1_{X/Y}\langle D+C\rangle & \xrightarrow{\ \delta^1\ } & R^1f_*\mathcal{O}_X \otimes \Omega^1_Y\langle S\rangle \end{array}$$

δ is the Kodaira-Spencer class and, since f is not isotrivial, both δ and δ' are non zero. If we tensorize 4.3 with $\omega_{X/Y}$ and apply Rf_* again we get

(4.5)

$$\begin{array}{ccc} f_*(\omega_{X/Y} \otimes \Omega^1_{X/Y}\langle D\rangle) & \xrightarrow{\ \gamma\ } & R^1f_*\omega_{X/Y} \otimes \Omega^1_Y\langle S\rangle \\ \downarrow & & \downarrow{=} \\ f_*(\omega_{X/Y} \otimes \Omega^1_{X/Y}\langle D+C\rangle) & \xrightarrow[\gamma']{} & R^1f_*\omega_{X/Y} \otimes \Omega^1_Y\langle S\rangle = \Omega^1_Y\langle S\rangle \end{array}$$

Since the left vertical arrow is injective γ' cannot be zero if γ is non zero.

LEMMA 4.6. *There is a commutative diagram*

$$\begin{array}{ccc} f_*\omega_{X/Y} \otimes f_*\Omega^1_{X/Y}\langle D\rangle & \xrightarrow{\ id\times\delta\ } & f_*\omega_{X/Y} \otimes R^1f_*\mathcal{O}_X \otimes \Omega^1_Y\langle S\rangle \\ \downarrow{m} & & \downarrow{\cup\otimes id} \\ f_*(\omega_{X/Y} \otimes \Omega^1_{X/Y}\langle D\rangle) & \xrightarrow[\gamma]{} & R^1f_*\omega_{X/Y} \otimes \Omega^1_Y\langle S\rangle \end{array}$$

where m is multiplication and $\cup$ *the cup product.*

Proof. One has a natural map $f^*f_*\omega_{X/Y} \to \omega_{X/Y}$ and taking the tensor product with the first row of 4.3 we get a commutative diagram

$$\begin{array}{ccccccccc} 0 \to & f^*f_*\omega_{X/Y} \otimes f^*\Omega^1_Y\langle S\rangle & \to & f^*f_*\omega_{X/Y} \otimes \Omega^1_X\langle D\rangle & \to & f^*f_*\omega_{X/Y} \otimes \Omega^1_{X/Y}\langle D\rangle & \to 0 \\ & \downarrow & & \downarrow & & \downarrow & \\ 0 \to & \omega_{X/Y} \otimes f^*\Omega^1_Y\langle S\rangle & \longrightarrow & \omega_{X/Y} \otimes \Omega^1_X\langle D\rangle & \longrightarrow & \omega_{X/Y} \otimes \Omega^1_{X/Y}\langle D\rangle & \longrightarrow 0 \end{array}$$

Applying Rf_* we obtain the diagram in 4.6.

Proof of 4.2. It is enough to show that γ is non zero. Let U be some open subvariety of Y such that $\Omega^1_Y\langle S\rangle$ is generated by a differential form α. Since δ is non zero we may choose $s \in f_*\Omega^1_{X/Y}\langle D\rangle(U)$ with $\delta(s) \neq 0$. For some

$0 \neq \lambda \in R^1 f_* \mathcal{O}_X(U)$ we can write $\delta(s) = \lambda \otimes \alpha$. Since $\bigcup$ induces a perfect pairing there is some $\mu \in f_* \omega_{X/Y}(U)$ with $\mu \bigcup \lambda \neq 0$. Then $\mu \bigcup \lambda \otimes \alpha \neq 0$ and by 4.6 this is the same as $\gamma(m(\mu \otimes s))$.

(4.7) Of course, 4.2 also implies that for non isotrivial families

$$\mathrm{sld}(f_*(\omega_{X/Y} \otimes \Omega^1_{X/Y}\langle D\rangle)) \leqslant 2q - 2 + s.$$

If f is semistable and minimal this together with 3.2 shows (see [9] for better bounds):

COROLLARY 4.8. *If f is semistable, relatively minimal and non isotrivial and if q is the genus of Y and s the number of degenerate fibres of f, then for $N > 1$*

$$(2q - 2 + s)(N \cdot (2g - 2) + 1)(2N - 1)(g - 1) \geqslant \deg(f_* \omega^N_{X/Y})$$

and

$$(2q - 2 + s)(2g - 2)^2 \geqslant \tfrac{1}{2} c_1(\omega_{X/Y})^2 > 0.$$

If f is not semistable let $\tau: Y' \to Y$ be a non singular Galois cover such that the ramification index of $s \in Y$ is divisible by the multiplicities of $f^{-1}(s)$. Let $\delta: X' \to X \times_Y Y'$ be the normalization, $f' = pr_2 \circ \delta: X' \to Y'$ and $\tau = pr_1 \circ \delta: X' \to X$. If the fibres of f are normal crossing divisors, then X' has at most rational Gorenstein singularities. Especially $\omega_{X'}$ is invertible and, if $\tau': X'' \to X'$ is a desingularization, then $\tau'_* \omega_{X''} = \omega_{X'}$. Moreover X' is non singular in a neighbourhood of a section C'. In fact, under the assumptions made, we may choose X'' such that $X'' \to Y'$ is semistable.

LEMMA 4.9. *Let $\mathcal{L}$ be an invertible sheaf on X and $\mathcal{L}' = \tau^* \mathcal{L}$. Then there is an inclusion $f'_*(\mathcal{L}' \otimes \omega_{X'/Y'}) \to \tau^* f_*(\mathcal{L} \otimes \omega_{X/Y})$ isomorphic in the general point of Y'.*

Proof. As in the proof of 2.4 one obtains by duality theory an inclusion

$$\delta_*(\mathcal{L}' \otimes \omega_{X'/Y'}) \to pr_1^*(\mathcal{L} \otimes \omega_{X/Y}).$$

4.9 follows by flat base change.

COROLLARY 4.10. *If $f: X \to Y$ is not semistable, then the bound for $h(\sigma(Y))$ given in Theorem 1 can be improved to*

$$h(\sigma(Y)) < 2 \cdot (2g - 1)^2 \cdot (2q - 2 + s) + s.$$

Sketch of proof. If $f: X \to Y$ is any morphism, not necessarily relatively minimal, and $C=\sigma(Y)$, then $h(C)\leqslant c_1(\omega_{X/Y})\cdot C$. Let $\tau\colon Y'\to Y$ be a nonsingular Galois cover with Galois-group G, such that $X\times_Y Y'\to Y'$ is birational to a semistable family of curves $\varphi\colon T\to Y'$, relatively minimal over Y'. Let B be the image of the section of φ lifting σ. The action of G on Y' extends to an action of G on T, such that X is birational to T/G.

CLAIM 4.11. There exists a sequence of at most $s\cdot\mathrm{ord}(G)$ blowing ups of points $\eta\colon X''\to T$ such that G acts on X'' and such that the quotient X''/G is non singular in some neighbourhood of C''/G, where C'' is the proper transform of B.

Using the notations from 4.11 one has $c_1(\omega_{X''/Y'})\cdot C''\leqslant s\cdot\deg(\tau)+h(B)$. We choose X to be a desingularisation of X''/G, isomorphic to X''/G near C''/G, such that $f\colon X\to Y$ has fibres with normal crossings. As above, let $f'\colon X'\to Y'$ be a minimal desingularization of $X\times_Y Y'\to Y'$, $\tau'\colon X'\to X$ the induced map, $C'=\tau'^{-1}(C)$ and $D'=\tau'^{-1}(D)$. One has $\tau'^*\Omega^1_{X'/Y'}\langle D'+C'\rangle=\Omega^1_{X'/Y'}\langle D'+C'\rangle$ and, since f' has a semistable model, both $f'_*\omega_{X'/Y'}(C')$ and

$$f'_*(\Omega^1_{X'/Y'}\langle D'+C'\rangle\otimes\omega_{X'/Y'})$$

are independent of the model chosen for f'. Especially both sheaves coincide. Using 3.5, 4.9, 1.2(b) and 4.2 one obtains

$$\begin{aligned}\tfrac{1}{2}(2g-1)^{-2}\cdot h(C')=\tfrac{1}{2}(2g-1)^{-2}\cdot h(B)&\leqslant \mathrm{sld}(f'_*\omega^2_{X'/Y'}(C))\\ &\leqslant \mathrm{sld}(f_*\omega_{X/Y}\otimes\Omega^1_{X/Y}\langle D+C\rangle)\cdot\deg(\tau)\\ &\leqslant (2q-2+s)\cdot\deg(\tau).\end{aligned}$$

On the other hand, near C the sheaves $\Omega^1_{X/Y}\langle D\rangle$ and $\omega_{X/Y}$ coincide and, using the notations introduced above, one has

$$\begin{aligned}h(C)&\leqslant c_1(\omega_{X/Y})\cdot C=c_1(\Omega^1_{X/Y}\langle D\rangle)\cdot C=\frac{1}{\deg(\tau)}c_1(\omega_{X'/Y'})\cdot C'\\ &\leqslant\frac{1}{\deg(\tau)}c_1(\omega_{X''/Y'})\cdot C''\leqslant s+\frac{h(B)}{\deg(\tau)}\leqslant 2\cdot(2g-1)^2(2q-2+s)+s.\end{aligned}$$

Proof of 4.11. The question is local in Y and we may replace G by the ramification group of some $p\in Y'$. Hence we assume Y to be a small disk and $G=\langle\sigma\rangle$ to be cyclic of order N. Let $Q=B\cap\varphi^{-1}(P)\in T$. We can find local coordinates x and y near Q, such that x is the pullback of a coordinate on Y', and such that the zero set of y is B. Moreover we can assume that $\sigma(x)=e\cdot x$ and $\sigma(y)=e^{\mu}\cdot$, where e is a primitive Nth root of unit and $O\leqslant\mu<N$. Blowing up Q we obtain T' and local coordinates x' and y' near $Q'=\varphi'^{-1}(p)\cap B'$ with $x'=x$ and $y'=y/x$. Therefore $\sigma(x')=e\cdot x'$ and $\sigma(y')=e^{\mu'}\cdot y'$ for $\mu'=\mu-1$. After at most N blowing ups we may assume that $\mu'=0$. Then, however, the quotient is non singular.

COROLLARY 4.12. *Assume that f is not isotrivial. Let q be the genus of Y and s be the number of degenerate fibres of f. Then one has*

$$0 < \mathrm{sld}(f_*(\omega_{X/Y} \otimes \Omega^1_{X/Y}\langle D\rangle)) \leqslant 2q - 2 + s.$$

4.12 implies the well known fact that a non isotrivial family of curves of genus $g \geqslant 2$ over Y must have at least three degenerate fibres, if $Y = \mathbb{P}^1$, and at least one degenerate fibre, if Y is an elliptic curve.

(4.13) If $f: X \to Y$ is a family of higher dimensional canonically polarized manifolds with degenerate fibres one can consider the iterated Kodaira–Spencer map $f_*\Omega^{n-1}_{X/Y}\langle D\rangle \xrightarrow{\delta^{n-1}} (\Omega^1_Y\langle S\rangle)^{n-1} \otimes R^{n-1}f_*\mathcal{O}_X$ and

$$f_*(\omega_{X/Y} \otimes \Omega^{n-1}_{X/Y}\langle D\rangle) \xrightarrow{\gamma^{(n-1)}} (\Omega^1_Y\langle S\rangle)^{n-1}.$$

With 3.1 and the same arguments we used above one obtains

PROPOSITION 4.14. *Assume that δ^{n-1} is non trivial then* $(2q - 2 + s) > 0$.

However, we do not know any reasonable criterion implying that $\delta^{n-1} \neq 0$.

References

1. H. Esnault and E. Viehweg, Dyson's lemma for polynomials in several variables (and the theorem of Roth). *Invent. Math.* 78 (1984) 445–490.
2. H. Esnault and E. Viehweg, Logarithmic De Rham complexes and vanishing theorems. *Invent. Math.* 86 (1986) 161–194.
3. T. Fujita, On Kähler fibre spaces over curves. *J. Math. Soc. Japan* 30 (1978) 779–794.
4. H. Grauert, Mordell's Vermutung über rationale Punkte auf algebraischen Kurven und Funktionenkörper. *Publ. Math. IHES* 25 (1965) 131–149.
5. Y. Kawamata, A generalization of Kodaira–Ramanujam's vanishing theorem. *Math. Ann.* 261 (1982) 57–71.
6. Yu. I. Manin, Rational points on an algebraic curve over function fields. *Trans. Amer. Math. Soc.* 50 (1966) 189–234.
7. S. Mori, Classification of higher-dimensional varieties. Algebraic Geometry. Bowdoin 1985. *Proc. of Symp. in Pure Math.* 46 (1987) 269–331.
8. A.N. Parshin, Algebraic curves over function fields *I. Math. USSR Izv.* 2 (1968) 1145–1170.
9. L. Szpiro, Séminaire sur les pinceaux de courbes de genre au moins deux. *Astérisque* 86 (1981).
10. E. Viehweg, Vanishing theorems. *J. Reine Angew. Math.* 335 (1982) 1–8.
11. E. Viehweg, Weak positivity and the additivity of the Kodaira dimension for certain fibre spaces. *Adv. Stud. Pure Math.* 1 (1983) 329–353 North-Holland.
12. E. Viehweg, Vanishing theorems and positivity in algebraic fibre spaces. *Proc. Intern. Congr. Math.*, Berkeley 1986, 682–687.
13. E. Viehweg, Weak positivity and the stability of certain Hilbert points. *Invent. Math.* 96 (1989) 639–667.
14. P. Vojta, Mordell's conjecture over function fields. Preprint 1988.
15. A.N. Parshin, Algebraic curves over function fields. *Soviet Math. Dokl.* 9 (1968) 1419–1422.

Compositio Mathematica **76**: 87–100, 1990.

Eine Klassenzahlformel für singuläre Moduln der Picardschen Modulgruppen

JAN FEUSTEL

Akademie der Wissenschaften der DDR, Karl-Weierstraß-Institut für Mathematik, Mohrenstraße 39, DDR-1086 Berlin

Received 3 December 1988; accepted in revised form 6 February 1990

Es ist altbekannt, daβ die elliptische Invariante $J(\tau)$ den Körper der Modulfunktionen auf der oberen Halbebene bzgl. $\mathrm{Sl}_2(Z)$ erzeugt und auf den Punkten $\tau \in H$, die imaginär-quadratische Zahlkörper erzeugen (und damit Fixpunkte von Elementen aus $\mathrm{Gl}_2^+(Z)$ sind), algebraische Werte annimmt. Durch Zuordnung dieser Punkte τ mit $Q(\tau) = K$ zu Idealen in Ordnungen des imaginär-quadratischen Zahlkörpers K erhalten wir mit der 'Klassengleichung' eine algebraische Gleichung für $J(\tau)$ mit rationalen Koeffizienten. In diesem Zusammenhang sei erinnert, daβ $J(\tau)$ zusammen mit der Weierstraβschen $\wp$-Funktion der zugehörigen elliptischen Kurve E_τ alle Abelschen Erweiterungen von $Q(\tau)$ erzeugt ('Kroneckers Jugendtraum').

Hecke verallgemeinerte in seiner Dissertation [3] diese Theorie auf die Wirkung der Hilbertschen Modulgruppe $\mathrm{Sl}_2(o_k)$ (k reell-quadratischer Zahlkörper) auf das Produkt zweier oberer Halbebenen (siehe Kapitel III).

Für die komplexe Kugel B und die Picardsche Modulgruppe $U((2,1), o_K) = \Gamma_K$ (wobei K ein imaginär-quadratischer Zahlkörper ist) lagen bis vor kurzer Zeit keine einschlägigen Ergebnisse vor. In seiner Arbeit [4] bestimmte Holzapfel erstmals den Ring der automorphen Formen bzgl. Γ_K und einiger Untergruppen für $K = Q(\sqrt{-3})$.

Der Ring dieser automorphen Formen von Γ_K wird erzeugt durch Formen G_i vom Gewicht i ($i = 2, 3, 4$) (exakter: der Ring der automorphen Formen von Γ_K bzgl. eines Twistes mit einem Charakter). Dabei sind die G_i die elementarsymmetrischen Funktionen $\sum 1 \leqslant j_1 < \cdots < j_i \leqslant 4\ X_{j_1} \cdots X_{j_i}$ für gewisse automorphe Formen einer Kongruenzuntergruppe von Γ_K (genauer für $\Gamma_K(\sqrt{-3}) = \{\gamma \in \Gamma_K \mid \det\gamma = 1, \gamma \equiv \mathrm{Id} \bmod \sqrt{-3}\}$), die sich nach den Ergebnissen von Shiga [10] und dem Autor [1] als Summen von Theta-Nullwerten auf $\mathscr{H}^3$ (obere Siegelsche Halbebene) mit einer automorphen Einlagerung $\eta\colon B \hookrightarrow \mathscr{H}^3$ und explizit bekannter Charakteristik darstellen lassen (exakt: $X_1 = \varphi_1^3 + \varphi_2^3 + \varphi_3^3$, $X_j = X_1 - 4\varphi_{j-1}^3$ für $2 \leqslant j \leqslant 4$, φ_j Theta-Nullwert).

Shiga zeigt nun auf analytischem Weg, daß für alle τ, für die ein

$$\gamma \in \left\{ \gamma' \in M_3(o_K) \mid {}^t\gamma' \begin{bmatrix} 1 & & 0 \\ & 1 & \\ 0 & & -1 \end{bmatrix} \bar{\gamma}' = p^2 \begin{bmatrix} 1 & & 0 \\ & 1 & \\ 0 & & -1 \end{bmatrix}, p \text{ träge in } K/Q \right\}$$

existiert, so daß τ isolierter Fixpunkt von γ ist, die Modulfunktionen

$$\delta_1 = \frac{G_3^2}{G_3^2} \quad \text{und} \quad \delta_2 = \frac{G_2^2}{G_4}$$

von Γ_K algebraische Werte annehmen [11]. Holzapfel komplettiert das Resultat für alle singulären Moduln z, d.h. $\{z \in B \mid \exists \gamma \in U((2, 1)m\, K)$: z isolierter Fixpunkt von $\gamma\}$; der Beweis arbeitet algebraisch-geometrisch und ist ‘unkonstruktiv’, d.h. liefert keinerlei Aussagen, in welchem Zahlkörper die Werte der Modulfunktionen liegen.

Das stellt natürlich erst einmal folgendes Programm auf: Zuerst wird eine arithmetische Charakterisierung aller singulärer Moduln für beliebiges imaginär-quadratisches K gesucht (Kapitel I).

Zweitens soll eine Zuordnung von gewissen singulären Moduln und Untergruppen von Idealklassengruppen von CM-Körpern $L \supset K$ mit $[L:K] = 3$ gegeben werden (Kapitel III). Die Klassifizierungsaufgabe der singulären Moduln würde für Picardsche Modulgruppen von der analogen aufgabe für Hilbertsche Modulgruppen, die in der Klassengleichung auftritt, abgelesen.

Von Interesse sind die automorphen Formen φ_i^3 bzw. G_i auch dadurch, daß sie die Umkehrung von Picards Abbildung ([7], [8]) vermitteln (siehe [5]):

Fur

$$\Lambda = \{(a'_1 : a'_2 : a'_3) | a'_i \neq 0,\ a'_i \neq a'_j \forall\, i, j,\ i \neq j\} \subset P_2(C)$$

und

$$\xi = (a_1 : a_2 : a_3) \in \Lambda \text{ sei } C_\xi\colon Y^3 = X \prod_{i=1}^{3} (X - a_i)$$

die zugehörige Picard-Kurve (vom Geschlecht 3 und mit K-Multiplikation).

Die Periodenmatrix von C bzgl. einer bis auf Monodromie von Λ fixierten Basis liefert eine mehrdeutige Abbildung in $\mathscr{H}^3$; bzgl. der oben erwähnten automorphen Einlagerung η: $B \hookrightarrow \mathscr{H}^3$ liegt das Bild in B, und die ‘Monodromie-Gruppe’ is gerade $\Gamma_K(\sqrt{-3})$, d.h. $\Phi\colon \Lambda \to B/\Gamma_K(\sqrt{-3})$ ist eindeutig, wobei $\operatorname{Im}\Phi =$
$B/\Gamma_K(\sqrt{-3}) \setminus D \cdot \Gamma_K/\Gamma_K(\sqrt{-3})$ für $D = \{(a, 0) \| a \backslash < 1\} \subset B$.

Die Umkehrung von Φ wird durch die drei Modulformen φ_i^3 vermittelt, und bei geeigneter Normierung der Nullstellen von $C_{\Phi^{-1}(\mathfrak{z})}$ gilt

$$C_{\Phi^{-1}(\mathfrak{z})}: Y^3 = \prod_{i=1}^{4} (X - X_i(\tau)) = X^4 + G_2X^2 + G_3X + G_4, \qquad \sum_{i=1}^{4} X_i(\mathfrak{z}) = 0,$$

d.h. die oben genannten G_i treten als Koeffizienten auf.

Im zweiten Kapitel wird nun nach Aussagen über $C_{(\varphi_1^3(\mathfrak{z}):\varphi_2^3(\mathfrak{z}):\varphi_3^3(\mathfrak{z}))}$ gefragt, speziell über die Einfachheit der Jacobischen.

0. Definitionen

Sei $B = \{(c_1, c_2) | c_i \in C, |c_1|^2 + |c_2|^2 < 1\} \subset C^2$ die zwei-dimensionale komplexe Einheitskugel, so wirkt $U((2,1), C)$ 'gebrochen rational' auf B, d.h. sei C^2 in $P_C^2 = PC^3$ eingelagert durch $\varphi(c_1, c_2) = (c_1 : c_2 : 1)$, so wirkt $U((2, 1), C)$ linear auf P_C^2, und für $\tau \in B$, $\gamma \in U((2, 1), C)$ gilt $\gamma \cdot \tau = \varphi^{-1} \cdot \gamma \cdot \varphi(\tau)$. Sei auf C^3 bzgl. Einer fixierten Basis die hermitesche Form $\langle\ ,\ \rangle$ durch die Matrix

$$\begin{bmatrix} 1 & & 0 \\ & 1 & \\ 0 & & -1 \end{bmatrix}$$

gegeben, d.h. $U((2, 1), C) = U(\langle\ ,\ \rangle, C)$. Sei K imaginär-quadratischer Zahlkörper und K^3 in C^3 bzgl. dieser Basis eingelagert, so ist o_K^3 ein hermitesches, unimodulares Gitter bzgl. $\langle\ ,\ \rangle$ und $U((2, 1), o_K) = \Gamma_K$ seine Automorphismengruppe (Isometrien), eine arithmetische Untergruppe von $U((2, 1), C)$, die auf B eigentlich diskontinuierlich wirkt.

Für $z = (z_1, z_2) \in B$ ist $\tilde{z} = {}^t(z_1, z_2, 1) \in C^3$, d.h. $\varphi z = \varphi\tilde{z}$.

Sei $z = (z_1, z_2) \in B$ algebraisch, d.h. $Q(z_1, z_2)$ algebraischer Zahlkörper und σ ein Automorphismus eines normalen Zahlkörpers über $Q(z_1, z_2)$, so ist $\tilde{z}^\sigma = {}^t(z_1^\sigma, z_2^\sigma, 1)$. $z \in B$ heißt K-singulärer Modul, falls ein $\gamma \in U((2, 1), K) = U_K$ existiert, für das z isolierter Fixpunkt ist.

I. Arithmetische Charakterisierung singulärer Moduln

LEMMA 1. *Sei L/K ein CM-Körper, K imaginär-quadratisch, so existiert ein $v \in L$ mit $v\bar{v} = 1$, $K(v) = L$.*

Beweis. Sei $m \in L$ mit $L = K(m)$, $m \neq \bar{m}$ (falls $m = \bar{m}$, so gehen wir zu

$m' = m + \kappa$, $\kappa \in K$, $\kappa \neq \bar{\kappa}$ über) und $\exists r \in \bar{N}$:

$$K\left(\frac{m}{\bar{m}}\right) = K\left(\frac{m + r}{\bar{m} + \varphi}\right)$$

(da nur endlich viele Zwischenkörper existieren). Wir nehmen an, daß aus $v \in L$, $v\bar{v} = 1$ folgt $K(v) \neq L$. Sei

$$v_1 = \frac{m}{\bar{m}}, \qquad v_2 = \frac{m + r}{\bar{m} + r},$$

so gilt $v_i\,\bar{v}_i = 1$ $(i = 1, 2)$ und $K(v_1) = K(v_2) = M \nsubseteq L$. Also existieren $n_1, n_2 \in M$ mit $m = n_1\bar{m}$, $m + r = n_2(\bar{m} + r)$ und damit $m(1 - (n_1/n_2)) = ((n_1/n_2) - n_1)r$. Da $m \notin M$, ist $n_2 = 1$, $n_2 = n_1$ und damit $m = \bar{m}$, ein Widerspruch. □

THEOREM 1. *Sei $z = (z_1, z_2) \in B$, so ist z ein K-singulärer Modul genau dann, wenn für $L_z = K(z_1, z_2)$ und M_z, den Galoisabschluß von L über K, gilt*

(1) $[L_z : K] \leqslant 3$
(2) *L_z ist CM-Körper*
(3) *Für alle $\sigma \in \mathrm{Gal}(M_z/K)$, $\sigma|_L \neq Id$ gilt $\langle \tilde{z}, \tilde{z}^\sigma \rangle = 0$.*

Beweis. ⇒Sei $\gamma \in U_K$ mit $\gamma z = z$, z isolierter Fixpunkt von γ, so ist $\gamma\tilde{z} = \lambda\tilde{z}$, λ erfüllt das charakteristische Polynom von γ und $\lambda\bar{\lambda} = 1$.

Daraus folgt sofort wegen der Einfachheit des Eigenwertes λ von γ (z ist isolierter Fixpunkt von γ!) $L_z = K(\lambda)$ und damit $[L_z : K] \leqslant 3$.

Sei $\sigma \in \mathrm{Gal}(M_z/K)$ wie in 3) gewählt, so ist λ^σ Eigenwert von γ mit Eigenvektor $\tilde{z}^\sigma$, und damit gilt $\lambda^\sigma\bar{\lambda}^\sigma = 1$. Die λ^σ dieser Form erzeugen M_z, und es gilt $\bar{\lambda}^\sigma = 1/\lambda^\sigma = (1/\lambda)^\sigma = (\bar{\lambda})^\sigma$ Damit kommutiert die komplexe Konjugation mit allen Einlagerungen von L_z in C, und es ist L_z CM-Körper nach dem Kriterium aus [6] S.6.

Weiterhin ist $\langle \tilde{z}, \tilde{z}^\sigma \rangle = \langle \gamma\tilde{z}, \gamma\tilde{z}^\sigma \rangle = \lambda\bar{\lambda}^\sigma\langle \tilde{z}, \tilde{z}^\sigma \rangle = \lambda/\lambda^\sigma\langle \tilde{z}, \tilde{z}^\sigma \rangle$, und da wegen $\sigma|_{L_z} \neq \mathrm{Id}$ $\lambda \neq \lambda^\sigma$ ist, gilt $\langle \tilde{z}, \tilde{z}^\sigma \rangle = 0$.

⇐Wir müssen Fallunterscheidungen machen:

(a) $L_z = K$: Sei γ gegeben durch $\gamma\tilde{z} = \tilde{z}$, $\gamma|_{\tilde{z}}\ \perp\ = -\mathrm{Id}$, so ist $\gamma K^3 = K^3$ und $\langle \delta\tilde{c}, \gamma\tilde{c} \rangle = \langle \tilde{c}, \tilde{c} \rangle$ für alle $\tilde{c} \in K^3$, also ist $\gamma \in U_K$ und $\tilde{z}$ isolierter Fixpunkt von γ.

(b) $[L_z : K] = 2$. Damit ist L_z normal über K und $\mathrm{Gal}(L_z/K) = (1, \sigma)$. Es gilt offensichtlich $\{\tilde{z}, \tilde{z}^\sigma\}^\perp = \tilde{z}^\perp \cap \tilde{z}^{\sigma\perp} = C\tilde{v}$, $\tilde{v} \in K^3$. Sei γ gegeben durch $\gamma\tilde{z} = \lambda\tilde{z}$, $\gamma\tilde{z}^\sigma = \lambda^\sigma\tilde{z}^\sigma$ und $\gamma\tilde{v} = \tilde{v}$, wobei λ nach Lemma 1) gewählt ist mit $\lambda\bar{\lambda} = 1$, $L_z = K(\lambda)$. Man sieht unmittelbar, daß $(L_z\tilde{z} \oplus L_z\tilde{z}^\sigma) \cap K^3 = \{\rho\tilde{z} + \rho^\sigma\tilde{z}^\sigma | \rho \in L_z\}$ und damit γ. $K^3 = K^3$, außerdem gilt wegen $\langle \gamma\tilde{c}, \gamma\tilde{c} \rangle = \langle \tilde{c}, \tilde{c} \rangle$ für alle $\tilde{c} \in C^3$ $\gamma \in U_K$ (da $\tilde{z}$, $\tilde{z}^\sigma$ und $\tilde{v}$ eine Basis von C^3 bilden), und z ist isolierter Fixpunkt von γ.

(c) $[L_z : K] = 3$. Es ist $[M_z : K] = 6$ oder 3, denn M_z ist Zerfällungskörper eines Polynoms vom Grade 3 über K. Sei $\sigma \in \mathrm{Gal}(M_z/K)$ gegeben mit $\mathrm{ord}\,\sigma = 3$ und damit $\sigma|_{L_z} \neq \mathrm{Id}$. Sei $\lambda \in L_z$ wieder nach Lemma 1) gewählt, so sei γ gegeben durch $\gamma\tilde{z}^{\sigma j} = \lambda^{\sigma j} \cong \sigma^j$ ($j = 0, 1, 2$). Dabei ist $K^3 = \{\sum_{j=0}^{2} \rho^{\sigma^j} \tilde{z}^{\sigma^j} | \rho \in L_z\}$, ergo $\gamma K^3 = K^3$ und $\langle \gamma\tilde{c}, \gamma\tilde{c}\rangle = \langle \tilde{c}, \tilde{c}\rangle$ für alle $\tilde{c} \in C^3$, d.h. $\gamma \in U_K$ und z ist isolierter Fixpunkt von γ (denn wegen $\langle \tilde{z}^{\sigma^j}, \tilde{z}^{\sigma^i}\rangle = 0$ für $i \neq j$ sind die z^j eine Basis von C^3). □

PROPOSITION 1: *Sei L CM-Körper mit $K \subset L$, $[L:K] \leqslant 3$, so existiert ein K-singulärer Modul z mit $L_z \simeq L$.*

Beweis. $[L:K] = 2$: Da L CM-Körper ist, ist L biquadratisch, $L = K(\sqrt{\kappa})$, $\kappa \in Q$, $\kappa > 0$. Sei $\omega = \sqrt{-d} + \sqrt{\kappa} + 1$ für $K = Q(\sqrt{-d})$, sei weiterhin $\lambda = \omega/\bar{\omega}^\sigma$ sowie φ: $L \hookrightarrow C$ als Einlagerung so vorgegeben, daß $\sqrt{\kappa} < 0$; es ist $\omega\bar{\omega} - \bar{\omega}^\sigma\omega^\sigma = 4\sqrt{\kappa} \neq 0$ für $\sigma \in \mathrm{Gal}(L/K)$. Es gilt wegen $\lambda\bar{\omega}^\sigma = \lambda(1 - \sqrt{-d}) - \lambda\sqrt{\kappa} = 1 + \sqrt{-d} + \sqrt{\kappa} = \omega$ sichtlich $\lambda \notin K$, also $L = K(\lambda)$. Man sieht unmittelbar, daß für $\tilde{z} = (\frac{3}{5}\lambda, \frac{4}{5}\lambda, 1) \in C^3$ gilt $\langle \tilde{z}, \tilde{z}\rangle = \lambda\bar{\lambda} - 1 < 0$, $\langle \tilde{z}, \tilde{z}^\sigma\rangle = 0$, d.h. z ist K-singulärer Modul.

$[L:K] = 3$: Hier wird der Beweis komplizierter, da sich keine explizite Erzeugung von L/K angeben läßt, die 'handhabbar' ist.

Sei $\{z_i\}$ eine K-Basis von L (als Vektorraum), $i = 1, 2, 3$, so setzen wir $H = (z_j^{\sigma^{(i-1)}}) \in M_3(L)$ für $\sigma \in \mathrm{Gal}(M/K)$, ord $\sigma = 3$. (M sei wieder der Galoisabschluß von L/K).

Sei $k = L \cap R$, so setze ich für

$$\rho' \in k H_{\rho'} = \begin{bmatrix} \rho' & 0 & 0 \\ 0 & \rho'^{\sigma} & 0 \\ 0 & 0 & \rho'^{\sigma^2} \end{bmatrix} \in M_3(L).$$

Es existiert ein $\rho \in k$ mit $\mathrm{Tr}\,\rho = 0$, o.B.d.A. sign $H_\rho = (2, 1)$ Wir setzen nun $\rho^* = \rho(-N_{k/Q}(\rho)^{-1} \det H \overline{\det H})$.

Es gilt für $N = {}^t\bar{H} H_{\rho^*}^{-1} H$:

(1) $N = (\mathrm{Tr}_{L/K}(\bar{z}_i z_j/\rho^*)) \in M_3(K)$

(2) sign N = sign $M_{\rho^*} = (2, 1)$

(3) $\det N = -(N_{k/Q}(\rho)/(\det H \overline{\det H}))^2 \equiv -1 \mod N_{K/Q}(K)$ (denn $\det H \overline{\det H} = \det {}^t H\bar{H} = \det(\mathrm{Tr}_{L/K} \bar{z}_i z_j) \in K \cap R = Q$.)

Nun sieht man aber sofort, daß dieselben Eigenschaften (1)–(3) auch für $N^{-1} = H^{-1} H_{\rho^*} {}^t\bar{H}^{-1}$ gelten, und nach dem Satz von Landherr (siehe [14], Kap. 10, §1, 1.6: (iv)). gilt nun

$$N^{-1} \underset{\mathrm{Gl}_3(K)}{\simeq} \begin{bmatrix} 1 & & 0 \\ & 1 & \\ 0 & & -1 \end{bmatrix}, \qquad \text{d.h. } N^{-1} = A \begin{bmatrix} 1 & & 0 \\ & 1 & \\ 0 & & -1 \end{bmatrix} {}^t\bar{A},$$

$A \in \mathrm{Gl}_3(K)$. Sei nun $(z_1, z_2, z_3)\ A = r(z_1', z_2', 1)$, $r, z_1', z_2' \in L$, so ist $(z_1', z_2') \in B$ K-singulärer Modul nach Theorem 1). □

II. Singuläre Moduln und Abelsche Mannigfaltigkeiten

Shimura hat in seinen Arbeiten [12] und [13] gezeigt, wie jedem imaginär-quadratischen Zahlkörper K und jedem Punkt $z \in B$ eine Abelsche Mannigfaltigkeit A_z mit Polarisierung Θ_z und einer mit Θ_z verträglichen Darstellung $\varphi_z\colon K \hookrightarrow \mathrm{End}^0 A_z = (\mathrm{End}\, A) \otimes_Z Q$ zugeordnet wird. Diese Zuordnung ist eindeutig mod Γ_K.

Dies geschieht auf folgende Weise:

$$\text{Sei } M = o_K^3 \subset K^3, \qquad T = \frac{1}{w}\begin{bmatrix} 1 & & 0 \\ & 1 & \\ 0 & & -1 \end{bmatrix} \text{ mit } w\sqrt{-1} > 0 \text{ und } (w) = \partial_{K/Q}$$

(Differente von K/Q, wobei diese stets Hauptideal in $\mathfrak{I}(K)$ ist) sowie h eine komplexe Struktur auf $M \otimes_Z R$ (M als Z-Gitter der Dimension 6 aufgefaßt), so daß $T\bar{h}$ hermitesch positiv definit ist. h ist dabei durch seinen $(-i)$-Eigenraum eindeutig bestimmt, d.h. durch dessen normierte Basis ${}^t(z_1, z_2, 1)$, $(z_1, z_2) = z \in B$ (und somit $h = h_z$). Dann sei $(M \otimes_Z R)^{h_z}$ der dreidimensionale C-Vektorraum, dessen Struktur von h_z herrührt, und $A_z = (M \otimes_Z R)^{h_z}/M \simeq C^3/\Lambda_z$. Die Riemannsche Form (und damit die Polarisierung Θ_z) ist durch $\mathrm{Tr}^t_{K/Q} \alpha T \mathscr{C}$ ($\alpha, \mathscr{C} \in K^3 \subset C^3$) gegeben, φ_z durch die gewöhnliche skalare Multiplikation in K^3.

Man macht sich klar, daß $(A_z, \Theta_z, \varphi_z) \simeq (A_z', \Theta_z', \varphi_z')$ genau dann gilt, wenn ein $\gamma \in \Gamma_K$ existiert mit $\gamma z = z'$.

Im weiteren sei $(A_z, \Theta_z, \varphi_z)$ das zu $z \in B$ gehörige Shimura-Tupel genannt.

THEOREM 2. *Sei $z \in B$ K-singulärer Modul mit $[L_z : K] = 3$ und $(A_z, \Theta_z, \varphi_z)$ das zugehörige Shimura-Tupel.*

Dann ist A_z einfach, und es gilt $\mathrm{End}^0 A_z = L_z$.

Beweis. Sei $\gamma \in \mathrm{U}_K$ mit $\gamma z = z$, $\gamma \tilde{z} = \lambda \tilde{z}$, $L_z = K(\lambda)$. Da $\gamma z = z$, so kommutiert γ als lineare Abbildung auf C^3 mit h_z und liefert ergo eine lineare Abbildung auf $(M \otimes_Z R)^{h_z}$ mit den Eigenwerten $\bar{\lambda}$, λ^σ, λ^{σ^2} für $\sigma \in \mathrm{Gal}(M_z/K)$, ord $\sigma = 3$. Man sieht unmittelbar, daß $\gamma \in \mathrm{End}^0\, (A_z)$.

Nach Konstruktion hat $\varphi_z(\alpha)$ für $\alpha \in K$ die Eigenwerte $\bar{\alpha}, \alpha, \alpha$ auf $(M \otimes_Z R)^{h_z}$, wobei die Eigenräume für $\bar{\lambda}$ und $\bar{\alpha}$ sowie für α und die direkte Summe derjenigen von λ^σ und λ^{σ^2} übereinstimmen, da $\gamma\ \Theta_z$ erhält.

Also existiert eine Darstellung $\Psi_z\colon L_z \hookrightarrow \mathrm{End}^0 A_z$ mit $\Psi_z|_K = \varphi_z$ und $\Psi_z(\lambda) = \gamma$; A_z mit dieser Darstellung Ψ_z ist in der Terminologie von Lang [6] also vom CM-Typ $\{L_z\colon \rho, \sigma, \sigma^2\}$ (ρ die komplexe Konjugation in $\mathrm{Gal}(M_z/Q)$). Für $S =$

$\{\rho, \sigma, \sigma^2\}$ gilt aber $\{\eta \in \mathrm{Gal}(M_z/Q) | S\eta = S\} = \mathrm{Id}$, und damit (nach Theorem 3.5 aus [6]) ist A_z einfach, und es gilt $\mathrm{End}^0 A_z = L_z$.

KOROLLAR 1. *Sei L CM-Körper über dem imaginär-quadratischen Zahlkörper K mit $[L:K] \leqslant 3$, $D = \{(a, 0) \mid a < 1\}$, $D' = \Gamma_K \cdot D$, so existiert ein $z \in B \backslash D'$ mit $L_z \simeq L$; falls $[L:K] = 3$, so liegt auf D kein z mit $L_z \simeq L$.*

Beweis. Man macht sich unmittelbar klar, daß für $z \in D$ und damit auch für $z \in D' A_z$ in das Produkt zweier Abelscher Mannigfaltigkeiten zerfällt, denn auf $K^t(0, 1, 0)$ stimmen für $z \in D$ die skalaren Multiplikationen mit K, die von der natürlichen Struktur von K^3 bzw. von h_z herrühren, überein, also kann $C^t(0, 1, 0)/o_K{}^t(0, 1, 0)$ von A_z abgespalten werden. Nach Theorem 2) kann also für einen singulären Modul $z \in B$ mit $[L_z : K] = 3$ z nicht auf D' liegen.

Falls $[L:K] = 2$, so betrachten wir z aus dem Beweis von Proposition 1. Sei $z \in \tilde{D} = \{{}^t(a, 0, b) \mid a, b \in C\}$, so ist $\gamma\tilde{z}^\sigma = (\gamma\tilde{z})^\sigma$ ebenfalls in $\tilde{D}$, und $\gamma\tilde{z}$ und $\gamma\tilde{z}^\sigma$ spannen $\tilde{D}$ auf, d.h. $\tilde{D}^\perp = \langle \gamma\tilde{z}, \gamma\tilde{z}^\sigma \rangle^\perp = C\gamma^t(4, -3, 0)$. Es ist nun $\tilde{D}^\perp \cap o_K^3 = o_K^t(0, 1, 0)$, d.h. $\tilde{D}^\perp \cap o_K^3$ enthält einen Vektor der Länge 1.

Andererseits ist $C\gamma^t(4, -3, 0) \cap o_K^3 = (C^t(4, -3, 0) \cap o_K^3) = o_K\gamma^t(4, -3, 0)$, und dies enthält keinen Vektor der Länge 1, ein Widerspruch. □

KOROLLAR 2. *Sei $z \in B$, z singulärer Modul mit $[L_z : K] = 3$, so ist* $\mathrm{Jac}(C_{\Phi^{-1}(z)})$ einfach. (Zur *Terminologie siehe die Einleitung*!).

Beweis. Auf der Abelschen Mannigfaltigkeit $\tilde{A}_z = \mathrm{Jac}(C_{\Phi^{-1}(z)})$ ist eine kanonische Polarisierung $\tilde{\Theta}_z$ und eine kanonische Darstellung $\tilde{\varphi}_z$ von o_K in $\mathrm{End}\, A_z$ gegeben (die durch die Wirkung des Automorphismus f der Ordnung 3, $f(y) = \omega y$, $f(x) = x$, ω primitive 3. Einheitswurzel von $C_{\Phi^{-1}(z)}$ auf die Differentiale erster Art definiert ist).

In seiner Arbeit [1*a*] hat der Autor gezeigt, daß $(\tilde{A}_z, \tilde{\Theta}_z, \tilde{\varphi}_z)$ und $(A_z, \Theta_z, \varphi_z)$ übereinstimmen. Die Aussage des Korollars folgt nun unmittelbar aus Theorem 2. □

III. Singuläre Moduln und Ideal-klassengruppen

In diesem Abschnitt soll ein Zusammenhang hergestellt werden zwischen den singulären Moduln $z \in B$ mit $L_z \simeq L$ für ein fixiertes L und gewisse Idealklassengruppen in L. Der Sinn der folgenden Ergebnisse besteht darin, daß sie sozusagen erste Schritte auf dem Weg zur Aufstellung konkreter algebraischer Gleichungen für die Werte der speziellen Modulfunktionen δ_1, δ_2 (siehe Einleitung) darstellen, der sogenannten Klassengleichung.

Hier seien kurz die 'parallelen' Ergebnisse Heckes [3] für den Fall Hilbertscher Modulflächen rekapituliert. Sei k ein reell-quadratischer Körper, $H \times H$ das Produkt zweier oberer Halbebenen, auf die $\mathrm{Gl}_2^+(o_K) =$

$\{\gamma \in \mathrm{Gl}_2(o_k) \mid \det \gamma \gg 0\}$ wirkt, s: $(\xi, \xi') = (-\xi, -\xi')$ Abbildung von $H \times H$ auf sich, $S = \langle s, \mathrm{Gl}_2^+(o_k) \rangle$.

Hecke zeigt in seiner Dissertation, daß (ξ, ξ') k-singulärer Modul ist, d.h. ein $\gamma \in M_2^+(o_k) = \{\gamma \in M_2^+(o_k) \mid \det \gamma \gg 0\}$ existiert mit $\gamma(\xi, \xi') = (\xi, \xi')$ genau dann, wenn ein biquadratischer CM-Körper K existiert mit $\xi \in K$, $\xi' \in H$ Wurzel der Gleichung von über k mit in k konjugierten Koeffizienten. Jedem singulären Modul (ξ, ξ') sei der o_k-Modul $M_\xi = o_k \cdot \xi \oplus o_k \cdot 1$ in K zugeordnet, der Ideal bzgl. einer o_k-Ordnung r_f in o_k ist. Dabei sind M_ξ und M_{ξ^0} gleich genau dann, wenn ein $\gamma \in \mathrm{Gl}_2^+(o_k)$ existiert mit $\gamma(\xi, \xi') = (\xi^0, \xi^{0\prime})$, d.h. $\gamma\xi = \xi^0$. Die Ordnung r_f mit Führer $f \in \mathfrak{I}(k)$ wird dabei durch die "Primitivdiskriminante' $D_\xi = f^2 \partial_{K/k}^2$ ($\partial_{K/k}$ die Differente) von ξ bestimmt.

Das Polynom

$$\Phi(X) = \prod_{\substack{\xi \bmod \mathrm{Gl}_2^+(o_k) \\ \xi \in K,\, D_\xi = f^2 \partial_{K/k}^2}} (X - F(\xi, \xi'))$$

hat für bestimmte Modulfunktionen F bzgl. S (die sich aus speziellen Thetanullwerten bei der automorphen Einlagerung von $H \times H$ in die obere Siegelsche Halbebene $\mathscr{H}^2$ ergeben) rationale Koeffizienten (Heckes Beweis ist allerdings lückenhaft), $F(\xi, \xi')$ ist also algebraisch, und der Grad der dadurch erzeugten Körpererweiterung ist ablesbar (da $\Phi(X)$ als irreduzibel bewiesen wird); für $f = 1$ ist er z.B. $N_{K/k}^{-1}(\mathrm{Cl}_k(1))$, da die freien o_k-Moduln $M \subset K$ mit $M \in \mathfrak{I}(K)$ bis auf die Multiplikation mit einer Konstanten aus K (projektive Äquivalenz) den Idealklassen $N_{K/k}^{-1}(A)$ für gewisse $A \in \mathrm{Cl}(k)$ entsprechen ($A\,\mathrm{Cl}_k(\partial_{K/k}^2) = \mathrm{Cl}_k(1)$).

Dies Programm soll sozusagen als Vorbild dienen für das Kapitel, das die Grundlagen der oben skizzierten Theorie konstituieren soll, nämlich die Zuordnung gewisser singulärer Moduln z B Idealklassen in $\mathrm{Cl}(L)$. Der Autor setzt seine Hoffnung darauf, daß diese Zuordnung vielleicht auch zu einem polynomialen Gleichungstyp ähnlich der Klassengleichung führen kann, wenn wir als Nullstellen die Werte der Funktionen δ_i in diesen gewissen singulären Moduln ansetzen. Von diesem Gleichungstyp müßten wir den Grad kennen und zu beweisen vermögen, daß die Koeffizienten in einem bestimmten algebraischen Zahlkörper liegen.

Im weiteren seien der imaginär-quadratische Zahlkörper K und der CM-Körper L mit $[L:K] = 3$ sowie eine Einlagerung $\varphi: N \hookrightarrow C$ für den Galoisabschluß N von L über K fixiert sowie $k = L \cap R$.

DEFINITION 1. Sei M ein o_K-Modul in L, so gelte:

$$E_L^M = \{\varepsilon \in L^* \mid \varepsilon M = M\}$$
$$E_k^M = \{\varepsilon \in k^* \mid \varepsilon M = M\}$$
$$E_k^{M+} = \{\varepsilon \in E_k^M \mid \varepsilon \gg 0\}$$

DEFINITION 2. Sei $G_L = \{M | M$ freier dreidimensionaler o_K-Modul in L, versehen mit der hermiteschen Form $\langle m, n\rangle = \mathrm{Tr}_{L/K} m\bar{n}$, für die ein $\rho \in k$ existiert mit $\varphi(\rho) < 0$, $\varphi(\rho^\sigma) > 0$ für alle $\sigma \in \mathrm{Gal}(N/K)$, $\rho^\sigma \neq \rho$, so daß $\rho \cdot M = M^{\#}\}$.

THEOREM 3. *Es existiert eine surjektive Abbildung* $\Psi: \{z \in B \,|\, z$ *K-singulärer Modul*, $L_z = \varphi(L)\}/\Gamma_K \to G_L \mathrm{mod}\ L^* = PG_L$ *mit* $\#\Psi^{-1}(M) = [E_k^{M+}: N_{L/k}(E_L^M)]$.

Beweis. 1. Konstruktion der Abbildung.

Sei $z = (z_1, z_2) \in B$ K-singulärer Modul mit $L_z = \varphi(L)$. Nach Theorem 1 gilt für $H = (z_j^{\sigma^{(i-1)}})$ ($z_3 = 1$, $\sigma \in \mathrm{Gal}(N/K)$ wie oben) und

$$\Phi = \begin{bmatrix} 1 & & 0 \\ & 1 & \\ 0 & & -1 \end{bmatrix} \quad \text{ergo} \quad H\Phi{}^t\bar{H} = H_\rho = \begin{bmatrix} \varphi(\rho) & & 0 \\ & \varphi(\rho^\sigma) & \\ 0 & & \varphi(\rho^{\sigma^2}) \end{bmatrix}$$

für $\rho \in k$, $\varphi(\rho) < 0$, $\varphi(\rho^{\sigma'}) < 0$ für alle $\sigma' \in \mathrm{Gal}(N/K)$, $\rho^{\sigma'} \neq \rho$. Damit gilt $\Phi = H^{-1} H_\rho {}^t\bar{H}^{-1}$ und wegen $\Phi^{-1} = \Phi$ auch $\Phi = {}^t\bar{H} H_{\rho^{-1}} H$, d.h. $(\mathrm{Tr}_{L/K}(z_i z_j/\rho)) \in \mathrm{Gl}_3(o_K)$. Setzen wir $M = \bigoplus_{i=1}^3 o_K z_i = \Psi(z)$, so gilt also für $M' = \bigoplus_{i=1}^3 o_K r_i$, $r_i = z_i/\rho$ $(\langle z_i, r_i\rangle) \in \mathrm{Gl}_3(o_K)$, d.h. jedes lineare Funktional auf M in o_K läßt sich durch ein $x \in M'$ mit $\varphi(y) = \langle y, x\rangle$ für alle $y \in M$ darstellen, und damit ist $M' = M^{\#}$, also $M \in G_L$.

2. *Surjektivität*

Sei $M^{\#} = \rho M$ für ein ρ aus Definition 2), so ist für $M = \bigoplus_{i=1}^3 o_K z_i$ und $D_M = (\mathrm{Tr}_{L/K} \rho \bar{z}_i z_j)$ D_M eine hermitesche Matrix aus $\mathrm{Gl}_3(o_K)$, sign $(D_M) = (2, 1)$ und nach [O] 2.2. Lemma 4 $D_M \underset{\mathrm{Gl}_3(o_K)}{\simeq} \Phi$ und damit auch $D_M^{-1} \underset{\mathrm{Gl}_3(o_K)}{\simeq} \Phi$.

Sei $H = (z_j^{\sigma^{(i-1)}})$ für ein $\sigma \in \mathrm{Gal}(N/K)$ mit ord $\sigma = 3$ und

$$H_\rho = \begin{bmatrix} \varphi(\rho) & & 0 \\ & \varphi(\rho^\sigma) & \\ 0 & & \varphi(\rho^{\sigma^2}) \end{bmatrix},$$

so ist $D_M = {}^t\bar{H} H_\rho H$ und ergo $H^{-1} H_\rho^{-1} {}^t\bar{H}^{-1} \underset{\mathrm{Gl}_3(o_K)}{\simeq} \Phi$, d.h. es existiert ein $A \in \mathrm{Gl}_3(o_K)$ mit $HA\Phi{}^t\bar{A}{}^t\bar{H} = H_\rho^{-1} = H_{\rho^{-1}}$. Nach Theorem 1) ist nun für $HA = (z_j'^{\sigma^{(i-1)}})$ und $M' = \bigoplus_{i=1}^3 o_K z_i'$ also $z = (z_1'/z_3', z_2'/z_3') \in B$ K-singulärer Modul mit $\Psi(z) \equiv M$ mod L^*, denn $M' = M$.

3. $\Psi^{-1}(M) = [E_k^{M+} : N_{L/k}(E_L^M)]$

Man sieht aus dem Beweis von 2. unmittelbar, daß HA (d.h. auch z) modulo

Rechtsmultiplikation mit $\langle \Gamma_K, E_L^M \cdot \mathrm{Id} \rangle$ (und damit auch $z \bmod \Gamma_K$) eineindeutig durch $M \bmod L^*$ und $\rho \bmod N_{L/k}(E_L^M)$ bestimmt ist. □

DEFINITION 3. Sei M ein endlich dimensionaler (torsionsfreier) Modul über einem Dedekindschen Ring R, so ist $M = \otimes \alpha_i m_i$, $m_i \in M \bigotimes_R Q_R$, $\alpha_i \in \mathfrak{I}(R)$, wobei Q_R der Quotientenkörper von R sei. Wir setzen $\mathrm{Cl}_R(M) = \mathrm{Cl}(\Pi \alpha_i) \in \mathrm{Cl}(R)$.

LEMMA 2. *Seien M_1, M_2 zwei o_K-Gitter gleicher Dimension mit $M_1 \supset M_2$ (d.h. torsionsfreie o_K-Moduln) und $M_1/M_2 \simeq o_K/\mathfrak{p}$ für ein Ideal $\mathfrak{p} \in \mathfrak{I}(K)$, so ist $\mathrm{Cl}_{o_K}(M_2)\, \mathrm{Cl}_{o_K}^{-1}(M_1) = \mathrm{Cl}(\mathfrak{p}) \in \mathrm{Cl}(K)$.*

Beweis. Reiner [9] Theorem 1.2. (v) S. 162.

DEFINITION 4. Sei $\mathfrak{I}_L = \{z \in B \mid z$ *K-singulärer Modul*, $L_z = \varphi(L)$, $\bigoplus_{i=1}^2 o_K z_i \otimes o_K \in \mathfrak{I}(L)\}$.

KOROLLAR 3. $\# \mathfrak{I}_L/\Gamma_K = [o_k^{*+} \colon N_{L/k}(o_L^*)] \# \{\alpha/\alpha \in \mathrm{Cl}(L), N_{L/k}(\alpha)\, \mathrm{Cl}_k(\partial_{L/K} \cap k) = \mathrm{Cl}_k(1)$, *$\partial_{L/K} \in \mathfrak{I}(L)$ die Differente, $N_{L/K}(\alpha)\mathfrak{p} = \mathrm{Cl}_K(1)$ für $\mathfrak{p} = \mathrm{Cl}_{o_K}(o_L)$ und es existiert ein $\rho \in k$ mit $\varphi(\rho) < 0$, $\varphi(\rho^\sigma) > 0$ für alle $\sigma \in \mathrm{Gal}(N/K)$, $\rho^\sigma \neq \rho$ und ein $\alpha' \in \mathfrak{I}(L)$, $\mathrm{Cl}_L(\alpha') = \alpha$, so daß $N_{L/k}(\alpha')(\partial_{L/K} \cap k) = (\rho)\}$.*

Beweis. Nach Theorem 3) ist nur zu zeigen: Sei $\alpha' \in \mathfrak{I}(L)$, so gilt

(1) $\alpha'^{\#} = \{b \in L \mid \mathrm{Tr}_{L/K} b\bar{a} \in o_K \, \forall a \in \alpha'\} = (\bar{\alpha} \cdot \partial_{L/K})^{-1}$.
(2) α' freier o_K-Modul genau dann, wenn $N_{L/K}(\mathrm{Cl}_L(\alpha'))\mathfrak{p}$ Haupt ideal ist
(3) $(\partial_{L/K} \cap k) \bigotimes_{o_k} o_L = \partial_{L/K}$

(1) Ergibt sich sofort aus der Definition der Differente. (2) Folgt sofort aus Lemma 2). (3) Zuerst macht man sich klar, daß $\partial_{L/K} = \bar{\partial}_{L/K}$, denn da L CM-Körper ist, ist mit $\mathfrak{p} \in \mathfrak{I}(K)$ auch $\bar{\mathfrak{p}}$ in L/K verzweigt. Es genügt also zu zeigen, daß für $\mathfrak{p} \in \mathfrak{I}(L)$ prim, $\mathfrak{p}/\partial_{L/K}$, $\mathfrak{p} = \bar{\mathfrak{p}}$, $(\mathfrak{p} \cap k) \bigotimes_{o_k} o_L \neq \mathfrak{p}$ (d.h. $\mathfrak{p}$ total verzweigt in L/Q) der Exponent von $\mathfrak{p}$ in $\partial_{L/K}$ gerade ist. Falls $N(\mathfrak{p}) \neq 3$, folgt dies sofort aus dem Exponentensatz ([2] S.431) wegen $[L:K] = 3$. Falls $N(\mathfrak{p}) = 3$, d.h. 3 total verzweigt in L/Q, so ist nach dem Schachtelungssatz für die Differente ([2] S.430) $\partial_{L/K}\partial_{K/Q} = \partial_{L/k}\partial_{k/Q}$, und da der Exponent von $\mathfrak{p}$ in $\partial_{L/k}$ nach dem Exponentensatz 1 ist, in $\partial_{K/Q}$ 3 und in $\partial_{k/q}$ gerade, ist der Exponent von $\mathfrak{p}$ in $\partial_{L/K}$ auch gerade. □

THEOREM 4. *Es gilt* $\# \mathfrak{I}_L/\Gamma_K = (h(L)/h(K)h(k))\, [o_k^* \colon N_{L/k}(o_L^*)]/8$.

Beweis. Nach Korollar 3 ist nur Folgendes zu beweisen:

(a) Für alle $v \in \mathrm{Cl}(k)$, $w \in \mathrm{Cl}(K)$ existiert ein $z \in \mathrm{Cl}(L)$ mit $N_{L/K}(z) = w$, $N_{L/k}(z) = v$.

(b) Für jedes Tupel (i_1, i_2, i_3) mit $i_j \in \{0, 1\}$ existiert ein $\alpha \in \mathfrak{I}(L)$ mit $N_{L/k}(\alpha) = (\rho)$, $\varphi(\rho^{\sigma^{(j-1)}})(-1)^{i_j} > 0$ für ein fixiertes $\sigma \in \mathrm{Gal}(N/K)$, ord $\sigma = 3$.

Zu (a) ist erst einmal zu beweisen, daß $N_{L/K}$ und $N_{L/K}$ auf $\mathrm{Cl}(L)$ surjektive Abbildungen sind. Für $N_{L/k}$ folgt dies sofort daraus, daß L und $\tilde{k}$ (der

Hilbertsche Klassenkörper von k) linear disjunkt sind. Für $N_{L/K}$ ist zu zeigen, daß L nicht im Hilbertschen Klassenkörper $\tilde{K}$ von K liegt.

Sei $L \subset \tilde{K}$, so ist L/K abelsch und unverzweigt. Nach dem Schachtelungssatz für Diskrimanten gilt $N_{K/Q}(\mathscr{D}_{L/K})\mathscr{D}_{K/Q}^3 = N_{k/Q}(\mathscr{D}_{L/k})\mathscr{D}_{k/Q}^2$. Da k/Q abelsch mit $[k:Q] = 3$, so existiert ein p, das in k/Q verzweigt ist. Nach dem Exponentensatz teilt p^2 $\mathscr{D}_{k/Q}$; falls $p \neq 2$, teilt aber nach demselben Satz p höchstens in der Potenz 1 die Diskriminante $\mathscr{D}_{K/Q}$; also muß p $N_{K/Q}(\mathscr{D}_{L/K})$ teilen, d.h. L/K ist nicht unverzweigt. Falls $p = 2$, so teilt p mit dem Exponenten 2 die Diskriminante $\mathscr{D}_{k/Q}$, aber mit einem Exponenten $\geqslant 2$ die Diskriminante $\mathscr{D}_{K/Q}$, also teilt 2 auch $N_{k/Q}(\mathscr{D}_{L/k})$, 2 ist voll verzweigt in L/Q, also auch in L/K. In jedem Fall folgt $L \not\subset \tilde{K}$, also sind L und K linear disjunkt, und $N_{L/K}$ ist surjektiv.

Sei nun $r_w \in N_{L/K}^{-1}(w)$, $r_v \in N_{L/k}^{-1}(v)$, so gilt für $z = r_w^4 r_v^3/(wvN_{L/K}(r_v)N_{L/k}(r_w)^2)$ (wobei die Ideale in k bzw. K durch Tensorieren mit o_L als Ideale in L aufgefaßt werden):

$$N_{L/K}(z) = w^4 N_{L/K}(r_v)^3 w^{-3} N_{L/K}(r_v)^{-3} = w,$$
$$N_{L/k(z)} = v^3 N_{L/k}(r_w)^4 v^{-2} N_{L/k}(r_w)^{-4} = v.$$

Am weitaus kompliziertesten gestaltet sich der Beweis von (b).

PROPOSITION 2. *Sei $\sigma \in \mathrm{Gal}(N/K)$, ord $\sigma = 3$ gegeben. Falls ein $\alpha \in k$ existiert, so daß $\alpha \gg 0$ und α prim zu $\mathscr{D}_{L/K}$ ist und für $H_o = \{(\alpha) \mid \alpha \in k, \alpha \gg 0, \alpha \equiv 1 \bmod \mathscr{D}_{L/k}\}$ gilt $(\alpha) \notin N_{L/k}(\mathfrak{J}_{\mathscr{D}_{L/k}}(L))H_0$, so existiert zu jedem Tupel $\{i_j\}, j = 1, 2, 3, i_j \in \{0, 1\}$ ein $\beta \in k$ mit $(\beta) \in N_{L/k}(\mathfrak{J}(L))$ und $\varphi(\beta^{\sigma^{(j-1)}})(-1)^{i_j} > 0$, wobei $\mathfrak{J}_{\mathscr{D}_{L/k}}(L) = \{\alpha \in \mathfrak{J}(L) \mid \alpha$ prim zu $\mathscr{D}_{L/k}\}$.*

Beweis. Nach [2a] S. 22-24 gilt dür $\mathfrak{J}_{\mathscr{D}_{L/k}}(L)$ und $\mathfrak{J}_{\mathscr{D}_{L/k}}(k)$ (analog definiert) $[\mathfrak{J}_{\mathscr{D}_{L/k}}(k): N_{L/k}(\mathfrak{J}_{\mathscr{D}_{L/k}}(L))H_0] = 2$.

Falls ein solches α wie oben existiert, so ist für $H_1 = \{(\alpha) \mid \alpha \in k, \alpha$ prim zu $\mathscr{D}_{L/k}, \alpha \gg 0\}$ und $H_2 = \{(\alpha) \mid \alpha \in k, \alpha$ prim zu $\mathscr{D}_{L/k}\}$ $[N_{L/k}(\mathfrak{J}_{\mathscr{D}_{L/k}}(L))H_1 : N_{L/k}(\mathfrak{J}_{\mathscr{D}_{L/k}}(L))H_0] = 2$ und damit $[N_{L/k}(\mathfrak{J}_{\mathscr{D}_{L/k}}(L))\, H_2 : N_{L/k}(\mathfrak{J}_{\mathscr{D}_{L/k}}(L))H_1] = 1$.

Da H_2 sichtlich zu jedem Tupel $\{i_j\}$ ein β' enthält mit $\varphi(\beta'\sigma^{(j-1)})(-1)^{i}j > 0$, gilt die obige Behauptung. □

LEMMA 3. *Seien L, K, k wie oben gegeben, so gilt für eine Primzahl p: Falls p in L/Q nicht total verzweigt ist, so sind die p-Komponenten von $\mathscr{D}_{L/k}$ und $\mathscr{D}_{K/Q}$ identisch. Falls p in L/Q total verzweigt ist, $\mathfrak{p}$ Verzweigungsideal von p in $\mathfrak{J}(k)$, p teilt $\mathscr{D}_{K/Q}$ mit dem Exponenten n, so teilt $\mathfrak{p}$ die Diskriminante $\mathscr{D}_{L/k}$ mit dem Exponenten $3n-2$.*

Beweis. Wir benutzen Schachtelungs- und Exponentensatz für Diskriminanten und bemerken nur, daß ein $\mathfrak{p} \in \mathfrak{J}(k)$, $\mathfrak{p}$ Zerlegungsideal von p, $\mathfrak{p}$ verzweigt in L/k mit demselben Exponenten in $\mathscr{D}_{L/k}$ erscheint wie das Ideal $\mathfrak{p}^\sigma \mathfrak{p}^{\sigma^2}$. □

DEFINITION 5. Sei für $a \in Z_{\mathscr{D}_{K/Q}} = \{b \in Z \mid b \text{ prim zu } \mathscr{D}_{K/Q}\}$

$$\varepsilon_K(a) = \begin{cases} 1 & \text{falls } 2 \nmid \mathscr{D}_{K/Q} \\ (-1)^{(a-1)/2} & \text{falls } \mathscr{D}_{K/Q} \equiv 4 \bmod 8 \\ (-1)(((a-1)/2) + (((-1)^{(a-1)/2}a - 1)/4 & \\ \quad + (\mathscr{D}_{K/Q}/8) - 1)(a-1)/4) & \text{falls } \mathscr{D}_{K/Q} \equiv 0 \bmod 8 \end{cases}$$

und

$$f_K(a) = \prod_{\substack{p \mid \mathscr{D}_{K/Q} \\ p \neq 2}} \left(\frac{a}{p}\right) \varepsilon_K(a).$$

ANMERKUNG. Da $f_K(a)$ auf $Z_{\mathscr{D}_{K/Q}}$ multiplikativ ist, kann

$$f_K(a) \quad \text{auf} \quad Q_{\mathscr{D}_{K/Q}} = \left\{ a = \frac{a_1}{a_2} \,\middle|\, a_i \in Z_{\mathscr{D}_{K/Q}} \right\}$$

fortgesetzt werden. Man sieht sofort, daß $f_K(a)$ auf $Q_{\mathscr{D}_{K/Q}}$ mod $\mathscr{D}_{K/Q}$ definiert ist.

LEMMA 4. *Falls* $v \in k$, $v \equiv 1 \bmod \mathscr{D}_{L/k}$, so ist $f_K(N_{k/Q}(v)) = 1$. *Falls* $w \in L$, *w prim zu* $\mathscr{D}_{L/k}$, *so ist* $f_K(N_{L/Q}(w)) = 1$.

Beweis. Wir benutzen Lemma 3, nach dem gilt $\mathscr{D}_{L/k}^{\sigma} = \mathscr{D}_{L/k}$ für alle $\sigma \in \mathrm{Gal}(N/Q)$ (in $\mathfrak{I}(N)$) und somit $N_{k/Q}(v) \equiv 1 \bmod \mathscr{D}_{L/k}$.

Ebenfalls aus Lemma 3 folgt (wegen $3n - 2 > n - 1$ für $n \in N$) damit $N_{k/Q}(v) \equiv 1 \bmod \mathscr{D}_{K/Q}$ und somit nach obiger Anmerkung $f_K(N_{k/Q}(v)) = 1$.

Sei $w \in L$, $w' = N_{L/K}(w)$, w prim zu $\mathscr{D}_{K/Q}$ und o.B.d.A. $w \in o_L$, so ist $N_{L/Q}(w) = N_{K/Q}(w')$. Für alle $p \mid \mathscr{D}_{K/Q}$, $p \neq 2$ folgt sofort $(N_{K/Q}(w')/p) = 1$. Falls $\mathscr{D}_{K/Q} \equiv 4 \bmod 8$, so ist $N_{K/Q}(w') = a^2 - b^2(\mathscr{D}_{K/Q}/4)$, $a, b \in Z$, (a, b) prim zu 2, $2 \mid ab$, $-(\mathscr{D}_{K/Q}/4) \equiv 1 \bmod 4$ und damit $\varepsilon_K(N_{K/Q}(w')) = 1$. Falls $\mathscr{D}_{K/Q} \equiv 0 \bmod 8$, so ist – wie oben –

$$N_{K/Q}(w') \bmod 8 \in \begin{matrix}\{1, 3\} \\ \{1, 7\}\end{matrix} \text{ falls } \frac{|\mathscr{D}_{K/Q}|}{8} \equiv \begin{matrix}1 \\ 3\end{matrix} \bmod 4.$$

Wir erhalten wieder $\varepsilon_K(N_{K/Q}(w')) = 1$. □

PROPOSITION 3. *Für ein* $v \in k$, *v prim zu* $\mathscr{D}_{L/k}$ *gilt: Falls* $(v) \in N_{L/k}(\mathfrak{I}_{\mathscr{D}_{L/k}}(L))H_0$, *so ist* $f_K(N_{k/Q}(v)) = 1$.

Beweis. Nach Lemma 4) genügt es zu zeigen: Falls $(v) \in N_{L/k}(\mathfrak{I}_{\mathscr{D}_{L/k}}(L))$, so ist $f_K(N_{k/Q}(v)) = 1$. Nach dem Dichtigkeitssatz von Tschebotarjow existiert in jeder Idealklasse aus $\mathrm{Cl}(L)$ ein Zerlegungsideal ϕ einer total zerlegten Primzahl. Falls $v \in N_{L/k}(\mathfrak{I}_{\mathscr{D}_{L/k}}(L))$, so existiert also ein solches $\phi \in \mathfrak{I}(L)$ mit $N_{L/k}(\phi) = (vN_{L/k}(v'))$, $v' \in L$ prim zu $\mathscr{D}_{L/k}$ und damit nach Lemma 4

$$f_K(q) = f_K(N_{k/Q}(vN_{L/k}(v'))) = f_K(N_{k/Q}(v)) \quad \text{für} \quad q = N(\phi).$$

Aus der totalen Zerlegtheit von q in L/Q folgt nun

$$1 = \left(\frac{\mathscr{D}_{K/Q}}{q}\right) = (-1)^{(q-1)/2} \prod_{\substack{p \mid \mathscr{D}_{K/Q} \\ p \neq 2}} \left(\frac{p}{q}\right)(q)$$

$$= (-1)^{((q-1)/2) + \sum_{\substack{p \mid \mathscr{D}_{K/Q} \\ p \neq 2}} (p-1)(q-1)/4)} \prod_{\substack{p \mid \mathscr{D}_{K/Q} \\ p \neq 2}} \left(\frac{q}{p}\right) \varepsilon(q)$$

für

$$\varepsilon(a) = \begin{cases} (2/a) = (-1)^{(((-1)^{(a-1)/2}a - 1)/4)} & \text{falls } 8 \mid \mathscr{D}_{K/Q} \\ 1 & \text{anderenfalls} \end{cases}$$

Es gilt aber

$$\sum_{\substack{p \mid \mathscr{D}_{L/K} \\ p \neq 2}} \frac{(p-1)(q-1)}{4} \equiv \frac{(|\mathscr{D}_{K/Q}|/(8, |\mathscr{D}_{K/Q}|) - 1)(q-1)}{4} \bmod 2$$

und damit

$$(-1)^{\sum_{\substack{p \mid \mathscr{D}_{L/K} \\ p \neq 2}} (p-1)(q-1)/4} = \begin{cases} 1 & \text{falls } \mathscr{D}_{L/K} \equiv 4 \bmod 8 \\ (-1)^{(q-1)/2} & \text{falls } 2 \nmid \mathscr{D}_{L/K}. \end{cases}$$

Damit ist

$$1 = \left(\frac{\mathscr{D}_{K/Q}}{q}\right) = f_K(q) = f_K(N_{k/Q}(v)).$$ □

Nach dem chinesischen Restsatz existiert ein $a \in Z_{\mathscr{D}_{L/K}}$, $a > 0$ mit $f_K(a) = f_K(a^3) = f_K(N_{k/Q}(a)) = -1$.

Für dieses a gilt nun nach Proposition 3

$$(a) \notin N_{L/k}(\mathfrak{I}_{\mathscr{D}_{L/k}}(L))H_0,$$

das heißt nach Proposition 2 existiert ein $\beta \in k$ mit $(\beta) \in N_{L/k}(\mathfrak{I}(L))$ und $\varphi(\beta^{\sigma^{(j-1)}})(-1)^{i_j} > 0$ für ein beliebiges Tupel $\{i_j\}$, $i_j \in \{0, 1\}$.

Damit ist (b) bewiesen. Es folgt also unmittelbar Theorem 4). □

Literaturverzeichnis

[0] Feustel, J.-M., Holzapfel, R.-P.: Symmetry points and Chern invariants of Picard modular surfaces, *Math. Nachr.* 111 (1983), S. 7-40.

[1] Feustel, J.-M.: Representation of Picard modular forms by theta constants, *Rev. Roumaine Math. Pures Appl.* 33 (1988), S.275–281.

[1a] Feustel, J.-M.: Arithmetik und Geometrie Picardscher Modulflächen, Dissertation B, Akademie der Wissenschaften der DDR, Karl-Weierstrass-Institut für Mathematik (1987).

[2] Hasse, H.: Zahlentheorie,*Akademie Verlag*, Berlin (1963).

[2a] Hasse, H.: *Bericht über neuere Untersuchungen und Probleme aus der Theorie der algebraischen Zahlkörper*, Teil I, B. G. Teubner, Leipzig/Berlin (1930).

[3] Hecke, E.: Zur Theorie der Modulfunktionen von 2 Variablen und ihre Anwendung auf die Zahlentheorie, *Math. Annalen* 71, (1912), S. 1–37.

[4] Holzapfel, R.-P.: Geometry and arithmetic around Euler partial differential equations, Kluwer Academic Publishers, Dordrecht, Holland, (1986).

[5] Holzapfel, R.-P.: An arithmetic uniformization for arithmetic points of the plane by singular moduli, *J. Ramanujan Math. Soc.* 3(1), (1988), S. 35–62.

[6] Lang, S.: *Complex multiplication*, New York, Berlin, Heidelberg, Tokyo, Springer (1983).

[7] Picard, E.: Sur des fonctions de deux variables independentes analogues aux fonctions modulaires, *Acta Mathematica* 2 (1983), S. 114–135.

[8] Picard, E.: Sur les formes quadratiques rerneires indefinies et sur les fonctions hyperfuchsiennes, *Acta Mathematica* 5 (1884), S. 121–182.

[9] Reiner, I.: *A survey of integral representation theory*, Bulletin of the American Mathematical Society, Vol. 76, No. 2, (1970), S. 159–227.

[10] Shiga, H.: *On the representation of Picard modular function by Θ constants I–II*, Publ. RIMS, Kyoto Univ., 24 (1988), S. 311–360.

[11] Shiga, H.: *On the construction of algebraic numbers as special values of the Picard modular function*, Preprint, Chiba University.

[12] Shimura, G.: On analytic families of polarized Abelian varieties and automorphic functions, *Annals of Mathematic*, 78 (1963) No. 1, S. 149–192.

[13] Shimura, G.: Arithmetic of unitary groups, *Annals of Mathematic*, 79 (1964), S. 369–409.

[14] Scharlau, W.: *Quadratic and hermitian forms*, Springer, Berlin, Heidelberg, New York, Tokyo, (1985).

Compositio Mathematica **76**: 101–162, 1990.

Chow categories

J. FRANKE

Universität Jena, DDR-6900 Jena, Universitätshochhaus 17. OG, and Karl-Weierstraß-Institut für Mathematik, DDR-1086 Berlin, Mohrenstraße 39, Germany

Received 3 December 1988; accepted in revised form 15 March 1990

Introduction

This paper arose from an attempt to solve some questions which were posed at the seminar of A. N. Parchin when Deligne's program ([D]) was reviewed. These problems are related to hypothetical functorial and metrical versions of the Riemann-Roch-Hirzebruch theorem. One of the problems posed by Deligne is, for instance, the following construction:

Let a proper morphism of schemes $X \to S$ of relative dimension n and a polynomial $P(c_i(E_j))$ of absolute degree $n+1$ (where $\deg(c_i) = i$) in the Chern classes of vector bundles $E_1, \ldots, E_k$ be given. Construct a functor which to the vector bundles E_j on X associates a line bundle on S

$$I_{X/S} P(c_i(E_j)) \tag{1}$$

which is an 'incarnation' of $\int_{X/S} P(c_i(E_j)) \in CH^1(S)$. The functor (1) should be equipped with some natural transformations which correspond to well-known equalities between Chern classes (cf. [D, 2.1]). Further steps in Deligne's program are to equip the line bundles (1) with metrics, to prove a functorial version of the Riemann-Roch-Hirzebruch formula which provides an isomorphism between the determinant $\det(Rp_*(F))$ of the cohomology of a vector bundle F and a certain line bundle of type (1); and (finally) to compare the metric on the right side of the Riemann-Roch isomorphism and the Quillen metric on the determinant of the cohomology.

In [D], Deligne dealt with the case $n = 1$. He considered (1) as a closed expression. It is our strategy to give 'live' to each ingredient of (1). If one tries to do so, the ith Chern functor $\mathbf{c}_i(E)$ should take values in the ith Chow category $\mathbf{CH}^i(X)$. It is the aim of these notes to explain what we believe to be the best definition of the Chow category, and to define some of the basic functors between Chow categories.

Our proposal for (1) is the following expression:

$$\mathbf{p}_*(P(\mathbf{c}_i(E_j)), \tag{2}$$

where $p: X \to S$ is the morphism we have in consideration and $\mathbf{p}_*$ is a push-forward functor which will be introduced in §3. The most complicated ingredient of (2) is the Chern functor $\mathbf{c}_i(\cdot)$. Its construction has been outlined in [Fr1], and details are contained in the notes [Fr2] which I distributed in June 1988. We shall publish our results on Chern functors together with more considerations about the Riemann-Roch problem in a continuation of this paper.

One of the advantages of the approach to Deligne's program via Chow categories is that it allows us to state the functorial Riemann-Roch theorem in Grothendieck's form. Hence it should be possible to copy the standard proof for Riemann-Roch theorems. Our proposal for the Riemann-Roch-Grothendieck isomorphism is a canonical isomorphism

$$\mathbf{ch}(Rp_*E) \cup A \to \mathbf{p}_*(\mathbf{ch}(E) \cup Td(\Omega^{\bullet}_{X/S}{}^{\check{}}) \cup \mathbf{p}^!(A)) \tag{3}$$

for any local complete intersection p with relative cotangential complex $\Omega^{\bullet}_{X/S}$. The isomorphism (3) should be characterized by certain axiomatic properties.

To explain the ingredients of (3) further I mention that the Chern functor will not be a mere object of the Chow category but an intersection product

$$\mathbf{c}_i(E) \cup \cdot : \mathbf{CH}^k(X) \to \mathbf{CH}^{k+i}(X).$$

Therefore no regularity assumptions for X are necessary to define both sides of (3) as a functor with values in the quotient category $\mathbf{CH}^{\bullet}(S) \otimes Q$. The remaining ingredient of (3) is the Gysin functor $\mathbf{p}^!$. This is our first example of a non-trivial functor between Chow categories, and the most considerations of this paper are directly or indirectly devoted to its construction.

After recalling some basic properties of Quillen's spectral sequence in §1, we define the Chow categories and some of the basic functors in §2 and §3. §4 contains the construction of the Gysin functor. In §5 we use this Gysin functor to outline the construction of a functorial intersection product. As an example which lies outside Deligne's program, we apply the intersection product functor to construct a biextension between certain groups of algebraic cycles. This biextension generalizes the well-known autoduality of the Jacobian, and should be equivalent to a construction of Bloch.

I started my research on Chow categories while I was a postgraduate student in Moscow under the guidance of I. M. Gel'fand. I am much obliged to A. A. Beilinson, Ju. I. Manin, A. N. Parchin, V. V. Schechtman, and the participants of Parchin's seminar for many helpful discussions. In particular, Beilinson and Manin pointed out that the Chow category should provide an alternative construction of Bloch's biextension. Their proposal is carried out, at least partially, in §5.5.

Notations

Throughout this paper, schemes are assumed to be Noetherian, separated over Spec($\mathbb{Z}$) and universally catenary. Our notations of K-theory are as usual $K_i(X) = K_i(P(X))$ and $K'_i(X) = K_i(M(X))$, where $P(X)$ and $M(X)$ are the exact categories of vector bundles and of coherent $\mathcal{O}_X$-modules on X.

Products in K-theory are defined by Waldhausen's pairing $BQA \wedge BQB \to BQQC$ (cf. [W], [Gr]). The relation between the product and the boundary of the localization sequence is given by formula [Gr, Corollary (2.6)]. In particular, the boundary of the K-theoretic product of two invertible functions differs by a sign from the tame symbol.

1. The sheaves G_k

For a scheme X, denote by X_k the set of points of codimension k (i.e., of points x with dim $(\mathcal{O}_{X,x}) = k$) and by $X_{(k)}$ the set of points of X, equipped with the following topology. U is open in $X_{(k)}$ iff it is Zariski-open and for every $x \in X_l$ with $l < k$, we have either $x \in U$ or $x \notin \bar{U}$. In particular, $X_{(1)} = X_{\mathrm{Zar}}$. For a point $x \in X$, $k(x)$ is the residue field of x.

1.1. *Definition of G_k*

The descending filtration of $M(X)$ by $M_p(X) = \{$coherent sheaves on X with support in codimension $\geqslant p\}$ defines a spectral sequence (cf. [Q, (5.5)] or [G, p. 269]) with initial term

$$E_1^{p,q}(X) = K_{-p-q}(M_p/M_{p+1}) = \coprod_{x \in X_p} K_{-p-q}(k(x)) \tag{1}$$

converging to $K'_{-p-q}(X)$. In particular, $E_2^{p,q}(X)$ is the homology in the middle term of

$$\coprod_{x \in X_{p-1}} K_{1-p-q}(k(x)) \to \coprod_{x \in X_p} K_{-p-q}(k(x)) \to \coprod_{x \in X_{p+1}} K_{-p-q-1}(k(x)). \tag{2}$$

We are particularly interested in the groups $Z^k(X) = E_1^{k,-k}(X)$, $CH^k(X) = E_2^{k,-k}(X)$, and $G_k(X) = E_2^{k-1,-k}(X)$. By (2), they can be defined elementarily, using only K_0, K_1, and K_2 of fields.

The $E_k^{p,q}$ are presheaves on X_{Zar}. Furthermore, one checks easily that the restriction of G_k to $X_{(k)}$ is a sheaf. By (2), there is an exact sequence for U open in $X_{(k)}$

$$0 \to G_k(X - \bar{U})^0 \to G_k(X) \xrightarrow{\partial} G_k(U) \to Z^k(X, X - U) \to CH^k(X), \tag{3}$$

where $G_k(X - \bar{U})^0 = \text{image}(G_k(X) \to G_k(X - \bar{U}))$ and $Z^k(X, X - U) = \{x \in Z^k(X) \, \text{supp}(x) \subset X - U\}$. The arrow $G_k(X) \to \ker(\partial)$ has a natural splitting.

1.2. *Mayer-Vietoris and localization sequences*

If X is the union of its open subsets U and V, we have

$$0 \to E_1^{p,q}(X) \to E_1^{p,q}(U) \oplus E_1^{p,q}(V) \to E_1^{p,q}(U \cap V) \to 0$$

and hence

$$\begin{aligned} &\to E_2^{p-1,q}(U \cap V) \to E_2^{p,q}(X) \to E_2^{p,q}(U) \oplus E_2^{p,q}(V) \\ &\quad \to E_2^{p,q}(U \cap V) \to E_2^{p+1,q}(X) \to . \end{aligned} \tag{4}$$

Let $Z \subset X$ be closed. We call Z of pure codimension d if $X_k \cap Z = Z_{k-d}$ for $k \in \mathbb{Z}$. Then the exact sequence

$$0 \to E_1^{p-d,q+d}(Z) \to E_1^{p,q}(X) \to E_1^{p,q}(X - Z) \to 0$$

gives rise to

$$\begin{aligned} &\to E_2^{p-1,q}(X - Z) \to E_2^{p-d,q+d}(Z) \to E_2^{p,q}(X) \to E_2^{p,q}(X - Z) \\ &\quad \to E_2^{p+1-d,q+d}(Z) \to \cdots \end{aligned} \tag{5}$$

1.3. *Flat pull-back*

If $f\colon Y \to X$ is a flat morphism, it defines an exact functor $f^*\colon M(X) \to M(Y)$ which maps $M_k(X)$ into $M_k(Y)$. Consequently, we have a homomorphism $f^*\colon E_k^{p,q}(X) \to E_k^{p,q}(Y)$ which commutes with the differentials d_k, and hence preserves (3), (4) and (5).

1.4. *Proper push-forward*

Let $f\colon X \to Y$ be a morphism of finite type. We call f of constant relative dimension $d \in \mathbb{Z}$ if for every $x \in X_l$ such that $\dim(\overline{f(x)}) = \dim(\bar{x})$ we have $f(x) \in Y_{l-d}$. The proof of the following lemma is straightforward:

LEMMA. *Let*

$$\begin{array}{ccc} X' & \xrightarrow{g_X} & X \\ \downarrow{\scriptstyle f'} & & \downarrow{\scriptstyle f} \\ Y' & \xrightarrow{g} & Y \end{array}$$

be a Cartesian diagram in which f is of constant relative dimension d. In each of the following cases, f' is also of constant relative dimension d:

(i) *If g is flat.*

(ii) *If g and g_X are l.c.i. (local complete intersections) and for every $x \in X'$, $d_x(g_X) = d_{f'(x)}(g)$, where $d_x(g)$ is the relative dimension of the lci-morphism g at x* (cf. [FL, p. 89] *or* [SGA6, VIII.1.9.]).

Proof. Since the question is local, we may assume in (i) that X is a closed subscheme of A^n_Y. Then f is of relative dimension d if and only if X is of codimension $n - d$ in A^n_Y, and this condition remains valid after flat base change. By (i), (ii) is reduced to the case of a regular closed immersion f in which it is trivial. □

Now we assume that f: $X \to Y$ is a proper morphism of constant relative dimension d. Then we have exact functors

$$f_*: M_p(X)/M_{p+1}(X) \to M_{p-d}(Y)/M_{p+1-d}(Y) \tag{6}$$

defining

$$f_*: E_1^{p,q}(X) \to E_1^{p-d,q+d}(Y). \tag{7}$$

The following theorem is similar to results of Gillet and Schechtman:

THEOREM. (i) *The homomorphism* (7) *commutes with the differential d_1 of the Quillen spectral sequence. Hence it defines $f_*: E_2^{p,q}(X) \to E_2^{p-d,q+d}(Y)$*

(ii) *The homomorphism f_* on the E_2-terms is compatible with the localization sequence* (3), *i.e., if U is open in $X_{(k)}$ and $V = Y - f(X - U)$ then we have a commutative diagram*

$$\begin{array}{ccccccc} G_k(X) & \longrightarrow & G_k(U) & \longrightarrow & Z^k(X, X-U) & \longrightarrow & CH^k(X) \\ \downarrow{\scriptstyle f_*} & & \downarrow{\scriptstyle f_*} & & \downarrow{\scriptstyle f_*} & & \downarrow{\scriptstyle f_*} \\ G_{k-d}(Y) & \longrightarrow & G_{k-d}(V) & \longrightarrow & Z^{k-d}(Y, Y-V) & \longrightarrow & CH^{k-d}(Y) \end{array}$$

Proof of (i). It is possible to copy the proof in [G, 7.22]. It should also be possible to apply the results of [GN].

Proof of (ii). This follows from (1) and the definition of (3).

1.5. *Specialization*

This is a modification of [F, Remark 2.3.], cf. also [G, 8.6.]. Let $D \subset X$ be a regular embedding of codimension 1, and assume that f is a section of $\mathcal{O}_X$ in some Zariski-neighbourhood U of D generating the sheaf of ideals defining $D \subset U$.

The existence of f is a serious restriction to the embedding $D \subset X$, for instance it implies the triviality of the conormal bundle of the immersion, which means that we are in the situation described in [F, Remark 2.3.].

We define homomorphisms

$$sp_f\colon E_k^{p,q}(X-D) \to E_k^{p,q}(D), \qquad k \geqslant 1 \tag{9}$$

as follows.

The tensor product $P(U-D) \times M_p(U-D) \to M_p(U-D)$ defines

$$[f]\cup\cdot\colon E_k^{p,q}(U-D) \to E_k^{p,q-1}(U-D), \tag{10}$$

where $[f]$ is the class of $f\cdot$ in $K_1(U-D)$. Let

$$M'_p = \{\text{coherent } \mathcal{O}_U\text{-modules } F \text{ with } \operatorname{cod}_U(\operatorname{supp}(F)) \geqslant p \\ \text{and } \operatorname{cod}_D(D \cap \operatorname{supp}(F)) \geqslant p\}.$$

Then $M_p(U-D)/M_{p+k}(U-D) = (M'_p/M'_{p+k})/(M_p(D)/M_{p+k}(D))$, consequently we have

$$\partial\colon E_k^{p,q}(U-D) \to E_k^{p,q+1}(D). \tag{11}$$

If $k = 2$, ∂ is the boundary homomorphism in (5).

We define sp_g by the composition of

$$E_k^{p,q}(X-D) \xrightarrow{\text{restriction}} E_k^{p,q}(U-D) \xrightarrow{[f]\cup\cdot} E_k^{p,q-1}(U-D) \longrightarrow E_k^{p,q}(D). \tag{12}$$

On the line $p+q=0$, sp_f is independent of f, and we obtain the homomorphism i^* described in [F, Remark 2.3.]. The composition

$$E_k^{p,q}(X) \longrightarrow E_k^{p,q}(X-D) \xrightarrow{\mathrm{sp}_f} E_k^{p,q}(D) \tag{13}$$

is also independent of f.

1.6. *Compatibilities*

Let

$$\begin{array}{ccc} X' & \xrightarrow{g'} & X \\ \big\downarrow{\scriptstyle f'} & & \big\downarrow{\scriptstyle f} \\ Y' & \xrightarrow[g]{} & Y \end{array}$$

be a Cartesian square with g flat and f proper of constant relative dimension d. Then we have the base change identity

$$g^*f_* = f'_*g'^* \text{ in } \mathrm{Hom}(E_k^{p,q}(X), E_k^{p-d,q+d}(Y')), \quad k \in \{1, 2\}. \tag{14}$$

For the diagram of functors

$$\begin{array}{ccc} M_p(X)/M_{p+1}(X) & \xrightarrow{g'^*} & M_p(X')/M_{p+1}(X') \\ \big\downarrow{\scriptstyle f_*} & & \big\downarrow{\scriptstyle f'_*} \\ M_{p-d}(Y)/M_{p+1-d}(Y) & \xrightarrow{g^*} & M_{p-d}(Y')/M_{p+1-d}(Y') \end{array}$$

commutes up to a natural transformation.

Consider a fibre square

$$\begin{array}{ccc} D' & \subset & X' \\ \big\downarrow{\scriptstyle p_D} & & \big\downarrow{\scriptstyle p} \\ D & \subset & X \end{array}$$

in which $D \subset X$ and $D' \subset X'$ are regular embeddings of codimension 1. Let f be the same as in 1.5. If p is proper of constant relative dimension d, the lemma in 1.4 implies that p_D is of the same relative dimension d. We have

$$\mathrm{sp}_f p_* = p_{D*}\mathrm{sp}_{p^*(f)} \text{ in } \mathrm{Hom}(E_k^{p,q}(X'), E_k^{p-d,q+d}(D)), \qquad k \in \{1, 2\}. \tag{15}$$

If in the same fibre square p is flat, we have

$$p_D^* \mathrm{sp}_f = \mathrm{sp}_{p^*(f)} p^*. \tag{16}$$

(15) is a consequence of the commutative diagram

$$\begin{array}{ccccc}
E_1^{p,q}(X'-D') & \xrightarrow{[p^*(f)]\cup\cdot} & E_1^{p,q-1}(X'-D') & \longrightarrow & E_1^{p,q}(D') \\
\Big\downarrow p_* & \text{(A)} & \Big\downarrow p_* & \text{(B)} & \Big\downarrow p_* \\
E_1^{p-d,q+d}(X-D) & \xrightarrow{[f]\cup\cdot} & E_1^{p-d,q+1-d}(X-D) & \longrightarrow & E_1^{p-d,q+d}(D)
\end{array}$$

The commmutativity of (A) follows from the fact that the diagram of bilinear functors

$$\begin{array}{ccc}
P \times M & \longrightarrow & p^*(P) \otimes M \\
P(X-D) \times (M_p/M_{p+1})(X'-D') & \longrightarrow & (M_p/M_{p+1})(X'-D') \\
\Big\downarrow \mathrm{Id} \times p_* & & \Big\downarrow p_* \\
P(X-D) \times (M_{p-d}/M_{p+1-d})(X-D) & \xrightarrow{P \times M \to P \otimes M} & (M_{p-d}/M_{p+1-d})(X-D)
\end{array}$$

commutes up to a natural transformation, and (B) commutes because p_* maps sheaves on X' whose support is of codimension p and meets D' in codimension p to sheaves on X with the similar property, and hence defines a morphism between the quasi-fibrations used to define (11).

The proof of (16) is similar.

If we have a commutative diagram

$$\begin{array}{ccc}
D & \xrightarrow{i} & X \\
& {\scriptstyle g}\searrow \quad \swarrow{\scriptstyle h} & \\
& Z &
\end{array}$$

in which g and h are flat and D, X, and f satisfy the assumptions of 1.5, then

$$\mathrm{sp}_f h^* = g^*. \tag{17}$$

This can easily be reduced to the following general situation:

LEMMA. *Let the following objects be given:*

(i) *A sequence $\mathscr{A} \xrightarrow{a} \mathscr{B} \xrightarrow{b} \mathscr{C}$ of exact functors between exact categories such that $ab \simeq 0$ and $BQ\mathscr{A} \to BQ\mathscr{B} \to BQ\mathscr{C}$ is a fibration up to homotopy.*

(ii) *An exact functor $P \to P'$ between exact categories, an object Y of P and an endomorphism $f: Y \to Y$ in P which becomes an isomorphism in P'.*

(iii) *An exact category $\mathscr{D}$, biexact functors $\otimes: P \times \mathscr{D} \to \mathscr{B}$ and $P' \times \mathscr{D} \to \mathscr{C}$ such that*

$$\begin{array}{ccc} P \times \mathscr{D} & \xrightarrow{\otimes} & \mathscr{B} \\ \downarrow & & \downarrow b \\ P' \times \mathscr{D} & \xrightarrow{\otimes} & \mathscr{C} \end{array}$$

commutes up to a natural transformation, and an exact functor $G: \mathscr{D} \to \mathscr{A}$ such that there is a functorial exact sequence in $\mathscr{B}$:

$$0 \to Y \otimes A \xrightarrow{f \otimes \mathrm{Id}_A} Y \otimes A \xrightarrow{\Pi} G(A) \to 0 \quad (A \in \mathrm{Ob}(\mathscr{D})).$$

Let $[f] \in K_1(P')$ be the class of f viewed as an automorphism in P'. Then

$$\partial([f] \cup \xi) = G_*(\xi), \qquad \xi \in K_i(\mathscr{D}), \tag{18}$$

where ∂: $K_{i+1}(\mathscr{C}) \to K_i(\mathscr{A})$ is the boundary defined by the fibration (i), *$\cup: K_1(P') \times K_i(\mathscr{D}) \to K_{i+1}(\mathscr{C})$ is the pairing defined by $\otimes$, and $G_*: K_i(\mathscr{D}) \to K_i(\mathscr{A})$ is defined by G.*

To derive (17) *from* (18), *we put $\mathscr{A} = M_p(D)/M_{p+1}(D)$, $\mathscr{B} = M'_p/M'_{p+1}$* (cf. 1.5.) *$\mathscr{C} = M_p(X)/M_{p+1}(X)$, $\mathscr{D} = M_p(Z)/M_{p+1}(Z)$, $P = P(X)$, $P' = P(X - D)$, $Y = \mathcal{O}_X$, and f = multiplication by f. Furthermore we put $G = g^*$ and define $\otimes: P \times \mathscr{D} \to \mathscr{B}$ by $(M, E) \to h^*(M) \otimes E$.*

Proof of Lemma. The class $[f]$ is given by the homotopy class of the map $S^2 \to |BPQ'|$ defined by the diagram

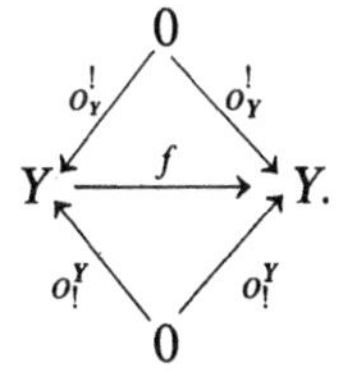

(19)

Here we use the usual notations for morphisms in QP', and $o_Y = Y \to 0$, $o^Y = 0 \to Y$. To get S^2 from the diagram (19), identify its left and right boundary.

Consequently, the homotopy class $\Sigma^2 |BQ\mathscr{D}| \to |BQQ\mathscr{C}|$ obtained by applying Waldhausen's pairing to $[f]$ can be defined by the geometric realization of the map which associates to $A \in \mathscr{D}$ the following diagram of vertical morphisms in $QQ\mathscr{C}$

$$\begin{array}{ccc} & 0 & \\ {}^{v}0^{!}_{Y\otimes A}\swarrow & & \searrow {}^{v}0^{!}_{Y\otimes A} \\ Y\otimes A & \xrightarrow{{}^{v}(f\otimes \mathrm{Id}_A)} & Y\otimes A \\ {}^{v}0_{!}^{Y\otimes A}\nwarrow & & \nearrow {}^{v}0_{!}^{Y\otimes A} \\ & 0 & \end{array} \qquad (20)$$

(the left superscript v denotes vertical morphisms in QQ) and to a morphism in $Q\mathscr{D}$ the similar diagram of bimorphisms in $QQ\mathscr{C}$. The diagram (20) has an obvious lifting to $QQ\mathscr{B}$:

$$\begin{array}{ccc} & 0 & \\ {}^{v}0^{!}_{Y\otimes A}\swarrow & & \searrow {}^{v}[\mathrm{im}(f)] \\ Y\otimes A & \xrightarrow{{}^{v}(f\otimes \mathrm{Id}_A)_!} & Y\otimes A \\ {}^{v}0_{!}^{Y\otimes A}\nwarrow & & \nearrow {}^{v}0_{!}^{Y\otimes A} \\ & 0 & \end{array} \qquad (21)$$

(${}^{v}[\mathrm{im}(f)]$ = vertical morphism from 0 to $Y\otimes A$ defined by the subobject $\mathrm{im}(f\otimes \mathrm{Id}_A) \subset Y\otimes A$). The diagram of bimorphisms corresponding to (20) has a lifting to $QQ\mathscr{B}$ which is similar to (21). Our task is now to compute the difference between the two homotopy classes $\Sigma DBQ\mathscr{D}| \to |BQQ\mathscr{B}|$ defined by the arrows on the left and the right boundary of (21). Because ${}^{v}[\mathrm{im}(f)]$ is equal to the composition

$$0 \xrightarrow{{}^{v}0_{!}^{G(A)}} G(A) \xrightarrow{{}^{v}\pi^{!}} Y\otimes A,$$

the map $\Sigma|BQ\mathscr{D}| \to |BQQ\mathscr{B}|$ defined by the vertical morphisms on the right boundary of (21) and the related bimorphisms is homotopic to the map defined by the diagram

$$\begin{array}{ccc} 0 & \xrightarrow{{}^{v}0_{!}^{G(A)}} & G(A) \\ & {}^{v}0_{!}^{Y\otimes A}\searrow & \downarrow {}^{v}\pi^{!} \\ & & Y\otimes A \end{array} \qquad (22)$$

and the similar diagram of bimorphisms. By the commutative diagram

$$\begin{array}{ccc} G(A) & \xleftarrow{{}^{v}0^{!}_{G(A)}} & \\ {}^{v}\pi^{!}\downarrow & & 0, \\ Y\otimes A & \xleftarrow{{}^{v}0^{!}_{Y\otimes A}} & \end{array}$$

(22) is homotopic to

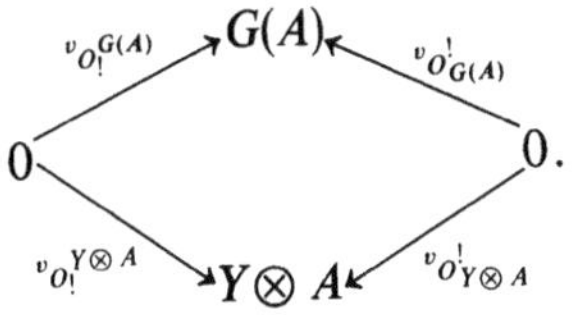

The bottom half of this diagram coincides with the left boundary of (21). By the very definition of the boundary operator ∂, we conclude that $\partial([f]\cup\cdot)$ is the homotopy class of the map $\Sigma|BQ\mathscr{D}| \to |BQQ\mathscr{A}|$ given by the diagram

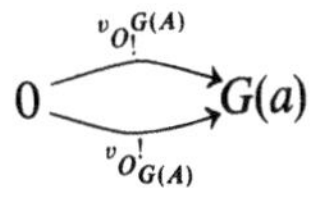

and the similar diagram of bimorphisms. This is, however, the composition of $G_*\colon |BQ\mathscr{D}| \to |BQ\mathscr{A}|$ with the map $\Sigma|BQ\mathscr{A}| \to |BQQ\mathscr{A}|$ defined in [W, p. 197]. The proof of the lemma and of (17) is complete.

1.7. *A relation between two specializations*

Let $D_i \subset X$ be regular immersions of codimension one such that the sheaf of ideals defining D_i is in some neighbourhood of D_i trivialized by f_i, $i \in \{1; 2\}$. We suppose also that $D_1 \cap D_2 \subset X$ is a regular immersion of codimension two, i.e., that $f_i|_{D_j}$ is not a zero-divisor if $i \neq j$. Then we have the identity

$$\mathrm{sp}_{f_2|_{D_1}}\mathrm{sp}_{f_1} = \mathrm{sp}_{f_1|_{D_2}}\mathrm{sp}_{f_2} \text{ in } \mathrm{Hom}(E_k^{p,q}(X - D_1 - D_2), E_k^{p,q}(D_1 \cap D_2)). \tag{23}$$

This follows from the commutative diagram

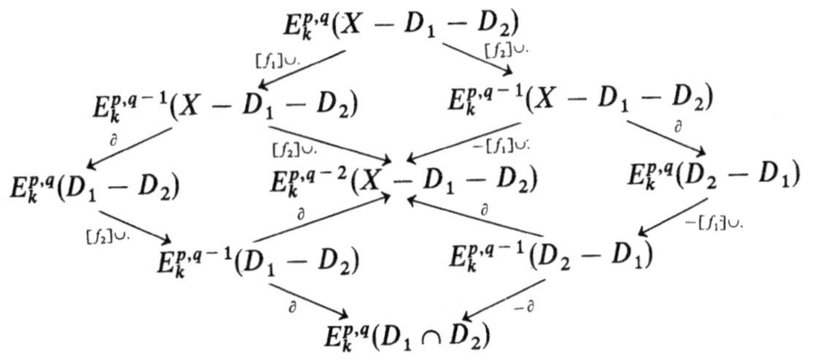

1.8. *A relation between specialization and restriction to closed subschemes*

Let X be a regular scheme satisfying Gersten's conjecture. Then we have an isomorphism

$$B\colon E_2^{p,q}(X) \to H^p(X, \mathcal{K}_{-q}) \tag{24}$$

defined by the well-known acyclic resolvent

$$\mathcal{K}_q \to \mathcal{E}_1^{\bullet,-q} \tag{25}$$

of the sheaf $\mathcal{K}_q$ associated to $U \to K_q(U)$. In (25), $\mathcal{E}_1^{p,q}$ is the Zariski-sheaf $U \to E_1^{p,q}(U)$.
If $i\colon Z \subset X$ is a closed regular subscheme of X, the composition

$$E_2^{p,q}(X) \xrightarrow{\ B\ } H^p(X, \mathcal{K}_{-q}) \xrightarrow{\text{restriction}} H^p(Z, \mathcal{K}_{-q}) \xrightarrow{\ B^{-1}\ } E_2^{p,q}(Z) \tag{26}$$

defines a homomorphism

$$i^*\colon E_2^{p,q}(X) \to E_2^{p,q}(Z). \tag{27}$$

PROPOSITION. *Let Z and D be closed regular subschemes of a regular scheme X satisfying Gersten's conjecture. We assume that $Z \cap D$ is regular and of codimension one in Z and that (X, D, f) satisfies the assumptions of* 1.5. *Then we have*

$$i_D^* \mathrm{sp}_f = \mathrm{sp}_f|_Z i^* \text{ in } \mathrm{Hom}(E_2^{p,q}(X - D), E_2^{p,q}(Z \cap D)), \tag{28}$$

where i and i_D are the inclusions $Z \subset X$ and $Z \cap D \subset D$.

Proof. This follows from the commutative diagram

$$\begin{array}{ccccccc}
E_2^{p,q}(X - D) & \xrightarrow{[f]\cup\cdot} & E_2^{p,q-1}(X - D) & & \xrightarrow{\ \partial\ } & & E_2^{p,q}(D) \\
\downarrow B & (A) & \downarrow B & & (B) & & \downarrow B \\
H^p(X - D, \mathcal{K}_{-q}) & \xrightarrow{(-1)^p[f]\cup\cdot} & H^p(X - D, \mathcal{K}_{1-q}) & \xrightarrow{\partial} & H_D^{p+1}(X, \mathcal{K}_{1-q}) & \xrightarrow{(a)} & H^p(D, \mathcal{K}_{-q}) \\
\downarrow & & \downarrow & & (C) & & \downarrow \\
H^p(Z - D, \mathcal{K}_{-q}) & \xrightarrow{(-1)^p[f]\cup\cdot} & H^p(Z - D, \mathcal{K}_{1-q}) & \xrightarrow{\partial} & H_{Z\cap D}^{p+1}(Z, \mathcal{K}_{1-q}) & \xrightarrow{(a')} & H^p(Z \cap D, \mathcal{K}_{-q}) \\
\downarrow B & (A') & \downarrow B & & (B') & & \downarrow B \\
E_2^{p,q}(Z - D) & \xrightarrow{[f]\cup\cdot} & E_2^{p,q-1}(Z - D) & & \xrightarrow{\ \partial\ } & & E_2^{p,q}(Z \cap D)
\end{array}$$

In this diagram, the arrow (a) and its symmetric counterpart (a′) are defined by the purity isomorphism

$$\mathcal{H}^p_D(X, \mathcal{K}_{q,X}) = \begin{cases} 0 \text{ if } p > 1 \\ \mathcal{K}_{q-1,0} \text{ if } p = 1, \end{cases} \tag{29}$$

where $\mathcal{H}^p_D$ is the derived functor of the sheaf of sections with support in D. The only non-zero isomorphism in (29) is normalized by the commutativity of

$$\begin{array}{ccccc} K'_q(U-D) = K_q(U-D) & \longrightarrow & H^0(U-D, \mathcal{K}_q) & \overset{\partial}{\searrow} & \\ \Big\downarrow \partial & & & & H^1_{U\cap D}(U, \mathcal{K}_q). \\ K'_{q-1}(U\cap D) = K_{q-1}(U\cap D) & \longrightarrow & H^0(U\cap D, \mathcal{K}_{q-1}) & \underset{(29)}{\nearrow} & \end{array} \tag{30}$$

The commutativity of the squares (B) and (B′) is therefore obvious. For the commutativity of (C), we denote by $\tilde{P}$ the category of sheaves F on U with the property

$$\mathrm{Tor}_i^{\mathcal{O}_X}(\mathcal{O}_Z, F) = 0 \quad \text{if } i > 0.$$

There is an obvious diagram

$$\begin{array}{ccccc} BQP(U\cap D) & \longrightarrow & BQ\tilde{P} & \longrightarrow & BQP(U-D) \\ \Big\downarrow & & \Big\downarrow & & \Big\downarrow \\ BQM(D\cap Z\cap U) & \longrightarrow & BQM(Z\cap U) & \longrightarrow & BQM((Z-D)\cap U) \end{array} \tag{31}$$

in which the rows are fibrations up to homotopy. Since the boundary homomorphism of the top row coincides with the left vertical arrow in (30), (31) implies the commutativity of

$$\begin{array}{ccc} K_q(U-D) & \longrightarrow & K_q(Z\cap U-D) \\ \Big\downarrow \partial & & \Big\downarrow \partial \\ K_{q-1}(U\cap D) & \longrightarrow & K_{q-1}(Z\cap U\cap D) \end{array}$$

for every Zariski-open U in X. By (30), this proves the commutativity of (C). The commutativity of (A) and its counterpart (A′) follows from the diagram of resolvents

$$\begin{array}{ccccccc}
\mathcal{K}_{q,X} & \longrightarrow & \mathcal{E}_1^{0,-q} & \longrightarrow & \mathcal{E}_1^{1,-q} & \longrightarrow & \cdots \\
\downarrow{\scriptstyle [f]\cup\cdot} & & \downarrow{\scriptstyle [f]\cup\cdot} & & \downarrow{\scriptstyle [f]\cup\cdot} & & \\
\mathcal{K}_{q+1,X} & \longrightarrow & \mathcal{E}_1^{0,-q-1} & \longrightarrow & \mathcal{E}_1^{1,-q-1} & \longrightarrow & \cdots
\end{array}$$

in which all the squares except the first one are anti-commutative. The commutativity of the other squares in the diagram is obvious. The proof of (28) is complete.

1.9. *Homotopy invariance*

Let $p\colon E \to X$ be the projection of a vector bundle to its base. Then $p^*\colon E_2^{p,q}(X) \to E_2^{p,q}(E)$ is an isomorphism.

Proof. By the localization sequence and the five lemma, we may reduce the assertion to the case that all connected components of X are irreducible and hence equidimensional. In this case the assertion follows from [G, Theorem 8.3].

2. Definition of the Chow category by means of cycles

On a normal locally factorial scheme X, every Weil divisor $B \in E_1^{1,-1}(X)$ defines a line bundle $O(D)$, and isomorphisms between $O(D)$ and $O(D')$ correspond to rational functions f with $\operatorname{div}(f) = D' - D$. We try to generalize this to higher codimension.

Let $\mathbf{CH}_z^i(X)$ be the following category. Objects of $\mathbf{CH}_z^i$ are cycles $z \in E_1^{i,-i}(X)$. Homomorphisms between z and z' are elements of the factor set

$$\mathrm{Hom}_{\mathbf{CH}_z^i(X)}(z, z') = \{f \in E_1^{i-1,-i}(X) \mid d_1(f) = z' - z\}/d_1E_1^{i-2,-i}(X). \tag{1}$$

The composition $\mathrm{Hom}(z, z') \times \mathrm{Hom}(z', z'') \to \mathrm{Hom}(z, z'')$ sends the equivalence classes of f and f' in (1) to the class of $f + f'$. It is easy to see that $\mathbf{CH}_z^i$ is a Picard category in the sense of [D, §4.1] if the sum is given by

$$z \oplus z' = z + z' \tag{2}$$

$$\text{class of } f \oplus \text{class of } f' = \text{class of } f + f'.$$

The commutativity and associativity law are simply identities between functors.

To admit also non-invertible arrows, we mention that $E_1^{i,-i}$ (being the free group generated by X_i) carries a natural ordering $\leqslant$, and define the extended Chow category $\mathbf{CH}_z^i(X)_e$ which has the same objects as $\mathbf{CH}_z^i(X)$ and

$$\mathrm{Hom}_{\mathbf{CH}_z^i(X)}(z, z') = \{f \in E_1^{i-1,-i}(X) \mid d_1(f) \geqslant z' - z\}/d_1 E_1^{i-2,-i}(X), \tag{3}$$

as morphisms between z and z'. The composition of arrows is defined by adding f and f'. (2) defines a sum in $\mathbf{CH}_z^i(X)_e$.

If X is normal and locally factorial, then $\mathbf{CH}_z^1(X)$ is (via $D \to O(D)$) equivalent to the category of line bundles and isomorphisms, while $\mathbf{CH}_z^1(X)_e$ is equivalent to the category of line bundles and inclusions of line bundles on X. The sum $\oplus$ corresponds to the tensor product of line bundles.

By the results of §1, there are flat pull-back, proper push-forward, and specialization functors between the categories $\mathbf{CH}_z^i$. If, for instance, $p\colon Y \to X$ is flat, the functor $\mathbf{p}^*$ sends the object z to $p^*(z)$ and the class of f in (1) to the class of $p^*(f)$. Using these functors, we could try to establish a functorial analogue of the usual intersection theory. We shall, however, prefer another definition of the Chow category which defines $\mathbf{CH}^i(X)$ as the category of principal homogeneous sheaves for G_i on $X_{(i)}$. We shall see in §3 that this definition is essentially equivalent to our previous definition. The advantages of the definition in §3 are that it is similar to the equivalence between line bundles and $\mathcal{O}_X^*$-principal homogeneous sheaves, that it is sometimes convenient to prove the commutativity of diagrams by computing images of so called 'rational sections', and that (in the case of manifolds over $\mathbb{C}$) it provides an easy definition of what a metric on an object of the Chow category should be.

3. The categories $\mathbf{CH}^k(X)$ and $\widetilde{\mathbf{CH}}^k(X)$

3.1. *Definition*

Let $k \geqslant 1$. Recall that $X_{(k)}$ is a topology on X consisting of sufficiently large Zariski open subsets. Let $\tilde{X}_{(k)}$ be the pretopology (cf. for instance [M]) on the category of open subsets in $X_{(k)}$ in which the U_i form a covering of U if and only if $U - U_i$ is of codimension $\geqslant k + 1$ in U. Then $G_k = E_2^{k-1,-k}$ is a sheaf on both $X_{(k)}$ and $\tilde{X}_{(k)}$. We recall from [M] that if G is a sheaf of groups over any site, then a G-principal homogeneous sheaf is a sheaf X of sets over this site which is equipped with a G-action such that the homomorphism

$$G \times X \to X \times X$$

$$(g, x) \to (x, gx)$$

is an isomorphism in category of sheaves of sets. A morphism in the category of G-principal homogeneous sheaves is a morphism in the category of sets which is compatible with the G-actions, such a morphism is automatically an isomorphism.

Let $\mathbf{CH}^k(X)$ (resp. $\widetilde{\mathbf{CH}}^k(X)$) be the category of G_k-principal homogeneous sheaves on $X_{(k)}$ (resp. $\tilde{X}_{(k)}$). If A is an object of one of these categories and if U is open in $X_{(k)}$, then the set of sections of A on U is denoted by $A(U)$. $\mathbf{CH}^k(X)$ is a full subcategory of $\widetilde{\mathbf{CH}}^k(X)$, and an object of $\widetilde{\mathbf{CH}}^k(x)$ belongs to $\mathbf{CH}^k(X)$ if and only if $X_{(k)}$ has a covering U_i such that $A(U_i)$ is not empty.

It is clear that the operation

$$A \oplus B = A \overset{G_k}{\times} B \tag{1}$$

defines the structure of a Picard category (in the sense of [D, §4]) on $\mathbf{CH}^k$ and $\widetilde{\mathbf{CH}}^k$. The commutativity law $A \oplus B \xrightarrow{\sim} B \oplus A$ sends $a \oplus b$ to $b \oplus a$, and the associativity law $(A \oplus B) \oplus C \xrightarrow{\sim} A \oplus (B \oplus C)$ sends $(a \oplus b) \oplus c$ to $a \oplus (b \oplus c)$ if a, b, and c are sections of A, B, and C on U. The zero object is G_k, and the isomorphism $G_k \oplus A \xrightarrow{\sim} A$ sends $g \oplus a$ to ga, where ga is the action of $g \in G_k(U)$ on $a \in A(U)$.

To admit also non-invertible arrows we define the following extended Chow category. Let $G_k^+(U)$ be the semi-group of self-homomorphisms of the zero object of $\mathbf{CH}_z^k(U)_e$. If $A \in \mathrm{Ob}(\widetilde{\mathbf{CH}}^k(X))$, put

$$A_e = G_k^+ \overset{G_k}{\times} A. \tag{2}$$

Homomorphisms from A to B in $\mathbf{CH}^k(X)_e$ (resp. $\widetilde{\mathbf{CH}}^k(X)_e$) are sheaf morphisms between A_e and B_e respecting the G_k^+-action.

Now we discuss the fundamental properties of these Chow categories.

3.2. *Relation to Line Bundles*

Since $X_{(1)} = X_{\mathrm{Zar}}$, the natural homomorphism $\mathcal{O}_X^* \to G_1$ defines a functor $\mathbf{c}_1$: (line bundles on X and isomorphisms) $\to \mathbf{CH}^1(X)$ and $\mathbf{c}_1$: (line bundles and $\mathcal{O}_X$-linear maps which are isomorphisms at the maximal points X_0) $\to \mathbf{CH}^1(X)_e$. This functor maps $\otimes$ between line bundles to $\oplus$ in $\mathbf{CH}^1$. It is faithful if X is reduced and an equivalence if X is normal.

3.3. *Rational sections and their cycles*

Let A be an object of $\mathbf{CH}^k(X)$. We define its sheaf of rational sections by

$$A_r(U) = \cup A(V), \tag{3}$$

where the union is over all V which are open in $U_{(k)}$ and meet every irreducible component of U. It is easy to check that $A_r(X)$ is not empty.

Every $a \in A_r(U)$ defines a cycle $\mathbf{c}(a)$ as follows. Choose a representative $a' \in A(V)$ for a. There exist a covering U_j of $\tilde{U}_{(k)}$ and sections $b_j \in A(U_j)$. Then $a'|_{V \cap U_j} - b_j|_{V \cap U_j} = c_j \in G_k(V \cap U_j)$. Let $z_j = \partial(c_j) \in Z^k(U_j)$, cf. 1(3). Then $z_i|_{U_i \cap U_j} = z_j|_{U_i \cap U_j}$, consequently (since Z^k is a sheaf on $\tilde{X}_{(k)}$) there exists $z \in Z^k(U)$ with $z|_{U_j} = Z_j$. We put $\mathbf{c}(a) = z$.

We have

$$A(U) = \{a \in A_r(U) | \mathbf{c}(a) = 0\}$$
$$A_e(U) = \{a \in A_r(U) | \mathbf{c}(a) \geqslant 0\} \tag{4}$$

If X is irreducible, A_r is a constant sheaf.

We will often use (4) to construct objects of the Chow category by first constructing their sets of rational sections, then specifying the cycle map c on the set of rational cycles, and then defining the object itself by the first equation in (4).

An example is the group of rational sections of G_k. For an open $G_k(U) = \ker(E_1^{k-1,-k}(U) \to E_1^{k,-k}(U))/\mathrm{im}(E_1^{k-2,-k}(U) \to E_1^{k-1,-k}(U))$. Because points of X of codimension larger than k are elements of U, replacing X by U does not change $E_1^{k-1,-k}(U)$ or $E_1^{k-2,-k}(U)$. However, every element of $E_1^{k,-k}(X)$ vanishes on some open and dense subset U of $X_{(k)}$. Consequently, $(G_k)_r(X) = E_1^{k-1,-k}(X)/E_1^{k-2,-k}(X)$. It is easy to see that on this set c is given by the E_1-differential.

3.4. *Relations between the several definitions of CH^k*

Let $k > 0$. Then there is an equivalence of categories

$$\mathbf{O}(\cdot)\colon \mathbf{CH}_z^k(X) \to \widetilde{\mathbf{CH}}^k(X)$$
$$\mathbf{O}(z)(U) = \mathrm{Hom}_{\mathbf{CH}_z^k(U)}(0, z). \tag{5}$$

$G_k(U)$, being the automorphism group of any object of $\mathbf{CH}_z^k(U)$, acts on the right side of (5). It is clear that a homomorphism from z to z' in $\mathbf{CH}_z^k(X)$ defines a homomorphism from $\mathbf{O}(z)$ to $\mathbf{O}(z')$ in $\widetilde{\mathbf{CH}}^k(X)$, that $\mathbf{O}(\cdot)$ is compatible with $\oplus$, and that $\mathbf{O}(\cdot)$ defines an equivalence of $\mathbf{CH}_z^k(X)_e$ and $\widetilde{\mathbf{CH}}^k(X)_e$.

An inverse to $0(\cdot)$ may be constructed as follows: For every object A of $\widetilde{\mathbf{CH}}^k(X)$, fix a rational section a_A of A. The inverse functor associates the cycle $\mathbf{c}(a_A) \in E_1^{k,-k}(X)$ to A and the element $a_{A'} - \varphi(a_A) \in (G_k)_r(X) = E_1^{k-1,-k}(X)/E_1^{k-2,-k}(X)$ to a morphism $\varphi\colon A \to A'$.

To investigate the relation between **CH** and $\widetilde{\mathbf{CH}}$, consider the following assumption:

(LF_k) If $x \in X$, $CH^k(\mathcal{O}_{X,x}) = 0$.

It is clear that (LF_k) is true if the local rings of X satisfy Gersten's conjecture. If X is regular, it satisfies (LF_k) up to torsion by the result of [S], and (LF_1) by the Auslander-Buxbaum theorem. Let X satisfy (LF_k). We want to prove $\mathbf{CH}^k(X) = \widetilde{\mathbf{CH}}^k(X)$. For every object A of $\widetilde{\mathbf{CH}}^k(X)$, we have to find a covering U_i of X such that A has a section on U_i. By the above remark, it suffices to do this if $A = \mathbf{O}(z)$ for a codimension k cycle z on X. By (LF_k), for every $x \in X$ there exists $g_x \in E_1^{k-1,-k}(\operatorname{Spec} \mathcal{O}_{X,x})$ such that $(g_x) = z|_{\operatorname{Spec} \mathcal{O}_{X,x}}$. It is clearly possible to extend g_x to $g'_x \in E_1^{k-1,-k}(X)$. Let $U_x = X - \operatorname{supp}(z - \partial(g'_x))$. Then g'_x defines a section of $O(Z)$ on U_x. By our choice of g_x and g'_x, $X \in U_x$. Consequently, the U_x form a covering of $X_{(k)}$ on which $O(z)$ has sections.

3.5. *Convention*

For $k \leqslant 0$, we put $\mathbf{CH}^k(X) = \widetilde{\mathbf{CH}}^k(X) = \mathbf{CH}_z^k(X)$. If A is an object in $\mathbf{CH}^k(X)$, $k \leqslant 0$ and U open in $X_{(k)}$, $A_r(U)$ consists of a single element denoted by β. We put $\mathbf{c}(\beta) = A \in E_2^{0,0}(X) = E_1^{0,0}(X) = Z^0(X)$ if $k = 0$ and $\mathbf{c}(\beta) = 0$ if $k < 0$. $A(U)$ and $A_e(U)$ are defined by (4).

Note that $\mathbf{CH}^k(X) = \widetilde{\mathbf{CH}}^k(X)$ consists of only one zero object if $k < 0$.

3.6. *Definition of a fibred Picard Category*

Recall from [D, §4] that a commutative Picard category is a groupoid P together with a functor $\oplus: P \times P \to P$, an associativity law $a_{A,B,C}$. $(A \oplus B) \oplus C \xrightarrow{\sim} A \oplus (B \oplus C)$ and a commutativity law $c_{A,B}: A \oplus B \xrightarrow{\sim} B \oplus A$ satisfying the compatibilities [DM, (1.0.1) and (1.0.2)], such that the translation functor $X \oplus \cdot$ is an equivalence of categories for every $X \in \operatorname{Ob}(P)$. It follows that P has a zero object, which we assume to be fixed.

An additive functor is a functor $F: P \to P'$ between commutative Picard categories together with a functor-isomorphism $F(A \oplus B) \xrightarrow{\sim} F(A) \oplus F(B)$ satisfying the additive analogues of [DM, Definition 1.8]. An additive functormorphism is a natural transformation $F \to G$ satisfying the additive analogue of [DM, (1.12.1) and (1.12.2)].

Let K be a category. A fibred Picard category over K consists of:

(i) For every object $X \in K$, a commutative Picard category P_X.
(ii) For every homomorphism $f: X \to Y$ in K, an additive functor $f^*: P_Y \to P_X$.
(iii) For every pair of composable arrows f, g in K, an additive functormorphism

$\kappa_{f,g}\colon (fg)^* \xrightarrow{\sim} g^*f^*$ such that for every $X \xleftarrow{f} Y \xleftarrow{g} Z \xleftarrow{h} U$ in K, the diagram

$$\begin{array}{ccc}
(fgh)^*(A) & \xrightarrow{\kappa_{f,gh}} & (gh)^*(f^*(A)) \\
\downarrow{\scriptstyle \kappa_{fg,h}} & & \downarrow{\scriptstyle \kappa_{g,h}} \\
h^*(gf)^*(A) & \xrightarrow{h^*(\kappa_{f,g})} & h^*(g^*(f^*(A)))
\end{array} \tag{6}$$

commutes.

Let $\tilde{P}$ be a second fibred Picard category over K with pull-back functors $f^!$ and natural transformations $\tilde{\kappa}_{f,g}$ for composable arrows in K. An admissible functor of fibred Picard categories is a pair (F, φ), consisting of:

(i) For every $X \in \mathrm{Ob}(K)$, an additive functor $F_X\colon P_X \to \tilde{P}_X$

(ii) For an arrow $f\colon X \to Y$ in K, an additive functor-morphism $\varphi_f\colon F_X \circ f^* \xrightarrow{\sim} f^! \circ F_Y$ such that the diagram

$$\begin{array}{ccccc}
F_X \circ (gf)^* & & \xrightarrow{\varphi_{gf}} & & (gf)^! \circ F_Z \\
\downarrow{\scriptstyle F_X(\kappa_{g,f})} & & & & \downarrow{\scriptstyle \tilde{\kappa}_{g,f}} \\
F_X \circ f^* \circ g^* & & & & f^! \circ g^! \circ F_Z \\
& \searrow{\scriptstyle \varphi_f} & & \nearrow{\scriptstyle f^!(\varphi_g)} & \\
& & f^! \circ F_Y \circ g^* & &
\end{array} \tag{7}$$

commutes for every sequence $X \xrightarrow{f} Y \xrightarrow{g} Z$ in K.

If P and P' are fibred Picard categories over K and F, F' are admissible functors from P to P', an admissible functor-isomorphism from F to F' is a family $(\psi_X\colon F_X \to F'_X)_{X \in \mathrm{Ob}(K)}$ of additive functor-isomorphisms such that for every arrow $f\colon X \to Y$ in K the diagram

$$\begin{array}{ccc}
F_X f^* & \xrightarrow{\varphi_f} & f^! F_Y \\
\downarrow{\scriptstyle \psi_X} & & \downarrow{\scriptstyle f^!((\psi_Y)} \\
F'_X f^* & \xrightarrow{\varphi'_f} & f^! F'_Y
\end{array}$$

commutes.

A cofibred Picard category over K is a fibred Picard category over I^{op}. The definition of an admissible functor between cofibred Picard categories is similar to the fibred case.

3.7. *Flat pull-back*

For a continuous mapping $f: X \to Y$ we denote the pull-back functor from sheaves on Y to sheaves on X by f^+. If $f: X \to Y$ is flat, the map $f: X_{(k)} \to Y_{(k)}$ is continuous. In §1 we defined a flat pull-back morphism $f^+: f^*G_{k,Y} \to G_{k,X}$.

If $A \in \mathrm{Ob}(\mathbf{CH}^k(Y))$ (resp. $\widetilde{\mathbf{CH}}^k(Y)$), $f^+(A)$ is a $f^+(G_{k,Y})$-torser on $X_{(k)}$ (resp. on $\tilde{X}_{(k)}$). We define $\mathbf{f}^*(A) \in \mathrm{Ob}(\mathbf{CH}^k(X))$ (resp. $\widetilde{\mathbf{CH}}^k(X)$) to be the image of $f^+(A)$ under $f^*: f^+(G_{k,Y}) \to G_{k,X}$. This defines a functor from $\mathbf{CH}^k(Y)$ to $\mathbf{CH}^k(X)$ and from $\mathbf{CH}^k(Y)_e$ to $\mathbf{CH}^k(X)_e$, and similar for $\widetilde{\mathbf{CH}}$. Every $a \in A(U)$ (resp. $A_e(U)$, resp. $A_r(U)$) defines $f^*(a) \in (\mathbf{f}^*A)(f^{-1}(U))$ (resp. $f^*(a) \in (\mathbf{f}^*A)_e(f^{-1}(U))$, resp. $f^*(a) \in (\mathbf{f}^*A)_r(f^{-1}(U))$ with $\mathbf{c}(f^*(a) = f^*(\mathbf{c}(a)))$.

If $k \leqslant 0$, $\mathbf{f}^*$ is defined to be the functor introduced in §2, and we put $f^*(\beta) = \beta$ (cf. Convention 3.5).

Let $X \xrightarrow{f} Y \xrightarrow{g} Z$ be flat morphisms. There are a natural isomorphism $\mathbf{f}^*(A \oplus B) \xrightarrow{\sim} \mathbf{f}^*(A) \oplus \mathbf{f}^*(B)$ sending $f^*(a \oplus b)$ to $f^*(a) \oplus f^*(b)$, and a natural isomorphism $\mathbf{f}^*(\mathbf{g}^*(A)) \xrightarrow{\sim} (\mathbf{gf})^*(A)$ sending $f^*(g^*(a))$ to $(gf)^*(a)$. These data define on $\mathbf{CH}^k$ and $\widetilde{\mathbf{CH}}^k$ the structure of a fibred Picard category over (schemes, flat morphisms), and the functors

$$\mathbf{CH}^k_z \xrightarrow{\mathrm{O}(\cdot)} \widetilde{\mathbf{CH}}^k \xleftarrow{\text{inclusion}} \mathbf{CH}^k$$

can be extended to admissible functors of fibred Picard categories.

If $f: X \to Y$ is flat and $A \in \mathrm{Ob}(\mathbf{CH}^k(Y))$, we put $A(X) = (\mathbf{f}^*A)(X)$, $A_e(X) = (\mathbf{f}^*A)_e(X)$, and $A_r(X) = (\mathbf{f}^*A)_r(X)$. By the previous remarks, these are presheaves on Y_{fpqc}.

3.8. If $j: U \to X$ is an open immersion, we often write $|_U$ instead of j^*. With this notation, we have

PROPOSITION. *Let $k > 0$ and $A \in \mathrm{Ob}(\widetilde{\mathbf{CH}}^k(X))$. If X_{Zar} has a covering by open subsets U_i such that $A|_{U_i} \in \mathrm{Ob}(\mathbf{CH}^k(U_i))$, then $A \in \mathrm{Ob}(\mathbf{CH}^k(X))$.*

Proof. Let $r \in A_r(X)$ be a rational section. By replacing U_i by a covering of $(U_i)_{(k)}$, we may assume $A(U_i) \neq \varnothing$, with section $a_i \in A(U_i)$. Let $a_i = b_i + r|_{U_i}$ with $b_i \in (G_k)_r(U_j)$. Then b_i can be represented as the image of $c_i = (c_{i,x})_{x \in (U_i)_{k-1}} \in E_1^{k-1,-k}(U_i)$ in $(G_k)_r(U_i)$.

We define $c_i' = (c_{i,x}')_{x \in X_{k-1}} \in E_1^{k-1,-k}(X)$ by

$$c_{i,x}' = \begin{cases} 0 & \text{if } x \notin U_i \\ c_{i,x} & \text{if } x \in U_i \end{cases}.$$

Let b_i' be the image of c_i' in $(G_k)_r(X)$. Then $\mathbf{c}(b_i')|_{U_i} = \mathbf{c}(b_i)$, hence $\mathbf{c}(b_i' + r)|_{U_i} = 0$. If $V_i = X - \text{supp}(\mathbf{c}(b_i' + r))$, then V_i is open in $X_{(k)}$ and contains U_i, consequently the V_i form a covering of $X_{(k)}$ on which A is trivial.

3.9. PROPOSITION. *Let p: $E \to X$ be the projection of a vector bundle E to its base X. Then* $\mathbf{p}^*\widetilde{\mathbf{CH}}^k(X) \to \widetilde{\mathbf{CH}}^k(E)$ and $\mathbf{p}^*$: $\mathbf{CH}^k(X) \to \mathbf{CH}^k(E)$ are equivalences of categories.

Proof. In §1, we proved that p^* is an isomorphism between the E_2-terms of the Quillen spectral sequences for X and E. Since the group of isomorphism classes of objects of $\widetilde{\mathbf{CH}}^k$ is $E_2^{k,-k}$, and since the group of automorphisms of any object of $\widetilde{\mathbf{CH}}^k$ is $E_2^{k-1,-k}$, the result follows for $\widetilde{\mathbf{CH}}^k$.

To prove the proposition for $\mathbf{CH}^k$, it suffices to prove that $A \in \text{Ob}(\widetilde{\mathbf{CH}}^k(X))$ and $\mathbf{p}^*A \in \text{Ob}(\mathbf{CH}^k(E))$ implies $A \in \text{Ob}(\mathbf{CH}^k(X))$. By Proposition 3.8. We may assume E is trivial, i.e., $E = A_X^d$. By induction on d, we may also assume $d = 1$. Every $A \in \text{Ob}(\widetilde{\mathbf{CH}}^k(X))$ is isomorphic to $\mathbf{O}(z)$ for some $z \in Z^k(X)$. Let $x \in X$. If $\mathbf{p}^*(\mathbf{O}(z)) \in \text{Ob}(\mathbf{CH}^k(E))$, there exists a cycle $z' \in Z^k(E)$ with $(x, 0) \notin \text{supp}(z')$ and $[p^*z] = [z']$ in $CH^k(E)$. If t is the coordinate on A_X^1 and sp_t the homomorphism defined in §1, this implies $[\text{sp}_t(z')] = [\text{sp}_t(p^*(z))] = [z]$ in $CH^k(X)$. But $(x, 0) \notin \text{supp}(z')$, hence $x \notin \text{supp}(\text{sp}_t(z'))$, consequently $\mathbf{O}(z)$ is locally trivial.

3.10. *Proper Push-Forward*

Let f: $X \to Y$ be proper of constant relative dimension $d \in \mathbb{Z}$. We define push-forward functors $\mathbf{f}_*$: $\widetilde{\mathbf{CH}}^k(X) \to \widetilde{\mathbf{CH}}^{k-d}(Y)$ in the following manner. If $k - d < 0$ and $A \in \text{Ob}(\widetilde{\mathbf{CH}}^k(X))$, $\mathbf{f}_*(A)$ is the only object of $\widetilde{\mathbf{CH}}^{k-d}(Y)$, every arrow in $\widetilde{\mathbf{CH}}^k(X)$ is mapped to the identity arrow, and $f_*(a) = \beta$ for every $a \in A_r(X)$.

Let $k = d$. Then every object in $\widetilde{\mathbf{CH}}^k(\mathbf{X})$ defines its class $[A] \in E_2^{k,-k}(X)$. We define $\mathbf{f}_*(A)$ to be the object of $\widetilde{\mathbf{CH}}^0(Y)$ defined by $f_*([A]) \in E_2^{0,0}(Y) = E_1^{0,0}(Y)$. Every arrow in $\widetilde{\mathbf{CH}}^k(X)$ is mapped to the identity arrow, and $f_*(a) = \beta$ for $a \in A_r(X)$.

Let $k < 0$ and $k - d > 0$. For $a \in \text{Ob}(\widetilde{\mathbf{CH}}^k(X))$ (there is only one object, and only its identical arrow), we put $\mathbf{f}_*(A) = G_{k,Y}$ and $\mathbf{f}_*(\beta) = 0$. Finally we consider the case $k \geqslant 0$ and $k - d > 0$. For $g \in G_{k,X}(U)$, we have $f_*(g) \in G_{k-d,Y}(Y - f(X - U))$, hence f_*: $(G_{k,X})_r(X) \to (G_{k-d,Y})_r(Y)$. Also, we have f_*: $Z^k(X) \to Z^{k-d}(Y)$, and f_* is compatible with 1(3), hence with $\mathbf{c}$. Let $A \in \text{Ob}(\widetilde{\mathbf{CH}}^k(X))$. We define $(\mathbf{f}_*A)_r(Y)$ to be the set of equivalence classes of pairs (g, a) with $g \in (G_{k-d,Y})_r(Y)$ and $a \in A_r(Y)$. Two pairs are equivalent if they are of the form $(g + f_*(h), a)$ and $(g, a + h)$ for some $h \in (G_{k,X})_r(X)$. The group $(G_{k-d,Y})_r(Y)$ acts on $(\mathbf{f}_*A)_r(Y)$ by the rule h: $(g, a) \to (g + h, a)$. We define the cycle of an element of $(\mathbf{f}_*A)_r(Y)$ by $\mathbf{c}((g, a)) = \mathbf{c}(g) + f_*\mathbf{c}(a)$. Now the sheaves $(\mathbf{f}_*A)_e$ and $\mathbf{f}_*A$ can be defined by (4). For $a \in A_r(X)$, the class of $(0, a)$ in $(\mathbf{f}_*A)_r(Y)$ is denoted by $f_*(a)$. If $a \in A(U)$, then $f_*(a) \in (\mathbf{f}_*A)(Y - f(X - U))$. If φ: $A \to B$ is a morphism in $\widetilde{\mathbf{CH}}^k(X)$, then $\mathbf{f}_*(\varphi)$ sends $f_*(a)$ to $f_*(\varphi(a))$.

In the remaining part of this paper we use the abbreviations c.r.d. for the condition 'constant relative dimension' and $\mathbf{CH}^{\cdot}(X)$, $\widetilde{\mathbf{CH}}^{\cdot}(X)$ for the direct sums of categories

$$\mathbf{CH}^{\cdot}(X) = \bigoplus_{k=-\infty}^{\infty} \mathbf{CH}^k(X), \qquad \widetilde{\mathbf{CH}}^{\cdot}(X) = \bigoplus_{k=-\infty}^{\infty} \widetilde{\mathbf{CH}}^k(X). \tag{8}$$

It is easy to see that the functors $\mathbf{f}_*$ constitute the structure of a cofibred Picard category over (schemes, proper morphisms of c.r.d.) on $\widetilde{\mathbf{CH}}^{\cdot}$. The transformation $\mathbf{f}_*(A) \oplus \mathbf{f}_*(B) \to \mathbf{f}_*(A \oplus B)$ maps $f_*(a) \oplus f_*(b)$ to $f_*(a \oplus b)$, and the isomorphism $\mathbf{g}_*(\mathbf{f}_*(A)) \to (\mathbf{gf})_*(A)$ maps $g_*(f_*(a))$ to $(gf)_*(a)$. It is easy to verify that the axioms of a cofibred Picard category are satisfied.

In the following cases, $\mathbf{f}_*(A)$ belongs to $\mathbf{CH}^{k-d}(Y)$:

(i) If Y satisfies $(LF)_{k-d}$.
(ii) *If* $k > 0$, f is a closed immersion and A belongs to $\mathbf{CH}^k(X)$.

3.11. *Definition of a bifibred Picard category*

For a bicategory K, we denote by Hor(K) (resp. Ver(K)) the category of objects of K with horizontal (resp. vertical) morphisms of K as morphisms. A bifibred Picard category over K consists of the following data:

(i) For every object $X \in \mathrm{Ob}(K)$, a commutative Picard category P_X.

(ii) The structure of a cofibred Picard category over Hor(K) for the P_X. We denote the push-forward functor of a horizontal morphism g by g_*.

(iii) The structure of a fibred Picard category, with pull-back functors f^*, over Ver(K) for the P_X.

(iv) For every bimorphism A in K with boundary

$$\begin{array}{ccc} X' & \xrightarrow{g'} & Y' \\ {\scriptstyle f'}\downarrow & (A) & \downarrow{\scriptstyle f} \\ X & \xrightarrow{g} & Y \end{array},$$

an additive functor-isomorphism $\varphi_A\colon f^*g_* \to g'_*f'^*$. The following compatibility assumption must be satisfied: If A, B, C are bimorphisms fitting into a diagram

$$\begin{array}{ccccc} & & \tilde{X} & \xrightarrow{\tilde{g}} & \tilde{Y} \\ & & {\scriptstyle e'}\downarrow & (B) & \downarrow{\scriptstyle e} \\ Z' & \xrightarrow{h'} & X' & \xrightarrow{g'} & Y' \\ {\scriptstyle \tilde{f}}\downarrow & (C) & {\scriptstyle f'}\downarrow & (A) & \downarrow{\scriptstyle f} \\ Z & \xrightarrow{h} & X & \xrightarrow{g} & Y \end{array},$$

then the following diagrams commute:

$$
\begin{array}{ccccc}
f^*(gh)_* & \xrightarrow{\text{can.}} & f^*g_*h_* & \xrightarrow{\varphi_A} & g'_*f'^*h_* \\
\Big\downarrow{\scriptstyle \varphi_{A_h \circ C}} & & & & \Big\downarrow{\scriptstyle \varphi_C} \\
(g'h')_*\tilde{f}^* & & \xrightarrow{\text{can.}} & & g'_*h'_*\tilde{f}^*
\end{array}
\tag{9}
$$

and a similar diagram for the vertical composition $A_v \circ B$. In (9), can. denotes isomorphisms canonically defined by the datum (ii). If we are given a second bifibred Ricard category P over K with pull-back and push-forward functors $f^!$ and $g_!$ and base-change isomorphisms $\tilde{\varphi}_A$, an admissible functor of bifibred Picard categories consists of:

— For every object X, an additive functor F_X: $P_X \to P_X$.
— For every horizontal morphism $g: X \to Y$, an additive functor-isomorphism $\alpha_g: F_Y g_* \to g_! F_X$ such that F becomes an admissible functor between cofibred Picard categories over Hor(K).
— For a vertical morphism $f: X \to Y$, the isomorphism $\beta_f: F_X f^* \to f^! F_Y$ making F an admissible functor between fibred Picard categories over Ver(K).

The following compatibility between α and β must be satisfied: For every bimorphism

$$
\begin{array}{ccc}
X' & \xrightarrow{g'} & Y' \\
{\scriptstyle f'}\Big\downarrow & \text{A} & \Big\downarrow{\scriptstyle f} \\
X & \xrightarrow[g]{} & Y
\end{array},
$$

the following diagram commutes:

$$
\begin{array}{ccc}
F_{Y'} f^* g_* & \xrightarrow{F_{Y'}(\varphi_A)} & F_{Y'} g'_* f'^* \\
\Big\downarrow{\scriptstyle \beta_f} & & \Big\downarrow{\scriptstyle \alpha_{g'}} \\
f^! F_Y g_* & & g'_! F_{X'} f'^* \\
\Big\downarrow{\scriptstyle \alpha_g} & & \Big\downarrow{\scriptstyle \beta_{f'}} \\
f^! g_! F_X & \xrightarrow{\tilde{\varphi}} & g'_! f'^! F_X
\end{array}
$$

If $G: P \to \tilde{P}$ is another admissible functor between bifibred Picard categories, then the collection ψ of additive functor-isomorphisms $\psi_X: F_X \to G_X$ is called a biadmissible functor-isomorphism if it is admissible for fibred Picard categories over Hor(K) and cofibred Picard categories over Ver(K).

Let $F: P \to \tilde{P}$ be an admissible functor between bifibred Picard categories over K. If for every object X the functor F_X is an equivalence of categories, it has an inverse F_X^{-1} together with a natural transformation $F_X F_X^{-1} \to \mathrm{Id}$, and this transformation determines F_X^{-1} up to unique functor-isomorphism. There is a unique way of giving F^{-1} the structure of an admissible functor between bifibred Picard categories such that the transformation $FF^{-1} \to \mathrm{Id}$ is biadmissible. In a similar fashion, admissible functors between fibred or cofibred Picard categories which are equivalences may be inverted.

In this paper we restrict our attention to bicategories K of the following type. Let C be a category and S_*, S^* be two distinguished families of morphisms in C which contain the identify morphisms and are closed under composition. We also assume that for $g \in S_*$ and $f \in S^*$ the fibre product

$$\begin{array}{ccc} Y' & \xrightarrow{g'} & Y \\ {\scriptstyle f'}\downarrow & & \downarrow{\scriptstyle f} \\ X' & \xrightarrow{g} & X \end{array} \qquad f' \in S^*,\ g' \in S_*$$

exists. We denote by (C, S_*, S^*) the bicategory which has the same objects as C, elements of S_* (resp. S^*) as horizontal (resp. vertical) morphisms and precisely one bimorphism for every fibre square as above (with boundary given by g, $g' \in S_*$ and $f, f' \in S^*$). The composition of morphisms and bimorphisms is defined in the obvious manner.

3.12. *Base change*

Consider a cartesian diagram

$$\begin{array}{ccc} X' & \xrightarrow{f'} & Y' \\ {\scriptstyle g'}\downarrow & & \downarrow{\scriptstyle g} \\ Y & \xrightarrow{f} & X \end{array}$$

with f flat and g proper of constant relative dimension. For every $A \in \mathrm{Ob}(\widetilde{\mathbf{CH}}^k(Y'))$, there is a base-change isomorphism $\mathbf{f}^*\mathbf{g}_*A \xrightarrow{\sim} \mathbf{g}'_*\mathbf{f}'^*A$ sending $f^*(g_*(a))$ to $g'_*(f'^*(a))$ for $a \in A_r(X)$. That this definition is correct follows from the base change identity we proved for $E_2^{p,q}$ (and hence for G_k) in §1.

It is easy to check that these base-change isomorphisms constitute the structure of a bifibred Picard category (in the sense of 3.11.) on $\widetilde{\mathbf{CH}}^{\cdot}$ over (schemes, proper morphisms of c.r.d., flat morphisms).

3.13. *Specialization*

Let $D \subset X$ be a closed regular immersion of codimension 1, and that the ideal defining D is generated by f in some neighbourhood of D. We define a functor $\mathbf{sp}_f\colon \widetilde{\mathbf{CH}}^k(X - D) \to \widetilde{\mathbf{CH}}^k(D)$ in the following manner. Let $k \leqslant 0$. Then on the set of objects $\mathbf{sp}_f$ is defined using Convention 3.5 and the homomorphisms sp_f defined in §1. On the set of morphisms $\mathbf{sp}_f$ is trivial since there are only identity morphisms. Furthermore, put $\mathrm{sp}_f(\beta) = \beta$.

Let $k > 0$ and $A \in \mathrm{Ob}(\widetilde{\mathbf{CH}}^k(X - D))$. We define $(\mathbf{sp}_f(A))_r(D)$ as the set of equivalence classes of pairs (g, a) with $g \in (G_{k,D})_r(D)$ and $a \in A_r(X - D)$. Two pairs are equivalent iff they are of the form $(g + \mathrm{sp}_f(h), a)$ and $(g,\ h + a)$ with $g \in (G_{k,D})_r(D)$, $h \in (G_{k,X})_r(X - D)$, $a \in A_r(X - D)$. We put $\mathbf{c}((g, a)) = \mathbf{c}(g) + \mathrm{sp}_f(\mathbf{c}(a))$. The sheaves $(\mathbf{sp}_f A)_e$ and $\mathbf{sp}_f A$ are defined by (4). For $a \in A_r(X - D)$, we denote the equivalence class of $(0, a)$ by $\mathrm{sp}_f(a) \in (\mathbf{sp}_f(A))_r(D)$. We have $\mathbf{c}(\mathrm{sp}_f(a)) = \mathrm{sp}_f(\mathbf{c}(a))$. If $a \in A(U)$, then $\mathrm{sp}_f(a) \in (\mathbf{sp}_f(A))(D - (D \cap \overline{(X - U)}))$.

Let K_{sp} be the following category. Objects of K_{sp} are triples (D, X, f) satisfying the assumptions of 1.5. A morphism between (D, X, f) and (D', X', f') is a Cartesian diagram

$$\begin{array}{ccc} D' & \longrightarrow & X' \\ \Big\downarrow{\scriptstyle p_D} & & \Big\downarrow{\scriptstyle p} \\ D & \longrightarrow & X \end{array} \tag{10}$$

such that $f' = f \circ p = p^*(f)$. A morphism is called flat (proper) if p is flat (proper). If in the latter case p is of constant relative dimension, the lemma in 1.4. implies that p_D is of the same relative dimension.

As usual, we define an isomorphism $\mathbf{sp}_f(A \oplus B) \to \mathbf{sp}_f(A) \oplus \mathbf{sp}_f(B)$ which sends $\mathrm{sp}_f(a \oplus b)$ to $\mathrm{sp}_f(a) \oplus \mathrm{sp}_f(b)$. If in (10) p is proper and of constant relative dimension, we define an isomorphism $\mathbf{sp}_f(\mathbf{p}_*(A)) \to \mathbf{p}_{D*}(\mathbf{sp}_{f'}(A))$ which sends $\mathrm{sp}_f(p_*(a))$ to $p_{D*}(\mathrm{sp}_{f'}(a))$. If p is flat, the isomorphism $\mathbf{sp}_{f'}(\mathbf{p}^*(A)) \to \mathbf{p}_D^*(\mathbf{sp}_f(A))$ sends $\mathrm{sp}_{f'}(p^*(a))$ to $p_D^*(\mathrm{sp}_f(a))$. By (1.15) and (1.16), these definitions are correct.

By the results of 3.7, 3.10 and 3.12, the categories $\widetilde{\mathbf{CH}}^{\cdot}(X - D)$ and $\widetilde{\mathbf{CH}}^{\cdot}(D)$ are bifibred Picard categories over (K_{sp}, proper morphisms of c.r.d., flat morphisms). It is easy to check that $\mathbf{sp}_f$ is an admissible functor (in the sense of 3.11) between bifibred Picard categories over K_{sp}.

3.14. LEMMA. *Let $f\colon X \to Y$ be flat. For a Zariski-open $U \subset Y$, let $f_U\colon f^{-1}(U) \to U$ be the restriction of f. If for every Zariski-open subset $U \subset Y$ we are given an additive functor-automorphism φ_U of $\mathbf{f}_U^*\colon \widetilde{\mathbf{CH}}^k(U) \to \widetilde{\mathbf{CH}}^k(f^{-1}(U))$ such that the restriction of φ_U to $f^{-1}(V)$ is φ_V for every $V \subset U$, then $\varphi_U = \mathrm{Id}$ for every U.*

Proof. It suffices to prove $\varphi_Y = \mathrm{Id}$. Let $A \in \mathrm{Ob}(\widetilde{\mathbf{CH}}^k(Y))$, then there is a covering U_j of $\tilde{Y}_{(k)}$ such that the restriction of A to U_j is trivial. Since φ is additive, $\varphi_{U_j, A|_{U_j}} = \mathrm{Id}$. By our assumption, this implies that the action of $\varphi_{Y,A}$ on $\mathbf{f}^*(A)|_{f^{-1}(U_j)}$ is trivial. Since the $f^{-1}(U_j)$ form a covering of $\tilde{X}_{(k)}$, this proves the result.

3.15. *Some natural transformations involving* **sp**

If $(D, X, f) \in \mathrm{Ob}(K_{\mathrm{sp}})$ and if in the commutative diagram

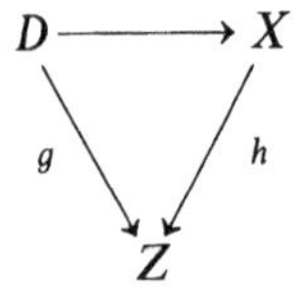

(11)

g and h are flat, then there is an additive functor-isomorphism

$$\tau_{h,X,D,f}\colon \mathbf{sp}_f \mathbf{h}^* \to \mathbf{g}^* \tag{12}$$

sending $\mathrm{sp}_f(h^*(a))$ to $g^*(a)$. By (1.17), this definition is correct. It is easy to check that τ satisfies the following compatibility with flat and proper base-changes $Z' \to Z$: It is clear that (since g is flat) the assumptions to (11) remain true after arbitrary base change $Z' \to Z$. In particular, $\widetilde{\mathbf{CH}}^{\cdot}(Z')$, $\widetilde{\mathbf{CH}}^{\cdot}(X \times_Z Z')$ and $\widetilde{\mathbf{CH}}^{\cdot}(D \times_Z Z')$ are bifibred Picard categories over (Z-schemes, proper morphisms of c.r.d., flat morphisms).

Both sides of (12) are admissible functors between bifibred Picard categories. The compatibility with base-changes $Z' \to Z$ is the fact that τ is a biadmissible transformation. By Lemma 3.14, τ is characterized uniquely by this compatibility. This implies a compatibility of τ with flat maps $Z \to Z'$ and flat base-changes $X' \to X$.

If in the diagram

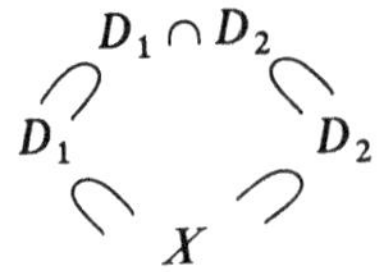

the triples (D_1, X, f_1), $(D_1 \cap D_2, D_2, f_1|_{D_2})$, (D_2, X, f_2), $(D_1 \cap D_2, D_1, f_2|_{D_1})$ belong to $\mathrm{Ob}(K_{\mathrm{sp}})$, we have a natural transformation

$$\omega_{f_1,f_2}\colon \mathbf{sp}_{f_1}|_{D_2}\mathbf{sp}_{f_2} \to \mathbf{sp}_{f_2}|_{D_1}\mathbf{sp}_{f_1} \tag{15}$$

sending $\mathrm{sp}_{f_1}(\mathrm{sp}_{f_2}(a))$ to $\mathrm{sp}_{f_2}(\mathrm{sp}_{f_1}(a))$. By (1.23), this is a correct definition. The following compatibilities are satisfied:

(*A*) *Compatibility with flat and proper base-changes.* Let K^b_{sp} be the category whose objects are 5-tuples (D_1, D_2, X, f_1, f_2) as in (14), and morphisms from $(D'_1, D'_2, X', f'_1, f'_2)$ to (D_1, D_2, X, f_1, f_2) are morphisms $X' \xrightarrow{h} X$ such that $f'_i = f_i \circ h$ and $D'_i = D_i \times_X X'$. In K^b_{sp} the fibre product of two morphisms exists if one of them is flat, hence the bicategory (K^b_{sp}, proper morphisms of c.r.d., flat morphisms) is well-defined. The Picard categories $\widetilde{\mathbf{CH}}^{\bullet}(X)$, $\widetilde{\mathbf{CH}}^{\bullet}(D_i)$, and $\widetilde{\mathbf{CH}}^{\bullet}(D_1 \cap D_2)$ are bifibred over (K^b_{sp}, proper morphisms of c.r.d., flat morphisms), and both sides of (15) are biadmissible functors. Then the transformation (15) is biadmissible.

(*B*). $\omega_{f_1,f_2}\omega_{f_2,f_1} = \mathrm{Id}$.

(*C*) *The Coxeter equality* $(\omega_{12}\omega_{23})^3 = \mathrm{Id}$. If $(D_i, X, f_i) \in \mathrm{Ob}(K_{\mathrm{sp}})$ for $i \in \{1; 2; 3\}$ and if for $i \neq j$ the embeddings $D_i \cap D_j \to D_j$ and $D_1 \cap D_2 \cap D_3 \to D_i \cap D_j$ are regular of codimension one, then the following diagram, in which each arrow is in an obvious manner constructed from the transformations (15), commutes:

$$\begin{array}{ccc}
\mathbf{sp}_{f_3}|_{D_1\cap D_2}\mathbf{sp}_{f_2}|_{D_1}\mathbf{sp}_{f_1} & \longrightarrow & \mathbf{sp}_{f_3}|_{D_1\cap D_2}\mathbf{sp}_{f_1}|_{D_2}\mathbf{sp}_{f_2} \\
\downarrow & & \downarrow \\
\mathbf{sp}_{f_2}|_{D_1\cap D_3}\mathbf{sp}_{f_3}|_{D_1}\mathbf{sp}_{f_1} & & \mathbf{sp}_{f_1}|_{D_2\cap D_3}\mathbf{sp}_{f_3}|_{D_2}\mathbf{sp}_{f_2} \\
\downarrow & & \downarrow \\
\mathbf{sp}_{f_2}|_{D_1\cap D_3}\mathbf{sp}_{f_1}|_{D_3}\mathbf{sp}_{f_3} & \longrightarrow & \mathbf{Sp}_{f_1}|_{D_2\cap D_3}\mathbf{sp}_{f_2}|_{D_3}\mathbf{sp}_{f_3}
\end{array} \tag{16}$$

(*D*) *Compatibility between* ω *and* τ. Let (Y, D, f) and (X, D_2, f_2) be objects of K_{sp}, and let $p\colon X \to Y$ be a flat projection whose restriction p_2 to D_2 remains flat. Then (D_1, D_2, X, f_1, f_2) belongs to K^b_{sp}, where $D_1 = p^{-1}(D)$ and $f_1 = f \circ p$. The following diagram commutes:

$$\begin{array}{ccccc}
\mathbf{sp}_{f_1}|_{D_2}\mathbf{sp}_{f_2}\mathbf{p}^* & \xrightarrow{\tau_p} & \mathbf{sp}_{f_1}|_{D_2}\mathbf{p}_2^* & \searrow & \\
\big\downarrow{\scriptstyle \omega_{f_1,f_2}} & & & & \mathbf{p}_{12}^*\mathbf{sp}_f, \\
\mathbf{sp}_{f_2}|_{D_1}\mathbf{sp}_{f_1}\mathbf{p}^* & \longrightarrow & \mathbf{sp}_{f_2}|_{D_1}\mathbf{p}_1^*\mathbf{sp}_f & \nearrow_{\tau_{p_1}} &
\end{array} \tag{17}$$

where p_1 and p_{12} are the restrictions of p to D_1 and $D_1 \cap D_2$. In the special case $(Y, D) = (D_2, D_1 \cap D_2)$, (such that $D_2 \to X$ is a section of p), p_2 is the identity, and (17) simplifies to

$$\begin{array}{ccc} \mathbf{sp}_{f_1}|_{D_2}\mathbf{sp}_{f_2}\mathbf{p}^* & \longrightarrow & \mathbf{sp}_{f_1}|_{D_2} \\ \downarrow & & \downarrow \\ \mathbf{sp}_{f_2}|_{D_1}\mathbf{sp}_{f_1}\mathbf{p}^* & \longrightarrow & \mathbf{sp}_{f_2}|_{D_1}\mathbf{p}_1^*\mathbf{sp}_{f_1}|_{D_2} \end{array} \tag{18}$$

4. Functoriality with respect to local complete intersections

For a better understanding of what follows, we suggest the reader to read the formulation of theorem 4.7. first.

4.1. LEMMA. *Let*

$$\begin{array}{ccc} D \cap Z & \subset & Z \\ \cap & & \cap \\ D & \subset & X \end{array} \tag{1}$$

be a Cartesian diagram in which the horizontal inclusions are regular embeddings of codimension one and the vertical inclusions are regular of codimension d. Let $X_{\widetilde{Z \cap D}}$ be the blow-up of X along $Z \cap D$. For a regular embedding $A \xrightarrow{j} B$, let N_A^B be the normal bundle of j. For a vector bundle E, let $\mathbb{P}(E)$ be its projective fibration.

(i) *We have $(X_{\widetilde{Z \cap D}}) \times_X D = D_{\widetilde{Z \cap D}} \bigcup_{\mathbb{P}(N_{Z \cap D}^D)} \mathbb{P}(N_{Z \cap D}^X)$, the components of the union are glued along the hyperplane $\mathbb{P}(N_{Z \cap D}^D)$ of $\mathbb{P}(N_{Z \cap D}^X)$. Let $(X_{\widetilde{Z \cap D}})^\circ = (X_{\widetilde{Z \cap D}}) - D_{\widetilde{Z \cap D}}$. Then the preimage of D in $(X_{\widetilde{Z \cap D}})^\circ$ can be identified with the vector bundle $N_{Z \cap D}^D \otimes (N_{Z \cap D}^Z)^{-1} = E$.*

(ii) *There exists a unique lifting $j: Z \to X_{\widetilde{Z \cap D}}$ of the embedding $Z \to X$. The immersion j is regular of codimension d, and its restriction to $Z \cap D$ is (in the notations of (i)) the zero-section of the bundle E.*

(iii) *Let $Z' \subset X$ be a regular inclusion of codimension $d - e$ containing Z, and suppose that $Z \subset Z'$ is regular of codimension e. If $\bar{Z}'$ is the closure of $Z' \cap (X - D)$ in $(X_{\widetilde{Z \cap D}})^\circ$, then $\bar{Z}' = (Z' \cap (X - D)) \cup N_{Z \cap D}^{Z' \cap D} \otimes (N_{Z \cap D}^Z)^{-1}$, and the inclusion $Z \to \bar{Z}'$ is regular of codimension e.*

(iv) *If $\hat{X} \to X$ is a morphism of schemes, we denote the base-change of objects on X to $\hat{X}$ by $\hat{}$. If the assumptions to the diagram (1) remain satisfied, with the same d, after base-change to $\hat{X}$, then $(\hat{X}_{\widetilde{\hat{Z} \cap \hat{D}}})^\circ = (X_{Z \cap D})^\circ \times_X \hat{X} (= ((X_{Z \cap D})^\circ)\hat{})$. If furthermore the assumption of (iii) remains satisfied, with the same e, after base-change to $\hat{X}$, then $\bar{\hat{Z}}' = (\bar{Z}')\hat{}$.*

(v) *If $Z \cap D$ and X are flat over a common base scheme S, then $\widetilde{X_{Z \cap D}}$ is flat over S. If furthermore in* (iii) *Z' is flat over S*, then so is $\bar{Z}'$.

Proof. The first part of (ii) follows from the universal property of the blow-up ([H, II.7.14]), since $Z \cap D \subset Z$ is (by assumption) regular of codimension one. The proofs of (i), (iii), (iv), and of the second part of (ii) are straightforward computations involving Micali's theorem (cf. [FL, §IV.2]).

To prove the first part of (v), it suffices to prove that J^k is flat over S, where J is the sheaf of ideals defining $Z \cap D$ on X. Since a locally free $\mathcal{O}_{Z \cap D}$-module is flat over S, this follows by induction on k from the sequence $0 \to J^{k+1} \to J^k \to S^k(N^X_{Z \cap D}) \to 0$. The proof of the second part of (v) is similar.

4.2. *Deformation to the normal bundle*

(cf. [F, §5.1]). Throughout 4.2–4.4, the symbol $Y \overset{d}{\subset} X$ denotes a regular embedding of codimension d.

LEMMA. *Let $n \geqslant 0$. To each sequence $X_0 \overset{d_1}{\subset} X_1 \overset{d_2}{\subset} \cdots \overset{d_n}{\subset} X_n$ of regular immersions, one can construct a commutative diagram*

$$\begin{array}{ccccccc} M_0 & \overset{d_1}{\subset} & M_1 & \overset{d_2}{\subset} & \cdots & \overset{d_n}{\subset} & M_n \\ \downarrow{\scriptstyle \pi_0} & & \downarrow{\scriptstyle \pi_1} & & & & \downarrow{\scriptstyle \pi_n} \\ \mathbb{P}^1_{X_0} & \overset{d_1}{\subset} & \mathbb{P}^1_{X_1} & \overset{d_2}{\subset} & \cdots & \overset{d_n}{\subset} & \mathbb{P}^1_{X_n} \end{array} \tag{2}$$

with the following properties:

(i) *On M_i, there is an action of the affine group $Aff_1(X_n) = \{g \in \Gamma(X_n, \mathcal{PGL}_2) \mid g(\infty) = \infty\}$ which is compatible with the action of $Aff_1(X_n)$ on $\mathbb{P}^1_{X_n}$.*

(ii) *The restriction of π_j to $\pi_j^{-1}(A^1_{X_j})$ is an isomorphism $\pi_j^{-1}(A^1_{X_j}) \xrightarrow{\sim} A^1_{X_j}$, and π_0 is an isomorphism everywhere: π_0: $M_0 \xrightarrow{\sim} \mathbb{P}^1_{X_0}$.*

(iii) *For a section $t \in \mathbb{P}^1(X_n)$, put $M_i^{(t)} = \pi_i^{-1}(t)$. Then $M_i^{(\infty)} \overset{1}{\subset} M_i$, and $M^{(\infty)}$ is the sequence*

$$X_0 \overset{d_1}{\subset} N^{X_1}_{X_0} \overset{d_2}{\subset} N^{X_2}_{X_0} \subset \cdots \overset{d_n}{\subset} N^{X_n}_{X_0}.$$

(iv) *If p: $X_n \to Z$ is a flat morphism such that the restriction of p to X_i is flat for every i, then M_i is flat over Z for every i.*

(v) *For a morphism $\tilde{X}_n \xrightarrow{\varphi} X_n$, we use $\tilde{}$ to denote base change to $\tilde{X}_n$. If we have $\tilde{X}_0 \overset{d_1}{\subset} \tilde{X}_1 \cdots \overset{d_n}{\subset} \tilde{X}_n$ (with the same d_i as in (2)!), then we can define an isomorphism M_φ: $M_j \times_{X_n} X_n \xrightarrow{\sim} M_j(\tilde{X}.)$ (the last symbol means the construction of M_i, applied to the sequence X.) On $M_j^{(\infty)} \times_{X_n} \tilde{X}_n$, M_φ is an* *isomorphism of vector bundles. If we apply a second base change* $\tilde{\tilde{X}}_n \xrightarrow{\psi} \tilde{X}_n \xrightarrow{\varphi} X_n$, *then the diagram*

$$\begin{array}{ccc} M_j \times_{X_n} \tilde{\tilde{X}}_n & \longrightarrow & (M_j \times_{X_n} \tilde{X}) \times_{\tilde{X}_n} \tilde{\tilde{X}}_n \\ \Big\downarrow M_{\varphi\psi} & & \Big\downarrow M_\varphi \times_{X_n} \tilde{\tilde{X}}_n \\ M_j(\tilde{\tilde{X}}.) & \xleftarrow{M_\psi} & M_j(\tilde{X}.) \times_{\tilde{X}_n} \tilde{\tilde{X}}_n \end{array}$$

commutes.

Proof. Let M_n be the blow-up of $\mathbb{P}^1_{X_n}$ along $\infty(X_0)$, where ∞ is the infinity-section of $\mathbb{P}^1_{X_n} \cdot M_0$ is the image of the immersion j constructed in 4.1(ii), and M_j for $0 < j < n$ is constructed by applying 4.1(iii) to $Z' = \mathbb{P}^1_{X_j}$ (of course, with $D = \infty$ and $Z = \mathbb{P}^1_{X_0}$). The Lemma follows from the results of 4.1.

4.3. *An awkward proposition*

Let $K_{d_1,\ldots,d_n}$ be the following category. Its objects are sequences

$$X^\cdot\colon X_0 \overset{d_1}{\subset} X_1 \overset{d_2}{\subset} X_2 \cdots \overset{d_n}{\subset} X_n.$$

A homomorphism from $X^\cdot$ to $Y^\cdot$ is a morphism from X_n to Y_n such that $X_i = Y_i \times_{Y_n} X_n$, it is called flat (proper) if $X_n \to Y_n$ is flat (proper). Since in $X^\cdot$ the fibre product of a flat and an arbitrary morphism exists, the bicategory ($K_{d_1,\ldots,d_n}$, proper morphisms of c.r.d., flat morphisms) is well-defined.

Let $K_{d_1,\ldots,d_n,\mathrm{sp}}$ be the following category. Objects are triples $(D^\cdot, X^\cdot, f)$ where $D^\cdot$ and $X^\cdot$. form a Cartesian diagram

$$\begin{array}{ccccccc} D_0 & \overset{d_1}{\subset} & D_1 & \overset{d_2}{\subset} & D_2 & \subset \cdots \overset{d_n}{\subset} & D_n \\ \cap 1 & & \cap 1 & & \cap 1 & & \cap 1 \\ X_0 & \overset{d_1}{\subset} & X_1 & \overset{d_2}{\subset} & X_2 & \subset \cdots \overset{d_n}{\subset} & X_n \end{array}$$

and f is a section of $\mathcal{O}_{X_n}$ in some neighbourhood of D_n such that $(D_i, X_i, f|_{X_i})$ belongs to $\mathrm{Ob}(K_{\mathrm{sp}})$ for every i. A morphism from $(D!, X!, f')$ to $(D^\cdot, X^\cdot, f)$ is a morphism X'_n to X_n such that $X'_i = X_i \times_{X_n} X'_n$, $D'_i = D_i \times_{X_n} X'_n$, and $f' =$ (pull-back of f to X'_n), it is flat (proper) if $X'_n \to X_n$ is flat (proper). The bicategory ($K_{d_1,\ldots,d_n,\mathrm{sp}}$, proper morphisms of constant relative dimension, flat morphisms) is well-defined.

By the results of §3, $\widetilde{\mathbf{CH}}^\cdot(X_i)$ (cf. (3.13)) are bifibred Picard categories over ($K_{d_1,\ldots,d_n}$, proper mor. of c.r.d., flat mor.). Over ($K_{d_1,\ldots,d_n,\mathrm{sp}}$, proper mor. of c.r.d., flat mor.), $\widetilde{\mathbf{CH}}^\cdot(X_i)$ and $\widetilde{\mathbf{CH}}^\cdot(D_i)$ are bifibred, and $\mathbf{sp}_f|_{X_i}\colon \widetilde{\mathbf{CH}}^\cdot(X_i) \to \widetilde{\mathbf{CH}}^\cdot(D_i)$ is biadmissible.

We are now ready to formulate our awkward proposition. To stimulate the reader's patience, we mention beforehand that the functor $F_{X\cdot}$ on K_{d_1} will play the role of the Gysin functor $\mathbf{i}^!$ for regular immersions of codimension d_1, that its uniqueness up to an additive functor-isomorphism on K_{d_1,d_2} will be used to construct the isomorphism $(\mathbf{ij})^! \to \mathbf{j}^!\mathbf{i}^!$, and that its uniqueness up to unique natural transformation on K_{d_1,d_2,d_3} will be used to verify (3.6).

PROPOSITION. *On* $K_{d_1,\ldots,d_n}$, *there exists an admissible functor between bifibred Picard categories* (*cf.* 3.11.) $F_{X\cdot}$: $\widetilde{\mathbf{CH}}^{\cdot}(X_n) \to \widetilde{\mathbf{CH}}^{\cdot}(X_0)$, *which preserves the graduation, together with the following data:*

(i) *If in the commutative diagram*

$$\begin{array}{ccccccc} X_0 & \overset{d_n}{\subset} & X_1 & \subset \cdots & \overset{d_n}{\subset} & X_n \\ & {\scriptstyle r_0}\searrow & \downarrow & & \swarrow{\scriptstyle r_n} & \\ & & Z & & & \end{array} \tag{3}$$

the projections r_i *are flat, then there is an additive functor-isomorphism*

$$\alpha_r \colon F_{X\cdot} \circ \mathbf{r}_n^* \to \mathbf{r}_0^* \tag{4}$$

which satisfies the following compatibility with base-changes $Z' \to Z$:

If $X_i' = X_i \times_Z Z'$ *and if* $r_i' \colon X_i' \to Z'$ *is the natural projection, then* (*since the* r_i *are flat*) $X_\cdot'$ *is in* $K_{d_1,\ldots,d_n}$, $\widetilde{\mathbf{CH}}^{\cdot}(X_i)$ *and* $\widetilde{\mathbf{CH}}^{\cdot}(Z')$ *are bifibred over* (*Z-schemes, proper mor. of* const. rel. dim., flat *morphisms*), *and the functors* $\mathbf{r}_i'^*$ *and* $F_{X'\cdot}$ *are biadmissible. The condition is that* $\alpha_{r'}$ *is a biadmissible transformation.*

(ii) *On* $K_{d_1,\ldots,d_n,\mathrm{sp}}$, *there is a biadmissible transformation*

$$\mathbf{sp}_f|_{X_0} F_{X\cdot} \to F_{D\cdot}\mathbf{sp}_f \qquad (\textit{denoted by } \beta_{f,X\cdot,D\cdot}) \tag{5}$$

with the following properties:

(ii.1) *Let* $(D_\cdot^{(1)}, X_\cdot, f_1)$ *and* $(D_\cdot^{(2)}, X_\cdot, f_2)$ *be objects of* $K_{d_1,\ldots,d_n,\mathrm{sp}}$ such that $(D_\cdot^{(12)}, D_\cdot^{(1)}, f_2|_{D_n^{(1)}})$ *and* $(D_\cdot^{(12)}, D_\cdot^{(2)}, f_1|_{D_n^{(2)}})$ *are both objects of* $K_{d_1,\ldots,d_n}$, *where* $D_i^{(12)} = D_i^{(1)} \cap D_i^{(2)}$. *Then the diagram*

$$\begin{array}{ccc} \mathbf{sp}_{f_1}|_{D_0^{(2)}}\mathbf{sp}_{f_2}|_{X_0}F_{X\cdot} & \xrightarrow{\mathbf{sp}_{f_1|D_0^{(2)}}(\beta_{f_2,X\cdot,D_\cdot^{(2)}})} & \mathbf{sp}_{f_1}|_{D_0^{(2)}}F_{D_\cdot^{(2)}}\mathbf{sp}_{f_2} \\ \downarrow{\scriptstyle \omega_{f_1|X_0,f_2|X_0}} & & \downarrow{\scriptstyle \beta_{f_1|D_n^{(2)},D_\cdot^{(2)},D_\cdot^{(12)}}} \\ \mathbf{sp}_{f_2}|_{D_0^{(1)}}\mathbf{sp}_{f_1}|_{X_0}F_{X\cdot} & & F_{D_\cdot^{(12)}}\mathbf{sp}_{f_1}|_{D_n^{(1)}}\mathbf{sp}_{f_2} \\ \downarrow{\scriptstyle \mathbf{sp}_{f_2|D_0^{(1)}}(\beta_{f_1,X\cdot,D_\cdot^{(1)}})} & & \downarrow{\scriptstyle F_{D_\cdot^{(12)}}(\omega_{f_1,f_2})} \\ \mathbf{sp}_{f_2}|_{D_0^{(1)}}F_{D_\cdot^{(1)}}\mathbf{sp}_{f_1} & \xrightarrow[\beta_{f_2|D_n^{(1)},D_\cdot^{(1)},D_\cdot^{(12)}}]{} & F_{D_\cdot^{(12)}}\mathbf{sp}_{f_2}|_{D_n^{(2)}}\mathbf{sp}_{f_1} \end{array}$$

commutes. The transformations ω *have been defined in* 3.15.

(ii.2) *If* $(D, Y, F) \in \mathrm{Ob}(K_{\mathrm{sp}})$, $X. \in \mathrm{Ob}(K_{d_1,\ldots,d_n})$, *and* $r_n\colon X_n \to Y$ *is a flat projection whose restriction* r_i *to* X_i *remains flat, we put* $D_i = X_i \times_Y D$ *and denote the projection of* D_i *to* D *by* $r_{D,i}$. *Then the following diagram commutes:*

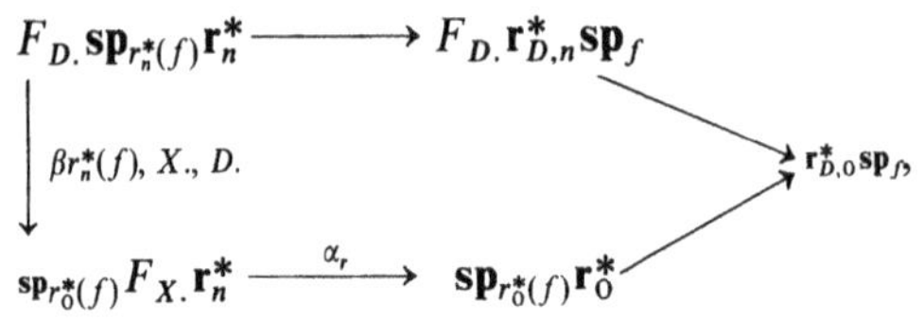

(7)

UNIQUENESS. The data (i) and (ii) determine F_X up to a unique biadmissible transformation between biadmissible functors over $K_{d_1,\ldots,d_n}$.

REMARK. In the special case $Y = X_0$, (7) simplifies to

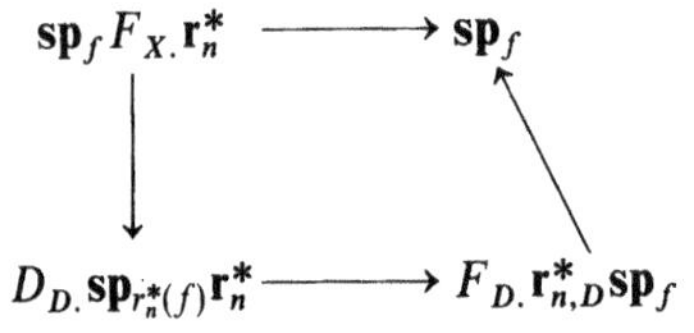

(8)

Proof of Proposition

4.4. *Construction of* $F_{X.}$. For $X_0 \subset \cdots \subset X_n$, let $M_0 \subset \cdots \subset M_n$ be the result of construction 4.2. The projections π_n have been defined in (2). We put

$$\begin{aligned} q_n &= \text{composition of } M_n \xrightarrow{\pi_n} \mathbb{P}^1_{X_n} \to X_n \\ q_{n,a} &= \text{restriction of } q_n \text{ to } \pi_n^{-1}(A^1_{X_n}) \end{aligned} \tag{9}$$

Furthermore, let

$$p_\infty\colon M_n^{(\infty)} \to M_0^{(\infty)} \tag{10}$$

be the bundle projection (cf. 4.2 (iii)). By 4.2 (v), the categories $\mathbf{CH}^\cdot(M_i^{(\infty)})$ are bifibred over $K_{d_1,\ldots,d_n}$, and $\mathbf{p}_\infty^*$ is a biadmissible functor. By 3.9, we know that it is also an equivalence of categories. Consequently it has a biadmissible inverse $(\mathbf{p}_\infty^*)^{-1}$ (cf. 3.11). Finally, let f_0 be a rational function on $\mathbb{P}^1_{\mathrm{Spec}(\mathbb{Z})}$ whose only zero is a simple zero along ∞ and whose pole does not intersect ∞. We denote the pull-back of f_0 to $\mathbb{P}^1_{X_n}$ by f and put

$$F_{X.} = (\mathbf{p}_\infty^*)^{-1} \mathbf{sp}_{\pi_n^*(f)} \mathbf{q}_{n,a}^*. \tag{11}$$

By 4.2(v) and our previous remarks, this is a biadmissible (over $K_{d_1,\ldots,d_n}$) functor.

Let u_n: $X_n \to Z$ be a flat morphism whose restriction u_i to X_i remains flat. 4.2(iv) proves that $M_i \to Z$ is flat, hence we can define α_u by

$$F_X\mathbf{u}_n^* = (\mathbf{p}_\infty^*)^{-1}\mathbf{sp}_{\pi_n^*(f)}\mathbf{q}_{n,a}^*\mathbf{u}_n^* \xrightarrow{\tau u_n q_{n,a}, M_n, M_n^{(\infty)}} (\mathbf{p}_\infty^*)^{-1}(\mathbf{u}_p\mathbf{p}_\infty)^*$$
$$\to (\mathbf{p}_\infty^*)^{-1}\mathbf{p}_\infty^*\mathbf{u}_0^* \to \mathbf{u}_0^*. \qquad (12)$$

It is biadmissible over (Z-schemes) since (12) contains only biadmissible transformations.

Let $(D., X., g) \in \mathrm{Ob}(K_{d_1,\ldots,d_n,\mathrm{sp}})$. We denote by $M.$ and $M_.^{(D)}$ the construction 4.2, applied to $X.$ and $D.$ respectively. The analogues for $D.$ of the morphisms (9) and (10) are denoted by a superscript (D). By 4.2(v), we have $M_i^{(D)} \simeq M_i \times_{X_n} D_n \subset M_i$. There is a unique biadmissible (over $K_{d_1,\ldots,d_n,\mathrm{sp}}$) transformation φ: $\mathbf{sp}_g(p_\infty^*)^{-1} \to (\mathbf{p}_\infty^{(D)*})^{-1}\mathbf{sp}_{p_\infty^*(g)}$ such that the diagram

$$\begin{array}{ccc}
\mathbf{p}_\infty^{(D)*}\mathbf{sp}_g(\mathbf{o}_\infty^*)^{-1} & \xrightarrow{\mathbf{p}_\infty^{(D)*}(\varphi)} & \mathbf{p}_\infty^{(D)*}(\mathbf{p}_\infty^{(D)*})^{-1}\mathbf{sp}_{p_\infty^*(g)} \\
\Big\downarrow 2 & & \Big\downarrow 2 \\
\mathbf{sp}_{p_\infty^*(g)}\mathbf{p}_\infty^*(\mathbf{p}_\infty^*)^{-1} & \longrightarrow & \mathbf{sp}_{p_\infty^*(g)}
\end{array}$$

commutes. We also mention that $p^*(g) = q_n^*(g)|_{M_n^{(\infty)}}$ and that the sheaf of ideals defining $D_n \subset M_n$ is in some neighbourhood of D_n trivialized by $q_n^*(g)$. We define the datum (ii) by

$$\mathbf{sp}_g F_X = \mathbf{sp}_g(\mathbf{p}_\infty^*)^{-1}\mathbf{sp}_{\pi_n^*(f)}\mathbf{q}_{n,a}'^* \xrightarrow{\varphi} (\mathbf{p}_\infty^{(D)*})^{-1}\mathbf{sp}_{q_n^*(g)|M_n^{(\infty)}}\mathbf{sp}_{\pi_n^*(f)}\mathbf{q}_{n,a}^* \to$$
$$\xrightarrow{\omega_{f,g}} (\mathbf{p}_\infty^{(D)*})^{-1}\mathbf{sp}_{\pi_n^{(D)*}(f)}\mathbf{sp}_{q_n^*(g)}\mathbf{q}_{n,a}^* \to (\mathbf{p}_\infty^{(D)*})^{-1}\mathbf{sp}_{\pi_n^{(D)*}(f)}\mathbf{q}_{n,a}^{(D)*}\mathbf{sp}_g = F_{D.}\mathbf{sp}_g.$$
$$(13)$$

Since (13) contains only biadmissible transformations, it defines a biadmissible (over $K_{d_1,\ldots,d_n,\mathrm{sp}}$) functor-isomorphism. It is straightforward to check that (6) follows from (3.16) (applied to the three functions $\pi_n^*(f)$, $q_n^*(g^{(1)})$ and $q_n^*(g^{(2)})$ on M_n) and (7) follows from (3.17) (applied to a fibre diagram of the form

$$\left(\begin{array}{ccc}
M_n^{(D)(\infty)} & \longrightarrow & M_n^{(\infty)} \\
\Big\downarrow & & \Big\downarrow \\
D & \longrightarrow & Y
\end{array}\right).$$

4.5. *Uniqueness of F*

Let G be a functor with the properties (i) and (ii), and let F be the functor constructed in 4.4. We use the notations of 4.4 and put $M_i^{(a)} = \pi_i^{-1}(A^1_{X_i})$. Since q_0 is flat and $M_0^{(\infty)} = X_0$, (3.12) provides us with a biadmissible (over $K_{d_1,\ldots,d_n}$) isomorphism

$$\psi\colon \mathbf{sp}_f|_{M_0}\mathbf{q}^*_{0,a}A \to A. \tag{14}$$

We define $\Phi_{X.}\colon G_{X.} \to F_{X.}$ by

$$\begin{aligned} G_{X.} &\to \mathbf{sp}_f|_{M_0}\mathbf{q}^*_{0,a}G_{X.} \xrightarrow{\text{(a)}} \mathbf{sp}_f|_{M_0}G_{M^{(a)}_.}\mathbf{q}^*_{n,a} \xrightarrow{\beta_{\pi^*_n(f),M.,M^{(\infty)}}} \\ &\to G_{M^{(\infty)}_.}\mathbf{sp}_{\pi^*_n(f)}\mathbf{q}^*_{n,a} \xrightarrow{\text{(b)}} G_{M^{(\infty)}_.}\mathbf{p}^*_\infty(\mathbf{p}^*_\infty)^{-1}\mathbf{sp}_{\pi^*_n(f)}\mathbf{q}^*_{n,a} \xrightarrow{\alpha_{p\infty}} \\ &\to (\mathbf{p}^*_\infty)^{-1}\mathbf{sp}_{\pi^*_n(f)}\mathbf{q}^*_{n,a} = F_{X.}. \end{aligned} \tag{15}$$

In (15), (a) is base-change with respect to the flat morphism $q_{n,a}$, and (b) is canonical. Since (15) contains only biadmissible transformations, $\Phi_{X.}$ is biadmissible (over $K_{d_1,\ldots,d_n}$). We have to check that it is compatible with the data (i) and (ii).

If we are given a commutative diagram (3), we define an additive functor-isomorphism $\chi_Z\colon \mathbf{r}^*_0 \to \mathbf{r}^*_0$ by

$$\mathbf{r}^*_0 \xrightarrow{(\alpha_r)^{-1}} G_{X.}\mathbf{r}^*_n \xrightarrow{\Phi_{X.}} F_{X.}\mathbf{r}^*_n \xrightarrow{\alpha_r} \mathbf{r}^*_0.$$

Since this definition contains only biadmissible (over Z-schemes transformations, we have $\chi_U = \chi_Z|_{r_0^{-1}(U)}$ for every Zariski-open $U \subset Z$. By 3.14, χ_Z is the identity, and Φ is compatible with the datum (i).

Diagram (16) implies that Φ is compatible with the datum (ii).

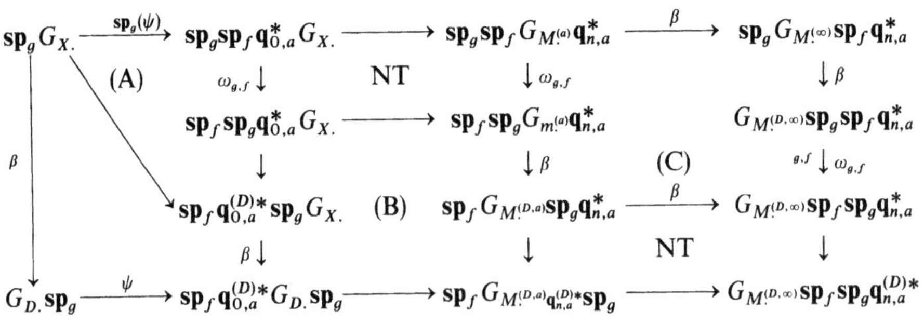

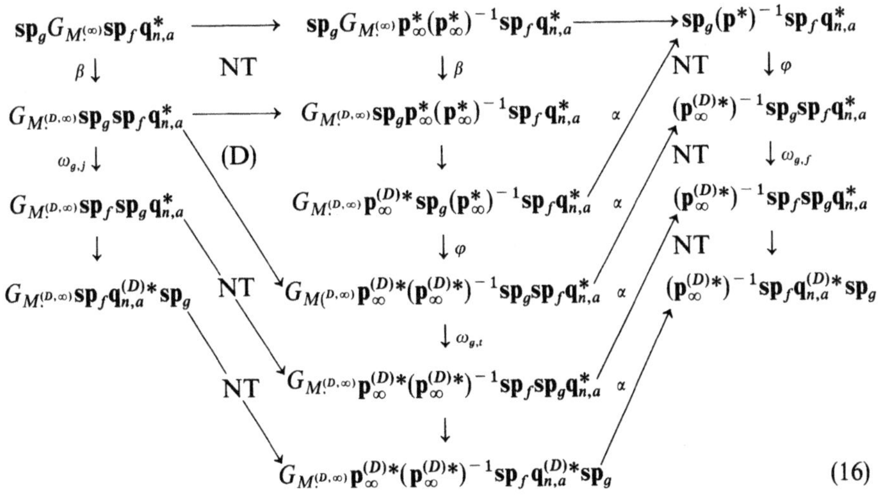

(16)

Glue the right boundary of the upper diagram with the left boundary of the lower diagram.

The notations in (16) are the same as in (13). To save space, the indices at the transformations α and β have been omitted, and the various pull-backs of f and g have been denoted by the same letter f or g. The vertical arrow at the left boundary of (16) is the isomorphism β for $G.$, and the outer right column of (16) is the sequence (13). The top and the bottom row of (16) are (15) for $X.$ and $D.$. The squares marked by 'NT' are commutative because they are of the form

$$\begin{array}{ccc} F(A) & \xrightarrow{\xi_A} & G(A) \\ {\scriptstyle F(\eta)}\downarrow & & \downarrow{\scriptstyle G(\eta)} \\ F(B) & \xrightarrow[\xi_B]{} & G(B) \end{array}$$

where $\xi\colon F \to G$ is a natural transformation between functors and $\eta\colon A \to B$ is a morphism.

The commutativity of (A) is (3.18). (B) commutes because β is admissible with respect to flat pull-backs in $K_{d_1,\ldots,d_n,\mathrm{sp}}$ (cf. the diagram at the end of 3.6). (C) is of type (6) and (E) of type (8). The commutativity of (D) is easily derived from the fact that **sp** is an admissible (with respect to flat pull-backs) functor.

4.6. *Uniqueness up to unique functor-isomorphism*

Suppose we are given a biadmissible automorphism Φ of F which is (in an obvious sense) compatible with the data (i) and (ii). From the compatibility with

(i) it follows tht $\Phi_{X.} = \mathrm{Id}$ if $X.$ is a sequence of vector bundles with base X_0. In particular, $\Phi_{M^{(\infty)}} = \mathrm{Id}$, where $M^{(\infty)}$ is the same as above. Since

$$\begin{array}{ccccccc}
F_{X.} & \longrightarrow & \mathbf{sp}_f \mathbf{q}^*_{0,a} F_{X.} & \longrightarrow & \mathbf{sp}_f F_{M^{(a)}} \mathbf{q}^*_{n,a} & \longrightarrow & F_{M^{(\infty)}} \mathbf{sp}_f q^*_{n,a} \\
\downarrow \Phi_{X.} & \mathrm{NT} & \downarrow \mathbf{sp}_f \mathbf{q}^*_{0,a}(\Phi_{X.}) & (\mathrm{A}) & \downarrow \mathbf{sp}_f(\Phi_{M^{(a)}}) & (\mathrm{B}) & \downarrow \Phi_{M^{(\infty)}} = \mathrm{Id} \\
F_{X.} & \longrightarrow & \mathbf{sp}_f \mathbf{q}^*_{0,a} F_{X.} & \longrightarrow & \mathbf{sp}_f F_{M^{(a)}} \mathbf{q}^*_{n,a} & \longrightarrow & F_{M^{(\infty)}} \mathbf{sp}_f \mathbf{q}^*_{n,a}
\end{array}$$

commutes, we have $\Phi_{X.} = \mathrm{Id}$, and the proof of 4.3 is complete. NT has the same meaning as in (16), (A) commutes since Φ is biadmissible, and (B) commutes since Φ is compatible with the datum (ii).

4.7. *The Gysin-functor*

We are now ready to prove the main theorem of §4. Throughout this paper, 'lci' will be an abbreviation for local complete intersection. A morphism $X \to Y$ is called a smoothable lci-morphism (abbreviated: slci-morphism) if it has a factorization $X \to S \to Y$ where $S \to Y$ is smooth. Then it follows ([SGA 6, Exp. VIII] or [FL, IV.3.10.]) that $X \to S$ is a regular immersion. The relative dimension of a lci-morphism at $x \in X$ has been defined in [SGA 6] and [FL].

Let K_{lci} be the following category: Objects are triples (f, X, Y) with $f: X \to Y$ a slci-morphism. A morphism from (f', X', Y') to (f, X, Y) is a Cartesian diagram (which we denote by (p_X, p_Y))

$$\begin{array}{ccc}
X' & \xrightarrow{f'} & Y' \\
\downarrow p_X & & \downarrow p_Y \\
X & \xrightarrow{f} & Y
\end{array}$$

such that for every $x \in X'$ $d_x(f') = d_{p_X(x)}(f)$, where $d_x(f)$ is the relative dimension of the lci-morphism f at the point x. A morphism in K_{lci} is called flat (resp. proper of c.r.d.) if so is p_Y (and hence p_X, cf. the lemma in 1.4). The bicategory (K_{lci}, proper morphisms of c.r.d., flat morphisms) is well-defined and will be denoted by $\mathbb{K}_{\mathrm{lci}}$.

Let $K_{\mathrm{lci,sp}}$ be the following category. Objects are 5-tuples (f, X, Y, D, λ) such that $f: X \to Y$ is a slci-morphism, $D \subset Y$ is a regular immersion of codimension one, λ is a section of $\mathcal{O}_Y$ in some neighbourhood of D which generates the sheaf of ideals defining D, and $D_X \to X$ is a regular immersion of codimension one, where $D_X = f^{-1}(D)$. A morphism in $K_{\mathrm{lci,sp}}$ from $(f', X', Y', D', \lambda')$ to (f, X, Y, D, λ) is a morphism (p_X, p_Y): $(f', X', Y') \to (f, X, Y)$ in K_{lci} such that $D' = p_Y^{-1}(D)$ and

$\lambda' = p_Y^*(\lambda)$. A morphism is said to be flat (proper), if so is p_Y. The bicategory ($K_{\mathrm{lci,sp}}$, proper morphisms of c.r.d., flat morphisms) is denoted by $\mathbb{K}_{\mathrm{lci,sp}}$.

Let $K_{\mathrm{lci,com}}$ be the category whose objects 5-tuples (f, g, X, Y, Z) where f: $X \to Y$ and g: $Y \to Z$ are lci-morphisms such that g and gf (and hence f too) are slci. A morphism from (f', g', X', Y', Z') to (f, g, X, Y, Z) is a triple (p_X, p_Y, p_Z), p_X: $X' \to X$, p_Y: $Y' \to Y$, p_Z: $Z' \to Z$ such that (p_Y, p_Z): $(g', Y', Z') \to (g, Y, Z)$ and (p_X, p_Y): $(f', X', Y') \to (f, X, Y)$ are morphisms in K_{lci}. It is flat (proper) if so is p_Z (and hence p_X and p_Y too). As usual, $\mathbb{K}_{\mathrm{lci,com}}$ refers to the bicategory ($K_{\mathrm{lci,com}}$, proper morphisms of c.r.d., flat morphisms).

The main result of §4 is the construction of an inverse image functor $f^!$ for local complete intersections f. Unlike the functors constructed in §3, it is no longer possible to define this functor directly. Instead, we describe it as a certain biadmissible functor, equipped with certain natural isomorphisms described in 4.7.1–4.7.3 which have to satisfy certain conditions explained in these paragraphs. The system of functors $f^!$ (for slci-morphisms f) is unique in the sense that, given another system of functors $f^?$ together with similar natural transformations satisfying the same conditions there exists a unique functor-isomorphism $f^! \to f^?$ respecting the natural transformations 4.7.1–4.7.3.

It should also be mentioned that the notions of a natural transformation and of a natural isomorphism are equivalent if applied to functors between groupoids, in particular to functors between Picard categories. Consequently, these two notions are used synonymously in the following text, and natural transformation is often abbreviated to transformations because confusions are impossible.

The main result of §4 is

THEOREM. *Let us denote objects of* $\mathbb{K}_{\mathrm{lci}}$ *by* (f, X, Y), *such that* $\widetilde{\mathbf{CH}}^{\cdot}(X)$ *and* $\widetilde{\mathbf{CH}}^{\cdot}(Y)$ *are bifibred Picard categories over* $\mathbf{K}_{\mathrm{lci}}$. *Then there is a biadmissible functor* $\mathbf{f}^!$: $\widetilde{\mathbf{CH}}^{i}(Y) \to \widetilde{\mathbf{CH}}^{i}(X)$ *between the bifibred Picard categories over* $\mathbb{K}_{\mathrm{lci}}\mathbf{CH}^{\cdot}(Y)$ *and* $\mathbf{CH}^{\cdot}(X)$, *together with the following data:*

4.7.1. For each flat morphism h: $Y \to Z$ such that hf is flat, we are given an isomorphism $\gamma_{f,h}$: $\mathbf{f}^!\mathbf{h}^* \to (\mathbf{hf})^*$ satisfying the following compatibility with flat and proper base changes $Z' \to Z$. For every $Z' \to Z$, f': $X' = Z' \times_Z X \to Z' \times_Z Y = Y'$ is slci (this is so because hf is flat), $\widetilde{\mathbf{CH}}^{\cdot}(X')$ $\widetilde{\mathbf{CH}}^{\cdot}(Y')$, $\widetilde{\mathbf{CH}}^{\cdot}(Z')$ are bifibred over (Z-schemes Z', proper morphisms of c.r.d., flat morphisms), and $\mathbf{f}'^!$, $\mathbf{h}'^*$ and $(\mathbf{h}'\mathbf{f}')^*$ are biadmissible functors between these categories. The condition is that $\gamma_{f',h'}$: $\mathbf{f}'^!\mathbf{h}'^* \to (\mathbf{h}'\mathbf{f}')^*$ is a biadmissible functor-isomorphism.

4.7.2. If we denote objects of $\mathbb{K}_{\mathrm{lci,sp}}$ by (f, X, Y, D, λ) and put $D_X = f^{-1}(D)$, then $\widetilde{\mathbf{CH}}^{\cdot}(X)$, $\widetilde{\mathbf{CH}}^{\cdot}(Y)$, $\widetilde{\mathbf{CH}}^{\cdot}(D)$, and $\widetilde{\mathbf{CH}}^{\cdot}(D_X)$ are bifibred Picard categories over $\mathbb{K}_{\mathrm{lci,sp}}$. The functors $\mathbf{sp}_\lambda$: $\widetilde{\mathbf{CH}}^{\cdot}(Y) \to \widetilde{\mathbf{CH}}^{\cdot}(D)$, $\mathbf{sp}_{f^*(\lambda)}$: $\widetilde{\mathbf{CH}}^{\cdot}(X) \to \widetilde{\mathbf{CH}}^{\cdot}(D_X)$, $\mathbf{f}^!$:

$\widetilde{\mathbf{CH}}^{\cdot}(Y) \to \widetilde{\mathbf{CH}}^{\cdot}(X)$, and $\mathbf{f}^!_D\colon \widetilde{\mathbf{CH}}^{\cdot}(D) \to \widetilde{\mathbf{CH}}^{\cdot}(D_X)$ are biadmissible (f_D is the restriction of f to $f^{-1}(D)$). The datum we require is a biadmissible functor-isomorphism

$$\delta_{\lambda,f}\colon \mathbf{f}^!_D \mathbf{sp}_\lambda \to \mathbf{sp}_{f^*(\lambda)} \mathbf{f}^!$$

which satisfies the following properties:

4.7.2.1. If $(f, X, Y, D_i, \lambda_i)$ ($i \in \{1, 2\}$) are objects of $K_{\mathrm{lci,sp}}$ such that the immersions $D_{12} = D_1 \cap D_2 \to D_i$ and $D_{X,12} = D_{X,1} \cap D_{X,2} \to D_{X,i}$ ($i \in \{1, 2\}$) are regular of codimension one, then the diagram similar to (6) commutes:

$$\begin{array}{ccccc} \mathbf{sp}_{\lambda_2}\mathbf{sp}_{\lambda_1}\mathbf{f}^! & \to & \mathbf{sp}_{\lambda_2}\mathbf{f}^!\mathbf{sp}_{\lambda_1} & \to & \mathbf{f}^!\mathbf{sp}_{\lambda_2}\mathbf{sp}_{\lambda_1} \\ \downarrow & & & & \downarrow \\ \mathbf{sp}_{\lambda_1}\mathbf{sp}_{\lambda_2}\mathbf{f}^! & \to & \mathbf{sp}_{\lambda_1}\mathbf{f}^!\mathbf{sp}_{\lambda_2} & \to & \mathbf{f}^!\mathbf{sp}_{\lambda_1}\mathbf{sp}_{\lambda_2} \end{array} \tag{17}$$

(For the sake of simplicity, the various pull-backs of λ_i and restrictions of f have been denoted by the same letters.)

4.7.2.2. Let $(D, Z, \lambda) \in \mathrm{Ob}(K_{\mathrm{sp}})$, $p\colon X \to Y$ a slci-morphism and $q\colon Y \to Z$ be a flat morphism such that qp is flat. We denote by D_X and D_Y the pre-images of D in X and Y, by p_D the restriction of p to D_X, and by q_D the restriction of q to D_Y. The condition is that the following diagram (which is similar to (7)) commutes:

$$\begin{array}{ccccc} \mathbf{p}^!_D\mathbf{q}^*_D\mathbf{sp}_\lambda & \to & \mathbf{p}^!_D\mathbf{sp}_{q^*(\lambda)}\mathbf{q}^* & \to & \mathbf{sp}_{(qp)^*(\lambda)}\mathbf{p}^!\mathbf{q}^* \\ \downarrow & & & & \downarrow \\ (\mathbf{q}_D\mathbf{p}_D)^*\mathbf{sp}_\lambda & & \longrightarrow & & \mathbf{sp}_{(qp)^*(\lambda)}(\mathbf{qp})^* \end{array} \ . \tag{18}$$

4.7.3. Let us denote objects of $\mathbb{K}_{\mathrm{lci,com}}$ by (f, g, X, Y, Z), such that $\widetilde{\mathbf{CH}}^{\cdot}(X)$, $\widetilde{\mathbf{CH}}^{\cdot}(Y)$, $\widetilde{\mathbf{CH}}^{\cdot}(Z)$ are bifibred Picard categories over $\mathbb{K}_{\mathrm{lci,com}}$, and $\mathbf{f}^!$, $\mathbf{g}^!$, and $(\mathbf{gf})^!$ are biadmissible functors between them. The datum we need is a biadmissible functor-isomorphism

$$\varepsilon_{f,g}\colon \mathbf{f}^!\mathbf{g}^! \to (\mathbf{gf})^!$$

subject to the following conditions:

4.7.3.1. If (f, g, X, Y, Z) is an object of $\mathbb{K}_{\mathrm{lci,com}}$ and $(D, Z, \lambda) \in \mathrm{Ob}(K_{\mathrm{sp}})$ such that $(g^{-1}(D), Y, g^*(\lambda))$ and $((gf)^{-1}(D), X, (gf)^*(\lambda))$ are objects of K_{sp}. Then the following diagram commutes:

$$\begin{array}{ccc} \mathbf{f}^!\mathbf{g}^!\mathbf{sp}_\lambda \to \mathbf{f}^!\mathbf{sp}_{g^*(\lambda)}\mathbf{g}_D^! & \to & \mathbf{sp}_{(gf)^*(\lambda)}\mathbf{f}_D^!\mathbf{g}_D^! \\ \big\downarrow & & \big\downarrow \\ (\mathbf{gf})^!\mathbf{sp}_\lambda & \longrightarrow & \mathbf{sp}_{(gf)^*(\lambda)}(\mathbf{g}_D\mathbf{f}_D)^! \end{array} \quad . \tag{19}$$

The restrictions of f and g to $(gf)^{-1}(D)$ and $g^{-1}(D)$ have been denoted by f_D and g_D.

4.7.3.2. The analogue of 3.(6), applied to lci-morphisms f, g, h such that h, hg, and hgf exist and are slci, commutes (of course, $*$ is replaced by $^!$).

4.7.3.3. If we have a Cartesian square

$$\begin{array}{ccc} X' & \xrightarrow{g'} & Y' \\ {\scriptstyle f'}\big\downarrow & & \big\downarrow{\scriptstyle f} \\ X & \xrightarrow{g} & Y \end{array}$$

with f smooth and g slci, then the following diagram is commutative:

$$\begin{array}{ccc} \mathbf{g}'^*\mathbf{f}^! & \xrightarrow{\text{base change}} & \mathbf{f}'^!\mathbf{g}^* \\ \big\downarrow{\scriptstyle \gamma_{g',\mathrm{Id}}{}^{-1}} & & \big\downarrow{\scriptstyle \mathbf{f}'^!(\gamma_{g,\mathrm{Id}})^{-1}} \\ \mathbf{g}'^!\mathbf{f}^! & \xrightarrow{\varepsilon_{g',f}} (\mathbf{fg}')^! = (\mathbf{gf}')^! \xleftarrow{\varepsilon_{f',g}} & \mathbf{f}'^!\mathbf{g}^! \end{array} \quad . \tag{20}$$

By 'base change' we mean the base change isomorphism defined by the coadmissible structure of $\mathbf{f}^!$ over $\mathrm{Ver}(\mathbb{K}_{\mathrm{lci}})$.

4.7.4. UNIQUENESS. The data 4.7.1-3, determine a biadmissible functor $\mathbf{f}^!$ over $\mathbb{K}_{\mathrm{lci}}$ up to unique biadmissible functor-isomorphism.

Proof. The proof will be carried out in steps 4.8–4.17.

4.8. $i^!$ *for regular closed immersions.* Let $i: X_0 \to X_1$ be a regular closed immersion. Since the codimension of i is locally constant, it suffices to construct $\mathbf{i}^!$ if i is of constant codimension d. Then $X. = (X_0 \subset X_1)$ is an object of K_d, and we put $\mathbf{i}^! = F_{X.}$ (cf. 4.3). The data 4.7.1 and 4.7.2 are given by the isomorphisms α and β in Proposition 4.3.

To construct $\varepsilon_{i,j}$: $\mathbf{i}^!\mathbf{j}^! \to (\mathbf{ji})^!$, we may assume that i and j are of constant codimensions d_1 and d_2. Then $X. = (X_0 \xrightarrow{i} X_1 \xrightarrow{j} X_2) \in \mathrm{Ob}(K_{d_1,d_2})$, and both $\mathbf{i}^!\mathbf{j}^!$

and $(\mathbf{ji})^!$ are candidates for $F_{X.}$. Hence, Proposition 4.3 implies that there is a unique biadmissible (over K_{d_1,d_2}) isomorphism $\varepsilon_{i,j}: \mathbf{i}^!\mathbf{j}^! \to (\mathbf{ji})^!$ which is compatible with the datum 4.7.2, i.e., which satisfies 4.7.3.1 (that such an isomorphism is compatible with the datum 4.7.1 follows from 3.14).

If we denote objects of K_{d_1,d_2,d_3} by $X. = (X_0 \xrightarrow{i} X_1 \xrightarrow{j} X_2 \xrightarrow{k} X_3)$, then $(\mathbf{kji})^!$ and $\mathbf{i}^!\mathbf{j}^!\mathbf{k}^!$ satisfy the conditions for $F_{X.}$, hence a 'good' isomorphism between them is unique. This is 4.7.3.2. in the case of regular closed immersions.

4.9. *The isomorphisms* $\varphi_{j,p}$

Let

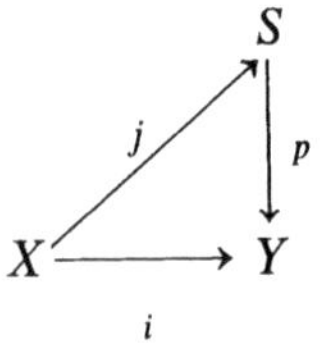

be a commutative diagram with regular closed immersions i and j and a smooth morphism p. We construct an isomorphism

$$\varphi_{j,p}: \mathbf{j}^!\mathbf{p}^* \to \mathbf{i}^! \tag{21}$$

as follows: Let $\tilde{S} = X \times_Y S$, $\tilde{p}: \tilde{S} \to X$ be the natural projection, and $\tilde{i}: \tilde{S} \to S$ be the base-change of i. The inclusion j determines a section $\tilde{j}$ of $\tilde{p}$:

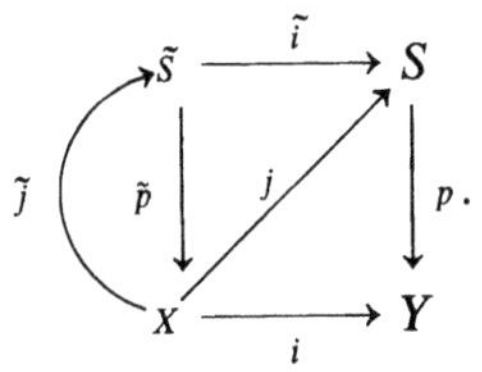

(22)

Then

$$\mathbf{j}^!\mathbf{p}^* \longrightarrow \tilde{\mathbf{j}}^!\tilde{\mathbf{i}}^!\mathbf{p}^* \longrightarrow \tilde{\mathbf{j}}^!\tilde{\mathbf{p}}^*\mathbf{i}^! \xrightarrow{\gamma_{\tilde{i},\tilde{p}}} \mathbf{i}^! \tag{23}$$

determines $\varphi_{j,p}$.

We need the following compatibilities:

(i) If $k: Z \to K$ is a regular immersion, we have

$$\varphi_{jk,p} = \mathbf{k}^!(\varphi_{j,p}) \tag{24}$$

(ii) If we have a commutative diagram

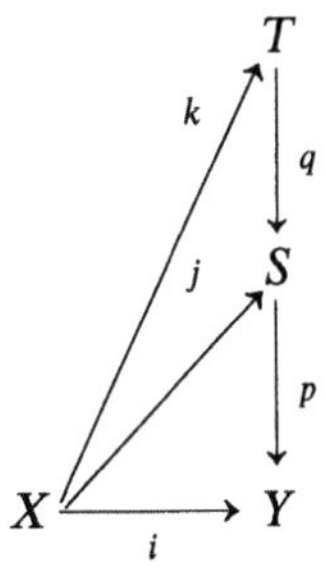

(25)

with regular imbeddings i, j, k and smooth morphisms p, q, then

$$\varphi_{k,pq} = \varphi_{j,p}\varphi_{k,q}. \tag{26}$$

(iii) Compatibility with flat and proper base changes $Y' \to Y$.

(iv) If we have a commutative diagram

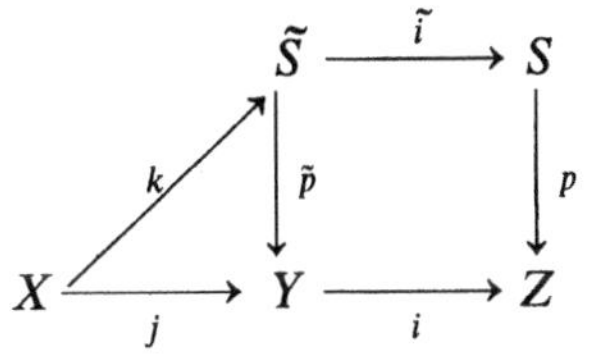

(27)

in which the square is Cartesian, i and j are regular immersions, and p is smooth, then the following diagram commutes:

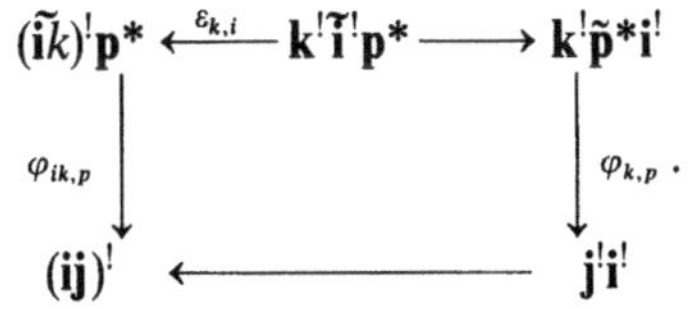

(28)

Since (23) contains only biadmissible transformations, (iii) is clear.

Proof of (i). We need the following

SUBLEMMA 1. *Let* $k: Z \to X$ *be a regular closed immersion,* $\tilde{S} \xrightarrow{\tilde{p}} X$ *a smooth morphism,* $\bar{S} \xrightarrow{\bar{p}} Z$ *the restriction of* $\tilde{S}$ to Z, $\tilde{j}: X \to \tilde{S}$ a section of $\tilde{p}$, $\bar{j}$ its restriction to Z:

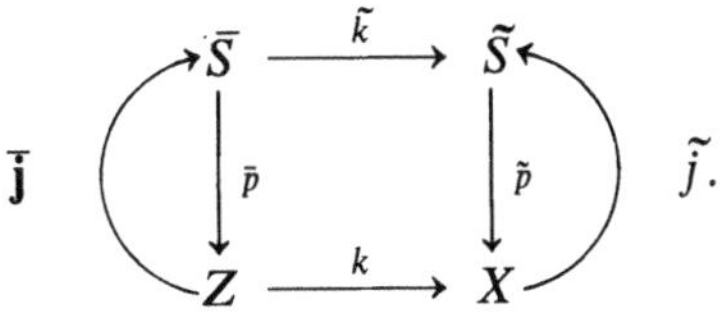

(29)

Then

$$\begin{array}{ccc} \mathbf{k}^! & \longrightarrow & \bar{\mathbf{j}}^!\bar{\mathbf{p}}^*\mathbf{k}^! \\ \downarrow & & \downarrow \\ \mathbf{k}^!\tilde{\mathbf{j}}^!\tilde{\mathbf{p}}^* & \longrightarrow & \bar{\mathbf{j}}^!\tilde{\mathbf{k}}^!\tilde{\mathbf{p}}^* \end{array} . \tag{30}$$

commutes.

Proof of Sublemma 1. For $A \in \mathrm{Ob}(\mathbf{CH}^i(X))$, let $\delta_{X,A}$ be the unique automorphism of $\mathbf{k}^! A$ making (30) commutative. We have to show that $\delta_{X,A} = 0$.

First we assume $k\colon Z \to X$ is the zero-section of a vector bundle. Then $\mathbf{k}^!$ is an equivalence of categories, hence $\delta_{X,A} = \mathbf{k}^!(\delta'_{X,A})$ for a unique automorphism $\delta'_{X,A}$ of A. Since (30) contains only biadmissible transformations, $(\delta'_{X,A})|_U = \delta'_{U,A}|_U$ for every open $U \subset X$. By 3.14 (applied to $f = \mathrm{Id}$), this implies $\delta_{X,A} = 0$.

Now we return to the general situation. Let $M_0 \to M_1$ be the deformation of $Z \to X$ to the normal bundle. We use the notations q_n and p_∞ as in (9) and (10), and also the isomorphism (14). By (14), we have $\delta_{X,A} = \mathbf{sp}_f \mathbf{q}_0^*(\delta_{X,A})$ as elements of $G_i(Z)$. Since the transformations in (30) are compatible with flat base change, $\mathbf{sp}_f \mathbf{q}_0^*(\delta_{X,A}) = \mathbf{sp}_f(\delta_{M_1^{(a)}, \mathbf{q}_{1,a}^* A})$. Of course, $\delta_{M^{(a)},B}$ and $\delta_{M^{(\infty)},B}$ are defined by means of the pull-backs of $\tilde{S}$ to $M_1^{(a)}$ and $M_1^{(\infty)}$. Since the transformations in (30) are compatible with specialization, we get

$$\delta_{X,A} = \mathbf{sp}_f(\delta_{M_1^{(a)}, \mathbf{q}_{1,a}^* \mathbf{A}}) = \delta_{M_1^{(\infty)}, \mathbf{sp}_f \mathbf{q}_{1,a}^* A} = 0$$

by our previous considerations, and the proof of the sublemma is complete.

We return to the proof of (i). The composite diagram of (22) and (29) is

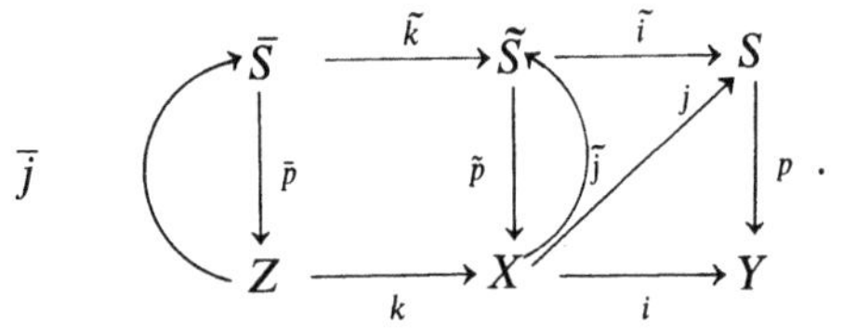

.

We have a diagram

$$\begin{array}{ccccccc} \mathbf{k}^!\mathbf{j}^!\mathbf{p}^* & \xrightarrow{(a)} & \bar{\mathbf{j}}^!\tilde{\mathbf{k}}^!\tilde{\mathbf{i}}^!\mathbf{p}^* & \xrightarrow{(b)} & \bar{\mathbf{j}}^!\tilde{\mathbf{k}}^!\tilde{\mathbf{p}}^*\mathbf{i}^! & \xrightarrow{(c)} & \bar{\mathbf{j}}^!\bar{\mathbf{p}}^*\mathbf{k}^!\mathbf{i}^! \\ & \overset{(a')}{\searrow} & \downarrow & \mathrm{NT} & \downarrow & & \downarrow (d) \\ & & \mathbf{k}^!\tilde{\mathbf{j}}^!\tilde{\mathbf{i}}^!\mathbf{p}^* & \xrightarrow{(b')} & \mathbf{k}^!\tilde{\mathbf{j}}^!\tilde{\mathbf{p}}^*\mathbf{i}^! & \xrightarrow{(c')} & \mathbf{k}^!\mathbf{i}^! \end{array} .$$

The commutativity of the triangle on the left is 4.7.3.2 for closed immersions, and the square on the right side is (30). It is clear that the composition of (a'), (b'), and (c') is $\mathbf{k}^!(\varphi_{j,p})$. Since $\bar{S} = Z \times_Y S$, the composition (d)(c)(b)(a) is $\varphi_{jk,p}$. The proof of (i) is complete.

Proof of (ii). We consider the diagram

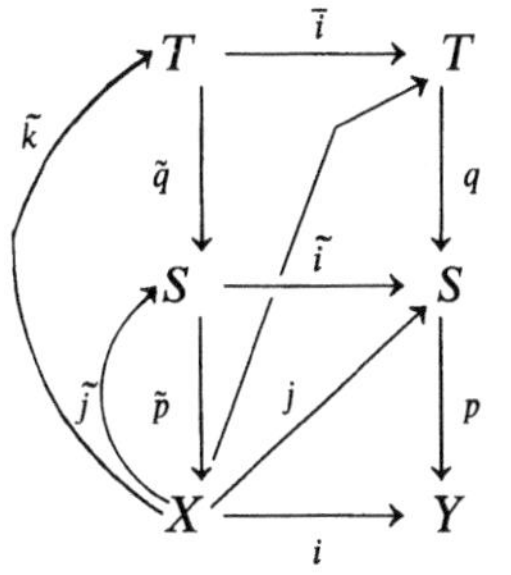

in which the squares are Cartesian. Our first aim is to show that the diagram

$$\begin{array}{ccc} \tilde{\mathbf{k}}^!(\tilde{\mathbf{p}}\tilde{\mathbf{q}})^* & \xrightarrow{\varphi_{\tilde{k},\tilde{p}\tilde{q}}} & \mathrm{Id} \\ \big\downarrow & & \big\downarrow \varphi_{\tilde{j},\tilde{p}} \\ \mathbf{k}^!\mathbf{q}^*\mathbf{p}^* & \xrightarrow{\varphi_{\tilde{k},\tilde{q}}} & \tilde{\mathbf{j}}^!\tilde{\mathbf{p}}^* \end{array} \qquad (31)$$

commutes. Since (31) contains only isomorphisms which are compatible with restriction to open subsets of X, this follows from 3.14.

Now we consider the diagram

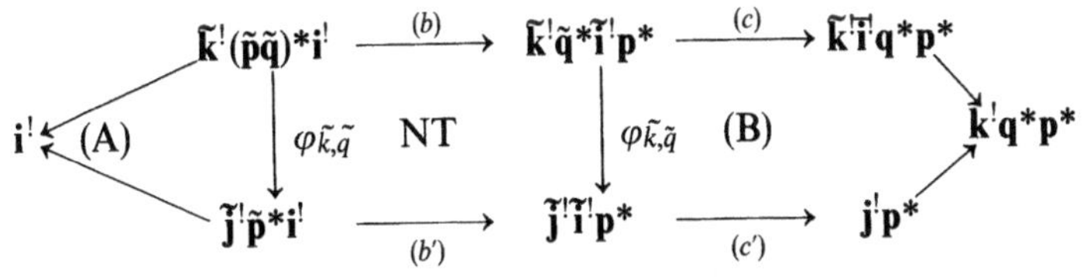

It is easy to check that $(a')(b')^{-1}(c')$ is $\varphi_{j,p}$ and that $(a)(b)^{-1}(c)^{-1}(d)^{-1}$ is $\varphi_{k,pq}$, such that (26) follows. It remains to prove that (32) commutes. For (A), this is an application of (31). For (B), form the fibre square

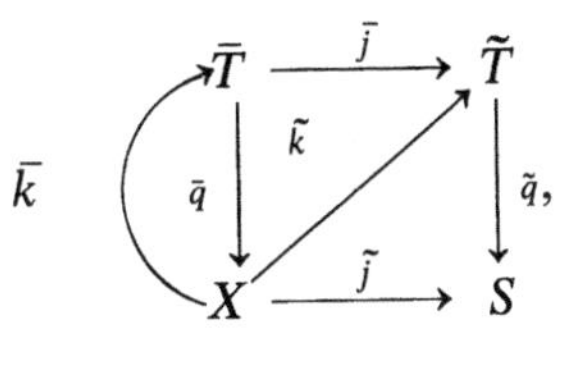

where $\bar{T} = X \times_S T = X \times_{\tilde{S}} \tilde{T}$ (with $X \to S$ and $X \to \tilde{S}$ given by j and $\tilde{j}$), and consider the isomorphisms

$$\begin{array}{l}
\mathbf{k}^!\mathbf{q}^* \longrightarrow \tilde{\mathbf{k}}^!\tilde{\mathbf{i}}^!\mathbf{q}^* \longrightarrow \tilde{\mathbf{k}}^!\tilde{\mathbf{q}}^*\tilde{\mathbf{i}}^! \xrightarrow{(a)} \tilde{\mathbf{k}}^!\tilde{\mathbf{j}}^!\tilde{\mathbf{q}}^*\tilde{\mathbf{i}}^! \\
\qquad \xrightarrow{(b)} \tilde{\mathbf{k}}^!\tilde{\mathbf{q}}^*\tilde{\mathbf{j}}^!\tilde{\mathbf{i}}^! \xrightarrow{(c)} \tilde{\mathbf{j}}^!\tilde{\mathbf{i}}^! \longrightarrow \mathbf{j}^!
\end{array}$$

Their composition is $\varphi_{k,q}$, and the composition $(c)(b)(a)$ is $\varphi_{\tilde{k},\tilde{q}}$, which proves the commutativity of (B).

Proof of (iv). With the same notations as in (27), put $S' = S \times_Z X \xrightarrow{p'} X$ and consider the diagram

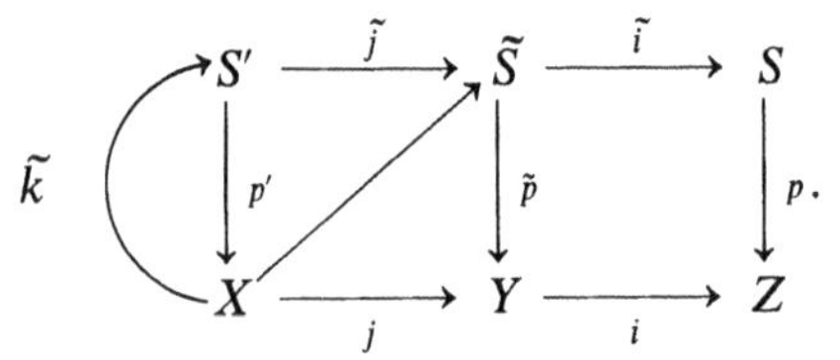

We have a commutative diagram

$$\begin{array}{ccccc}
(\mathbf{ik})^!\mathbf{p}^* & \longrightarrow & \mathbf{k}^!\mathbf{i}^!\mathbf{p}^* & \longrightarrow & \mathbf{k}^!\mathbf{p}^*\mathbf{i}^! \\
\downarrow & (A) & \downarrow & NT & \downarrow \\
\mathbf{k}^!(\mathbf{ij})^!\mathbf{p}^* & \longrightarrow & \mathbf{k}^!\mathbf{j}^!\mathbf{i}^!\mathbf{p}^* & \longrightarrow & \mathbf{k}^!\mathbf{j}^!\mathbf{p}^*\mathbf{i}^! \\
\downarrow & & (B) & & \downarrow \\
\mathbf{k}^!\mathbf{p}'^*(\mathbf{ij})^! & & \longrightarrow & & \mathbf{k}^!\mathbf{p}'^*\mathbf{j}^!\mathbf{i}^! \\
\downarrow & & NT & & \downarrow \\
(\mathbf{ij})^! & & \longrightarrow & & \mathbf{j}^!\mathbf{i}^!
\end{array}$$

(A) is 4.7.3.2., applied to the special case of regular immersions in which it has already been proved. (B) is the fact that $\varepsilon_{i,j}$ (cf. 4.7.3) is biadmissible at least in the special case of closed immersions. If we take the outer contour of this diagram and delete the vertices of the two middle rows, we get (28).

4.10. Now we are ready to construct $\mathbf{f}^!$. We choose a factorization

$$\delta\colon X \xrightarrow{\ i\ } S \xrightarrow{\ p\ } Y$$

of f, where p is smooth and i is a closed (and hence regular) immersion. We put

$$\mathbf{f}^!_\delta = \mathbf{i}^!\mathbf{p}^*. \tag{33}$$

Our task is to define the 'change of factorization'-isomorphism. First we assume we are given a new factorization δ' and a smooth morphism $r\colon S' \to S$ making

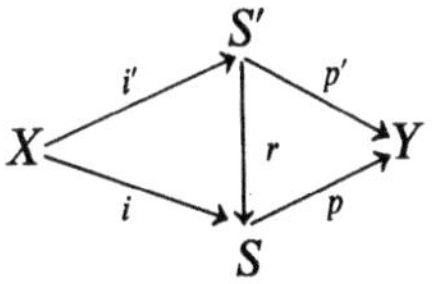

commutative. We define $\varphi_{\delta',\delta,r}$ by

$$\mathbf{i}'^!\mathbf{p}'^* \longrightarrow \mathbf{i}'^!\mathbf{r}^*\mathbf{p}^* \xrightarrow{\varphi_{i',r}} \mathbf{i}^!\mathbf{p}^*.$$

The following facts are consequences of (i)–(iv) in 4.9:

(i) If $k\colon Z \to X$ is a regular immersion and if σ_k is the factorization of fk by $Z \xrightarrow{ik} S \xrightarrow{p} Y$, then

$$\varphi_{\sigma'_k,\sigma_k,r} = \mathbf{k}^!(\varphi_{\sigma',\sigma,r}). \tag{34}$$

(ii) If $\tilde{\sigma}$ is a factorization of f sitting in the top row of the diagram

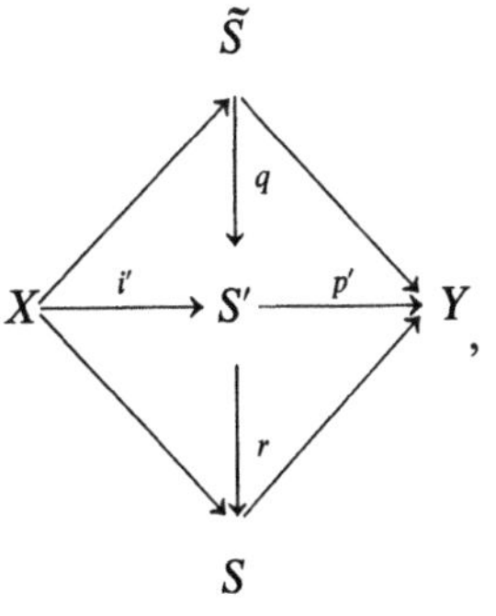

then we have

$$\varphi_{\tilde{\sigma},\sigma,rq} = \varphi_{\sigma',\sigma,r}\varphi_{\tilde{\sigma},\sigma',q}. \tag{35}$$

(iii) Compatibility with flat and proper base change $\tilde{Y} \to Y$.

If σ_1 and σ_2 are two factorizations of f, we denote by $\sigma_1 \times \sigma_2$ the factorization

$$X \to S_1 \times_Y S_2 \to Y,$$

by $r_{1,2}: S_1 \times_Y S_2 \to S_{1,2}$ the natural projections, and define $\varphi_{\sigma_1,\sigma_2}: \mathbf{f}^!_{\sigma_1} \to \mathbf{f}^!_{\sigma_2}$ by the composition of

$$\mathbf{f}^!_{\sigma_1} \xrightarrow{(\varphi_{\sigma_1\times\sigma_2,\sigma_1,r_1})^{-1}} \mathbf{f}^!_{\sigma_1\times\sigma_2} \xrightarrow{\varphi_{\sigma_1\times\sigma_2,\sigma_2,r_2}} \mathbf{f}^!_{\sigma_2}. \tag{36}$$

The identity $\varphi_{\sigma_1,\sigma_2}\varphi_{\sigma_2,\sigma_3} = \varphi_{\sigma_1,\sigma_3}$ follows from the diagram

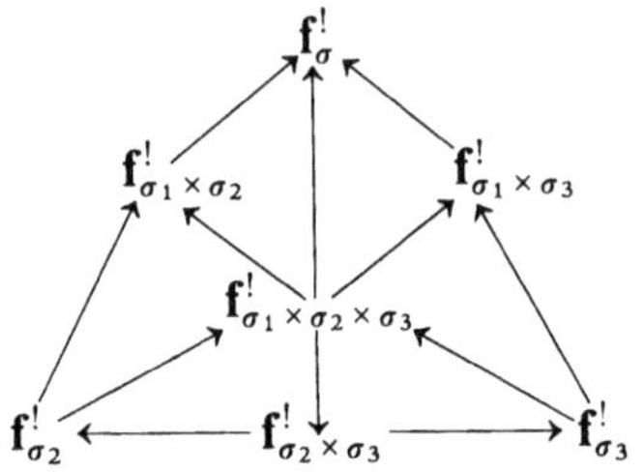

in which each arrow is in an obvious manner of the form $\varphi_{\sigma',\sigma,r}$. The commutativity of the small triangles follows from (35).

The isomorphisms (36) enable us to identify the functors $\mathbf{f}^!_\sigma$ with a unique functor $\mathbf{f}^!$. By fact (iii) above, this functor is biadmissible over $\mathbb{K}_{\mathrm{lci}}$.

The construction of the data 4.7.2. and 4.7.1 is easy. If $q: Y \to Z$ is flat and if qf is also flat, then for every factorization σ as in (33) we have an isomorphism $\mathbf{f}^!_\sigma\mathbf{q}^* \to \mathbf{i}^!(\mathbf{qp})^* \to (\mathbf{qf})^*$. It is easy to see that the isomorphism $\mathbf{f}^!\mathbf{q}^* \to (\mathbf{qf})^*$ defined this way is biadmissible in the sense explained in 4.7.1. By 3.14, such an isomorphism is unique, hence it is independent of the choice of the factorization.

Let $(f, X, Y, D, \lambda) \in \mathrm{Ob}(K_{\mathrm{lci,sp}})$. We choose a factorization $X \xrightarrow{i} S \xrightarrow{p} Y$ of f, put $D_X = f^{-1}(D)$, $D_S = p^{-1}(D)$, and denote by i_D and p_D the restrictions of i to D_X and p to D_S. By our assumptions to objects of $K_{\mathrm{lci,sp}}$, the restriction f_D of f to D_X is lci, and $f_D = p_D i_D$ is a factorization of f_D into a smooth morphism and a regular immersion. We define

$$\begin{aligned}\delta_{\lambda,f}: \mathbf{f}^!_D\mathbf{sp}_\lambda \to \mathbf{sp}_{f^*(\lambda)}\mathbf{f}^! \text{ by } & \mathbf{i}^!_D\mathbf{p}^*_D\mathbf{sp}_\lambda \\ & \to \mathbf{i}^!_D\mathbf{sp}_{p^*(\lambda)}\mathbf{p}^* \to \mathbf{sp}_{f^*(\lambda)}\mathbf{i}^!\mathbf{p}^*.\end{aligned}$$

Since our definition of the transformations $\varphi_{j,r}$, $\varphi_{\sigma,\sigma',r'}$, and $\varphi_{\sigma_1,\sigma_2}$ contains only isomorphisms compatible with specialization, $\delta_{\lambda,f}$ is independent of the factorization. The verification of 4.7.2.1 and 4.7.2.2 is easy.

4.11. Let $(f, g, X, Y, Z) \in \mathrm{Ob}(K_{\mathrm{lci,com}})$. For two Z-schemes A, B we denote $A \times_Z B \to B$ by $A_B \to B$. By our assumption, there exists a closed immersion k_0 of X into a smooth Z-scheme A. We have an induced immersion $k: X \to A_Y$ and a factorization $X \xrightarrow{k} A_Y \xrightarrow{\pi} Y$ of f, which we denote by α. We choose a factorization σ: $Y \xrightarrow{l} S \xrightarrow{p} Z$ of g and consider the diagram

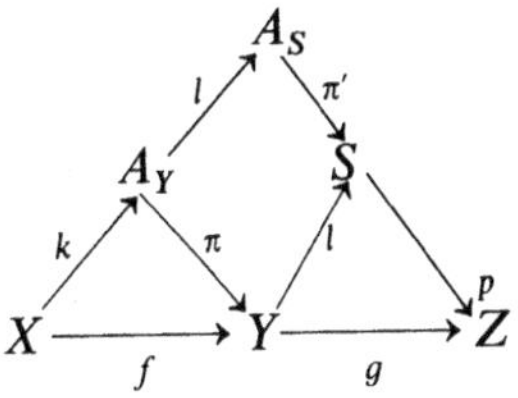

(37)

Let $\alpha * \sigma$ be the factorization of gf over A_S. We define

$$\varepsilon_{f,g,\alpha,\sigma}\colon \mathbf{f}^!_\alpha \mathbf{g}^!_\sigma = \mathbf{k}^!\boldsymbol{\pi}^*\mathbf{l}^!\mathbf{p}^* \to \mathbf{k}^!\mathbf{l}'^!\boldsymbol{\pi}'^*\mathbf{p}^* \to (\mathbf{l}'\mathbf{k})^!(\mathbf{p}\boldsymbol{\pi}')^* = (\mathbf{gf})^!_{\alpha*\sigma} \tag{38}$$

Let i_0 be another closed immersion of X into a smooth Z-scheme B, with induced factorization β: $X \xrightarrow{i} B_Y \xrightarrow{\rho} Y$ of f, and let τ: $Y \xrightarrow{j} T \xrightarrow{q} Z$ be another factorization of g. Consider the morphisms

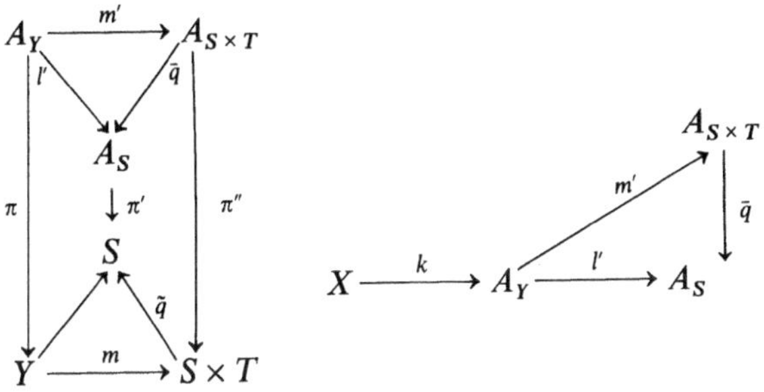

(39)

We get a commutative diagram

$$\begin{array}{ccccc}
\mathbf{k}^!\boldsymbol{\pi}^*\mathbf{m}^!\tilde{\mathbf{q}}^* & \longrightarrow & \mathbf{k}^!\mathbf{m}'^!\bar{\mathbf{q}}^*\boldsymbol{\pi}'^* & \longrightarrow & (\mathbf{m}'\mathbf{k})^!\bar{\mathbf{q}}^*\boldsymbol{\pi}'^* \\
\Big\downarrow {\scriptstyle k^!\pi^*(\psi_{m,q})} & \text{(A)} & \Big\downarrow & \text{(B)} & \Big\downarrow \\
\mathbf{k}^!\boldsymbol{\pi}^*\mathbf{l}^! & \longrightarrow & \mathbf{k}^!\mathbf{l}'^!\boldsymbol{\pi}'^* & \longrightarrow & (\mathbf{l}'\mathbf{k})^!\boldsymbol{\pi}'^*
\end{array} \quad . \tag{40}$$

(A) commutes by 4.9(iii) since in the left diagram (39) the upper triangle is the base change of the lower triangle by the flat map $A_{S\times T} \to S \times Y$, and (B) follows from 4.9(i), applied to the right side of (39).

If we apply (40) to $\mathbf{p}^*A$ for $A \in \mathrm{Ob}(\mathbf{CH}^\bullet(Z))$, we get a diagram whose outer contour is

$$\begin{array}{ccc}
\mathbf{f}^!_\alpha \mathbf{g}^!_{\sigma\times\tau} A & \longrightarrow & (\mathbf{fg})^!_{\alpha*(\sigma\times\tau)} \\
\Big\downarrow & & \Big\downarrow \\
\mathbf{f}^!_\alpha \mathbf{g}^!_\sigma A & \longrightarrow & (\mathbf{fg})^!_{\alpha*\sigma} A
\end{array} \quad , \tag{41}$$

where $(\mathbf{fg})^!_{\alpha^*(\sigma\times\tau)} \to (\mathbf{fg})^!_{\alpha^*\sigma}$ is defined by the commutative diagram of factorizations

$$
\begin{array}{ccccc}
 & & A_{S\times T} & & \\
 & \nearrow & \downarrow & \searrow & \\
X & \longrightarrow & S\times T & \longrightarrow & Z
\end{array} .
$$

Let us consider the morphisms ($A\times B$ means $A \times_Z B$)

$$
\begin{array}{ccccccccccc}
 & & (A\times B)_Y & \xrightarrow{m''} & (A\times B)_{S\times T} & & & & & & \\
 & {}^{n}\nearrow & \downarrow{\scriptstyle \rho'} & & \downarrow{\scriptstyle oA} & & & & & & \\
X & \xrightarrow[k]{} & A_Y & \xrightarrow[m']{} & A_{S\times T} & \xrightarrow[\pi'']{} & S\times T & \xrightarrow[\tilde q]{} & S & \xrightarrow[p]{} & Z
\end{array} .
$$

By 4.9(iv), the diagram

$$
\begin{array}{ccccc}
(\mathbf{m}''\mathbf{n})^! & \longrightarrow & \mathbf{n}^!\mathbf{m}''^!\boldsymbol{\rho}''^* & \longrightarrow & \mathbf{n}'^!\boldsymbol{\rho}^*\mathbf{m}'^! \\
\downarrow & & & & \downarrow \\
(\mathbf{m}'\mathbf{k})^! & & \longrightarrow & & \mathbf{k}^!\mathbf{m}^!
\end{array}
$$

commutes. Applying this to objects of the form $\boldsymbol{\pi}''^*\mathbf{q}^*\mathbf{p}^*A$, $A\in \mathrm{Ob}(\mathbf{CH}^{\boldsymbol{\cdot}}(Z))$, we get

$$
\begin{array}{ccc}
\mathbf{f}^!_{\alpha\times\beta}\mathbf{g}^!_{\sigma\times\tau}A & \longrightarrow & (\mathbf{gf})^!_{(\alpha\times\beta)*(\sigma\times\tau)}A \\
\downarrow & & \downarrow \\
\mathbf{f}^!_{\alpha}\mathbf{g}^!_{\sigma\times\tau}A & \longrightarrow & (\mathbf{gf})^!_{\alpha*(\sigma\times\tau)}A
\end{array} . \tag{42}
$$

Gluing (41) and (42) and using 4.10(ii), we get

$$
\begin{array}{ccc}
\mathbf{f}^!_{\alpha\times\beta}\mathbf{g}^!_{\sigma\times\tau} & \longrightarrow & (\mathbf{gf})^!_{(\alpha\times\beta)*(\sigma\times\tau)} \\
\downarrow & & \downarrow \\
\mathbf{f}^!_{\alpha}\mathbf{g}^!_{\sigma} & \longrightarrow & (\mathbf{gf})^!_{\alpha*\sigma}
\end{array} . \tag{43}
$$

Applying (43) another time, with the roles of α and σ and of β and τ interchanged, and using the definition (36), we arrive at the commutativity of

$$\begin{array}{ccc} \mathbf{f}^!_\alpha \mathbf{g}^!_\sigma & \longrightarrow & (\mathbf{gf})^!_{\alpha * \sigma} \\ \Big\downarrow {\scriptstyle \varphi_{\alpha,\beta}\mathbf{f}^!(\varphi_{\sigma,\tau})} & & \Big\downarrow \\ \mathbf{f}^!_\beta \mathbf{g}^!_\tau & \longrightarrow & (\mathbf{gf})^!_{\beta * \tau} \end{array} . \tag{44}$$

This proves that the transformations (38) fit together and define $\varepsilon_{f,g}\colon \mathbf{f}^!\mathbf{g}^! \to (\mathbf{gf})^!$.

4.12. We omit the proof of 4.7.3.1 since it is straightforward. To prove 4.7.3.2, we consider lci-morphisms $U \xrightarrow{f} X \xrightarrow{g} Y \xrightarrow{h} Z$ such that U, X, and Y admit closed immersions into smooth Z-schemes. We want to prove

$$\begin{array}{ccc} \mathbf{f}^!\mathbf{g}^!\mathbf{h}^! & \longrightarrow & \mathbf{f}^!(\mathbf{hg})^! \\ \Big\downarrow & & \Big\downarrow \\ (\mathbf{gf})^!\mathbf{h}^! & \longrightarrow & (\mathbf{hgf})^! \end{array} . \tag{45}$$

We choose closed Z-immersions of U, X, and Y into smooth Z-schemes A, B, and S. Then we have the following factorizations of f, g, and h:

$$\alpha\colon\ U \xrightarrow{\ i\ } A_X \xrightarrow{\ p\ } X$$

$$\beta\colon\ X \xrightarrow{\ j\ } B_Y \xrightarrow{\ q\ } Y$$

$$\sigma\colon\ Y \xrightarrow{\ k\ } S \xrightarrow{\ r\ } Z.$$

It suffices to prove

$$\begin{array}{ccc} \mathbf{f}^!_\alpha \mathbf{g}^!_\beta \mathbf{h}^!_\sigma & \longrightarrow & \mathbf{f}^!_\alpha(\mathbf{hg})^!_{\beta * \sigma} \\ \Big\downarrow & & \Big\downarrow \\ (\mathbf{gf})^!_{\alpha * \beta}\mathbf{h}^!_\sigma & \longrightarrow & (\mathbf{hgf})^!_{\alpha * \beta * \sigma} \end{array} \tag{46}$$

We consider the morphisms ($A \times B$ means $A \times_Z B$):

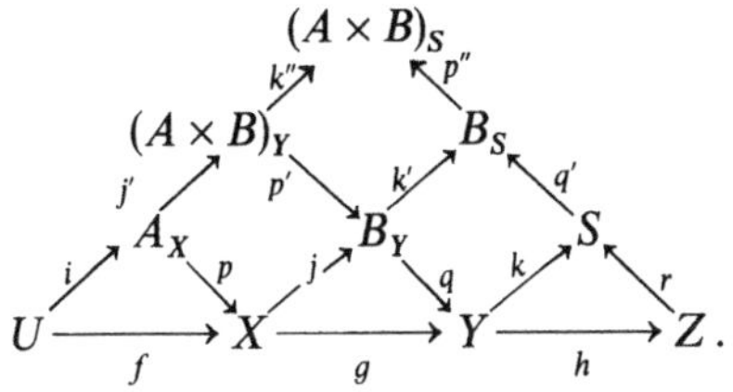

We have a commutative diagram

$$
\begin{array}{ccccc}
\mathbf{p}^*\mathbf{j}^!\mathbf{q}^*\mathbf{k}^!C & \longrightarrow & \mathbf{p}^*\mathbf{j}^!\mathbf{k}'^*\mathbf{q}'^!C & \longrightarrow & \mathbf{p}^*(\mathbf{k}'\mathbf{j})^!\mathbf{q}'^*C \\
\downarrow & \text{NT} & \downarrow & \text{(A)} & \\
\mathbf{j}'^!\mathbf{p}'^*\mathbf{q}^*\mathbf{k}^!C & \longrightarrow & \mathbf{j}'^!\mathbf{p}'^*\mathbf{k}'^!\mathbf{q}'^*C & & \\
 & & \downarrow & & \downarrow \\
 & \text{(B)} & \mathbf{j}'^!\mathbf{k}''^!\mathbf{p}''^*\mathbf{q}'^*C & \longrightarrow & (\mathbf{k}''\mathbf{j}')^!\mathbf{p}''^*\mathbf{q}'^*C \\
\downarrow & & \downarrow & \text{NT} & \downarrow \\
\mathbf{j}'^!(\mathbf{q}\mathbf{p}')^*\mathbf{k}^!C & \longrightarrow & \mathbf{j}'^!\mathbf{k}''^!(\mathbf{q}'\mathbf{p}'')^*C & \longrightarrow & (\mathbf{k}''\mathbf{j}')^!(\mathbf{q}'\mathbf{p}'')^*C
\end{array}
\tag{47}
$$

The commutativity of (A) belongs to the conditions which were used to characterize the isomorphism $\mathbf{j}^!\mathbf{k}'^! \to (\mathbf{k}'\mathbf{j})^!$ defined in 4.8, and (B) is of type 3(7) (applied to the biadmissible functor $F = \mathbf{k}^!$). If we insert $C = \mathbf{r}^*(.)$ in (46) and apply $\mathbf{i}^!$, we get a diagram whose outer contour can be identified with (46).

Now we prove 4.7.3.3. Since (20) is clear for a smooth morphism g, it suffices to consider the case of a regular closed immersion g. In this case, the proof consists of two parts:

SUBLEMMA 1. *If in* σ: $X \xrightarrow{i} Y \xrightarrow{p} Z$ *i is a regular immersion and p is smooth, then*

$$
\begin{array}{ccc}
\mathbf{i}^!\mathbf{p}^* & \xleftarrow{\mathbf{i}^!(\gamma_{p,\mathrm{Id}})} & \mathbf{i}^!\mathbf{p}^! \\
\| & & \downarrow{\scriptstyle \varepsilon_{i,p}} \\
(\mathbf{pi})^!_\sigma & \longrightarrow & (\mathbf{pi})^!
\end{array}
\tag{48}
$$

commutes.

SUBLEMMA 2. *We suppose that in a Cartesian diagram*

$$
\begin{array}{ccc}
X' & \xrightarrow{i'} & Y' \\
\downarrow & & \downarrow{\scriptstyle p} \\
X & \xrightarrow{i} & Y
\end{array}
$$

p is smooth and i is a regular immersion. Then $\sigma\colon X' \xrightarrow{i'} Y' \xrightarrow{p} Y$ is an admissible factorization of the lci-morphism ip'. With these notations, the diagram

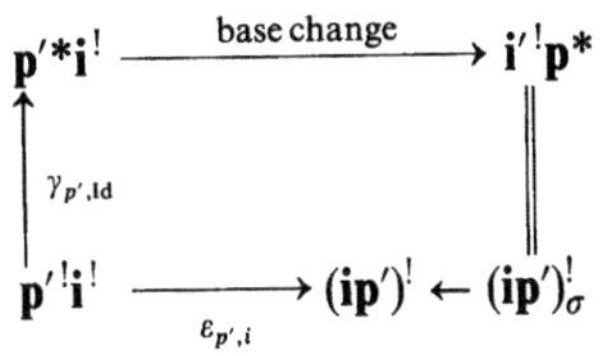

(49)

commutes.

It is clear that (20) *for a regular immersion g follows from* (48) *and* (49).

Proof of Sublemma 1. In 4.11, we choose for k_0 the immersion of X into the smooth Z-scheme Y, and put $S = Y$. Then (37) becomes

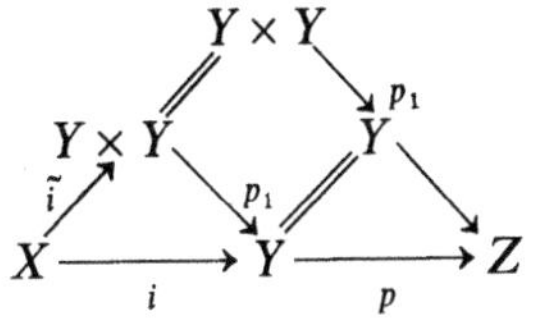

($Y \times Y = Y \times_Z Y$, and p_1 = projection to the first factor).

Hence $\varepsilon_{i,p}$ is

$$\begin{array}{l} \mathbf{i}^!\mathbf{p}^* \to \mathbf{i}^!\mathbf{p}_1^*\mathbf{p}^* \to \mathbf{i}^!(\mathbf{pp}_1)^* = (\mathbf{pi})^!_{\sigma\times\sigma} \to (\mathbf{pi})^! \\ \| \\ (\mathbf{pi})^!_\sigma \end{array} \tag{50}$$

By the definition made at the beginning of 4.10, the isomorphism $(\mathbf{pi})^!_\sigma \to (\mathbf{pi})^!_{\sigma\times\sigma}$ in (50) is $\varphi^{-1}_{\sigma\times\sigma,\sigma,p_1}$. So it remains to prove $\varphi_{\sigma\times\sigma,\sigma,p_1} = \varphi_{\sigma\times\sigma,\sigma}$. By (4.10(ii)), the following diagram commutes:

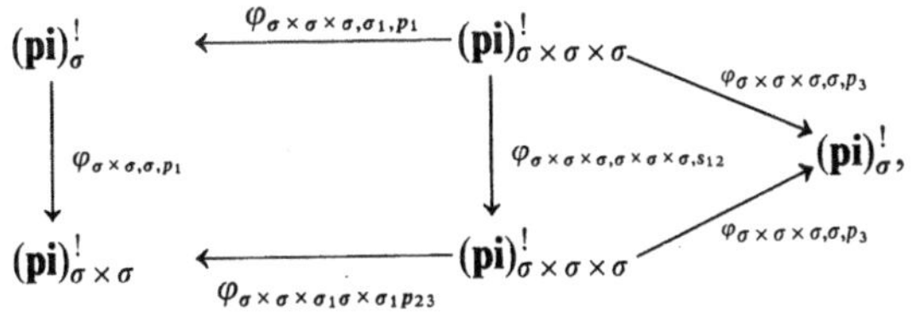

where $p_{23}\colon Y\times Y\times Y \to Y\times Y$ is projection to the last two factors and $s_{12}\colon Y\times Y\times Y \to Y\times Y\times Y$ interchanges the first two factors. Since the triangle on the right side commutes, the right vertical arrow is the identity. Using this and definition (36), we see that the commutativity of the square implies the desired equality $\varphi_{\sigma\times\sigma,\sigma,p_1} = \varphi_{\sigma\times\sigma,\sigma}$.

Proof of Sublemma 2. By a special choice of factorizations in the application of 4.11 to the composition $X' \xrightarrow{p'} X \xrightarrow{i} Y$ (namely, $A = Y'$ and $S = Y$), (37) becomes

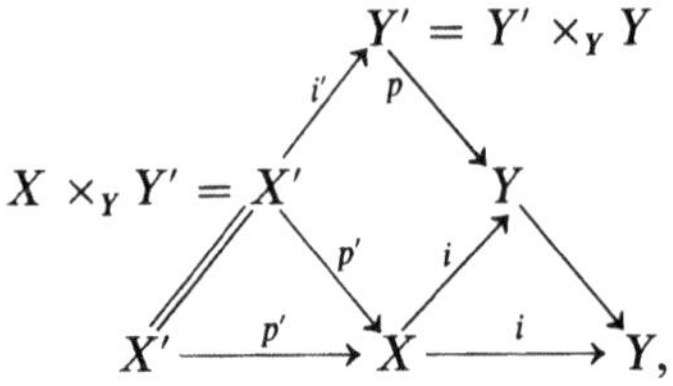

and (49) follows from definition (38).

4.13. Our previous considerations in 4.8–4.12 prove the existence of functors $\mathbf{f}^!$ with the properties required in 4.7.1–4.7.3. It remains to prove the uniqueness assertion 4.7.4. Let $\mathbf{f}^?$ be another collection of functors satisfying the same conditions. We proceed in several steps:

4.13.1. Let i be a regular closed immersion of codimension of codimension d, defining an object $X_0 \xrightarrow{i} X_1$ of K_d; we denote this object by $X_\cdot$. Then both $\mathbf{i}^!$ and $\mathbf{i}^?$ satisfy the conditions for $F_{X_\cdot}$ in 4.3. By the uniqueness result in Proposition 4.3, there is a unique biadmissible functor-isomorphism $\mathbf{i}^! \to \mathbf{i}^?$ which respects the datum 4.7.2. By 3.14, this isomorphism automatically respects 4.7.1. Since it was mentioned in 4.8 that the composition law $\mathbf{i}^!\mathbf{j}^! \to (\mathbf{ij})^!$ is the unique one which is biadmissible and respects the datum 4.7.2, it follows that our isomorphism $\mathbf{i}^! \to \mathbf{i}^?$ respects the composition law 4.7.3. Since the codimension of a regular immersion is locally constant, we get the isomorphism $\mathbf{i}^! \to \mathbf{i}^?$ for any regular closed immersion i. If in the next considerations an isomorphism $\mathbf{i}^! \to \mathbf{i}^?$ is used without any comment, it is supposed to be the isomorphism constructed here.

4.13.2. Let

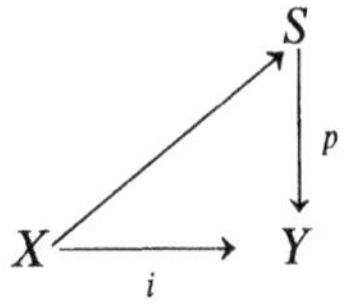

be a commutative diagram in which p is smooth and i is a regular closed immersion. We want to prove that the transformation

$$\mathbf{j}^!\mathbf{p}^* \to \mathbf{j}^?\mathbf{p}^* \xrightarrow{\mathbf{j}^?(\gamma_{p,\mathrm{Id}_X})^{-1}} \mathbf{j}^?\mathbf{p}^? \xrightarrow{\varepsilon_{j,p}} (\mathbf{pj})^? = \mathbf{i}^? \to \mathbf{i}^! \tag{51}$$

must be (21). Since confusions are impossible, the transformations 4.7.1–3 for $\mathbf{f}^!$ and $\mathbf{f}^?$ are denoted by the same letters γ, δ, ε. In the following diagram

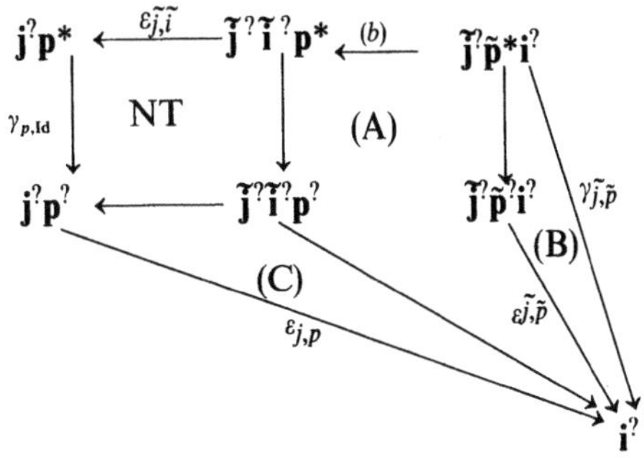

All notations are the same as in (22)!

The commutativity of the pentagon (A) is (20), commutativity of (B) follows from 3.14, and the commutativity of (C) follows from 4.7.3.2. If we identify $\mathbf{k}^?$ with $\mathbf{k}^!$ by 4.13.1, $\varepsilon_{j,p}\gamma_{\tilde{j},\mathrm{Id}}^{-1}$ becomes (51) and $\gamma_{\tilde{j},p}(b)\varepsilon_{\tilde{j},\tilde{i}}$ is (21), which proves our claim.

4.13.3. Let $f: X \to Y$ be a slci-morphism with a factorization $\sigma: X \xrightarrow{i} Z \xrightarrow{p} Y$ into a smooth projection and a regular immersion. We define $\Gamma_\sigma: \mathbf{f}^? \to \mathbf{f}^!$ by

$$\mathbf{f}^? \xrightarrow{\varepsilon_{i,p}^{-1}} \mathbf{i}^?\mathbf{p}^? \xrightarrow{\mathbf{i}^?(\gamma_{p,\mathrm{Id}})} \mathbf{i}^?\mathbf{p}^* \to \mathbf{i}^!\mathbf{p}^* = \mathbf{f}^!. \tag{52}$$

By the result of 4.13.2 and the definition of $\varphi_{\tilde{\sigma}r\sigma,r}$ in 4.10, if $\tilde{\sigma}: X \xrightarrow{\tilde{i}} \tilde{Z} \xrightarrow{\tilde{p}} Y$ is another factorization of f and $r: \tilde{Z} \to Z$ is smooth such that $r\tilde{i} = i$ and $pr = \tilde{p}$, then

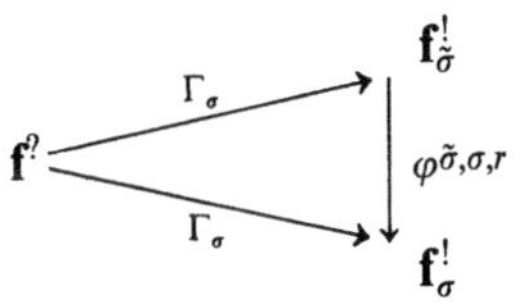

commutes. By definition (36), this implies $\varphi_{\sigma_1,\sigma_2}\Gamma_{\sigma_1} = \Gamma_{\sigma_2}$ for any two factorizations σ_1 and σ_2 of f. Consequently, (52) defines an unique isomorphism $\mathbf{f}^? \to \mathbf{f}^!$. This isomorphism is biadmissible over $\mathbb{K}_{\mathrm{lci}}$ since its definition contains only biadmissible transformations. In the case of a closed immersion we obtain the isomorphism constructed in 4.13.1. In paragraph 4.13.4, any isomorphism $\mathbf{f}^? \to \mathbf{f}^!$ will be the isomorphism constructed here.

4.13.4. It remains to prove that the isomorphism $\mathbf{f}^? \to \mathbf{f}^!$ is compatible with the isomorphisms 4.7.1–4.7.3. For 4.7.1, this follows from 3.14. To prove compatibility with the isomorphisms $\varepsilon_{p,q}$, i.e., with 4.7.3, it suffices by 4.7.3.2 to consider the following four cases:

(α) p and q are smooth. Then the assertion follows from 3.14.

(β) p and q are regular immersions. Then the compatibility follows from the considerations in 4.13.1.

(γ) $p = i$ is a regular closed immersion, and q is smooth. Then the assertion is a trivial consequence of definition (52).

(σ) $q = i$: $Y \to Z$ is a regular closed immersion, $A \xrightarrow{p'} Z$ is smooth, $X = A_Y = A \times_Z Y$, and $p: X \to Y$ is the natural projection:

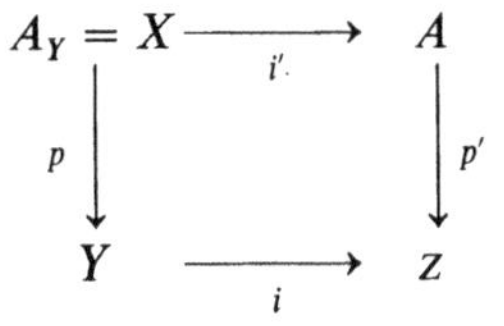

Case (δ) is a consequence of the following diagram:

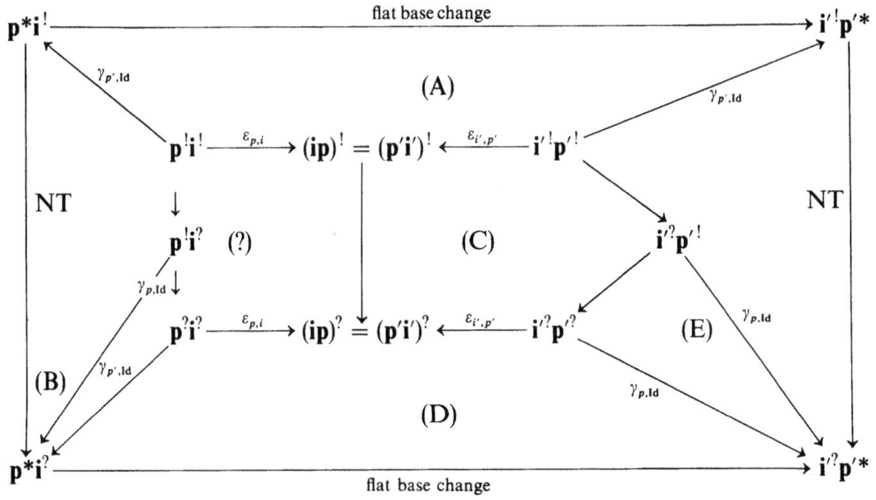

The outer contour commutes by the axiomatic description of $\mathbf{i}^! \hookrightarrow \mathbf{i}^?$ in 4.13.1 (biadmissibility). (A) and (D) are (20), (B) and (E) commute by 3.14, applied to p and p'. By case (γ), (C) commutes. It follows that (?) commutes, which is the desired result. The proof of case (δ) is complete.

It remains to prove the compatibility of $\mathbf{f}^! \to \mathbf{f}^?$ with the datum 4.7.2, i.e., with the isomorphisms $\delta_{\lambda,f}$. If f is smooth, this follows from 4.7.2.2. If f is a regular immersion, this is among the conditions characterizing the isomorphism defined in 4.13.1. By 4.7.3.1, the general case follows from these two cases since it has already been proved that our isomorphism commutes with $\varepsilon_{p,q}$.

4.14. *An alternative description of* $\mathbf{i}^! A$

Let $Z \subset X$ be a closed regular submanifold of a regular quasi-projective manifold X over an infinite field F. Since in this case Gersten's conjecture is known to be true, we have a homomorphism $i^*\colon E_2^{p,q}(X) \to E_2^{p,q}(Z)$ (cf. 1.8.). For $A \in \mathrm{Ob}(\mathbf{CH}^k(X)) = \mathrm{Ob}(\widetilde{\mathbf{CH}}^k(X))$ and $U \subset Z_{(k)}$ open we put

$$(\mathbf{i}^{!,0}\Lambda)(U) = \{(g, V, a)\}/\sim, \tag{53}$$

where:

- The triple (g, V, a) consists of $g \in G_k(U)$, an open subset $V \subset X_{\mathrm{Zar}}$ such that $V \cap Z = U$, and $a \in A(V)$.
- Two triples (g, V, a) and (g', V', a') are equivalent if and only if $i^*(a' - a) = g - g'$, where $a - a'$ is defined as an element of $G_k(V \cap V')$.

It is clear that (53) defines a separated presheaf on $Z_{(k)}$. By the moving lemma, every point $z \in Z_{(k)}$ has a neighbourhood U in $Z_{(k)}$ such that $(\mathbf{i}^{!,0}A)(U)$ is not empty. The group $G_k(U)$ acts on $(\mathbf{i}^{!,0}A)(U)$ by $\gamma\colon (g, V, a) \to (\gamma + g, V, a)$, and it acts simply transitive if $(\mathbf{i}^{!,0}A)(U)$ is not empty. For $a \in A(V)$, we denote the class of $(0, V, a)$ in (53) by $i^!(a)$.

Let $\mathbf{i}^! A$ be the sheaf associated to the presheaf $\mathbf{i}^{!,0}A$ on $Z_{(k)}$. From the previous remarks it is clear that $\mathbf{i}^! A$ is an object of $\mathbf{CH}^k(Z)$. We want to prove that this new functor $\mathbf{i}^!$ is canonically isomorphic to the old one constructed in Theorem 4.7. A careful examination of the proof of 4.3 shows that in 4.5 and 4.6 we do not leave the category of regular manifolds over a field if the closed immersions we consider belong to this category. Therefore it suffices to equip the functor $\mathbf{i}^!$ defined by (53) with the data 4.7.1–4.7.3:

4.14.1. Let

$$\begin{array}{ccc} Z & \xrightarrow{\ i\ } & X \\ & {\scriptstyle p}\searrow \quad \swarrow{\scriptstyle q} & \\ & Y & \end{array}$$

be a commutative diagram of regular manifolds over F, with a closed immersion i and flat morphisms p and q. If W is open in $Y_{(k)}$, then $U = p^{-1}(W)$ and $V = q^{-1}(W)$ are open in $Z_{(k)}$ and $X_{(k)}$ and satisfy the assumptions of (53). There is a unique isomorphism $\gamma: \mathbf{p}^*A \to \mathbf{i}^!\mathbf{q}^*A$ with the property that for every $a \in A(W)$ we have $\gamma(p^*(a)) = i^!(q^*(a))$.

4.14.2. Let $i: Z \to X$ be a closed immersion of codimension d between regular manifolds, $D \subset X$ a regular submanifold of codimension one such that $Z \cap D$ is also regular and of codimension one in Z, λ a section of $\mathcal{O}_X$ in some neighbourhood of D which trivializes the sheaf of ideals defining D.

Let $x \in D$ and $A \in \mathrm{Ob}(\mathbf{CH}^k(X - D))$. Since the restriction $CH^k(X) \to CH^k(X - D)$ is surjective, it is possible to extend A to an object of $\mathbf{CH}^k(X)$. Therefore the moving lemma implies the existence of an open $V \subset X - D$ such that $A(V)$ is not empty and $K =$ (closure of $X - D - V$ in X) meets $Z \cap D$ in codimension $\geqslant k$ and $x \notin K$. Let $U = Z - D - (K \cap Z)$, $W = D - (K \cap D)$, $Y = Z \cap D - (K \cap Z \cap D)$. If $a \in A(V)$, then

$$\begin{array}{ll} \mathrm{sp}_\lambda(a) \in (\mathbf{sp}_\lambda A)(W) & i_D^!(\mathrm{sp}_\lambda(a)) \in (\mathbf{i}_D^! \mathbf{sp}_\lambda A)(Y) \\ \mathbf{i}^!(a) \in (\mathbf{i}^{!,0} A)(U) \subset (\mathbf{i}^! A)(U) & \mathrm{sp}_\lambda|_Z(i^!(a)) \in (\mathbf{sp}_\lambda|_Z \mathbf{i}^! A)(Y), \end{array}$$

where i_D is the inclusion $Z \cap D \to D$.

There exists a unique isomorphism $\delta: \mathbf{i}_D^! \mathbf{sp}_\lambda A \to \mathbf{sp}_\lambda|_Z \mathbf{i}^! A$ with the property

$$(i_D^!(\mathrm{sp}_\lambda(a))) = \mathrm{sp}_\lambda|_Z(i^!(a)). \tag{54}$$

Indeed, it follows from Proposition 1.8 that the $G_k(Y)$-equivariant isomorphism $\delta: (\mathbf{i}_D^! \mathbf{sp}_\lambda A)(Y) \to (\mathbf{sp}_\lambda|_Z \mathbf{i}^! A)(Y)$ characterized by (54) is independent of $a \in A(V)$. Consequently, definition (54) is correct.

4.14.3. If $j: Y \to Z$ is another embedding satisfying our assumptions, then $\varepsilon_{i,j}: \mathbf{j}^! \mathbf{i}^! A \to (\mathbf{ij})^! A$ is characterized by $j^! i^!(a) \to (ij)^!(a)$.

It is easy to see that the isomorphisms 4.14.1–3 satisfy all assumptions of Theorem 4.7. If we replace the word 'slci-morphism' by 'locally closed embedding of regular manifolds over a field'. Consequently, (53) is an alternative description of the functors $\mathbf{i}^!$ defined in 4.7. We shall use this description in a forthcoming paper when metrics on objects of $\mathbf{CH}^k$ are investigated.

5. Intersection products. Bloch's biextension

Here we restrict our considerations to the case of regular manifolds over a field F. For the sake of simplicity we shall also assume that these manifolds are connected. By working with methods similar to [F, §20.2], we could also deal with the case of smooth schemes over a Dedekind domain. The smoothness assumption is absolutely essential only for 5.3.

5.1. *The cross product*

Let X and Y be manifolds and $p_{1,2}$ the projections of $X \times Y$ to X and Y. The biexact functor

$$\begin{aligned} M_p(X) \times M_q(Y) &\longrightarrow M_{p+q}(X \times Y) \\ (A; B) &\longrightarrow p_1^* A \otimes p_2^* B \end{aligned} \tag{1}$$

defines a product

$$\times\colon E_1^{p,q}(X) \times E_1^{s,t}(Y) \to E_1^{p+q,s+t}(X \times Y) \tag{2}$$

which satisfies

$$d_1(a \times b) = d_1(a) \times b + (-1)^{p+q} a \times d_1(b), \qquad a \in E_1^{p,q}. \tag{3}$$

Because of $E_1^{p+1,-p} = 0$, (3) implies that the following product is well-defined:

$$\begin{aligned} &\times\colon (E_1^{p-1,-p}(X)/d_1(E_1^{p-2,-p}(X)) \times E_1^{q,-q}(Y) \\ &\quad \to E_1^{p+q-1,-p-q}(X \times Y)/\mathrm{im}(d_1). \end{aligned}$$

In our terminology, this is a pairing

$$\times\colon (G_p)_r(X) \times Z^q(Y) \to (G_{p+q})_r(X \times Y). \tag{4}$$

(3) implies:

$$\mathbf{c}(g \times z) = \mathbf{c}(g) \times z. \tag{5}$$

In a similar manner, we get

$$\times\colon Z^p(X) \times (G_q)_r(Y) \to (G_{p+q})_r(X \times Y) \tag{6}$$

with the property

$$\mathbf{c}(z \times g) = z \times \mathbf{c}(g). \tag{7}$$

For $A \in \mathrm{Ob}(\mathbf{CH}^k(X))$, $B \in \mathrm{Ob}(\mathbf{CH}^l(Y))$, we define $A \boxtimes B \in \mathrm{Ob}(\mathbf{CH}^{k+l}(X \times Y))$ as follows:

$$(A \boxtimes B)_r(X \times Y) = \{(g, a, b)\}/\sim, \tag{8}$$

where:

— The entries of the triple are $g \in (G_{k+1})_r(X \times Y)$, $a \in A_r(X)$, $b \in B_r(Y)$.
— $(g, a, b) \sim (g', a', b')$ if and only if

$$\begin{aligned} g' - g &= \mathbf{c}(a) \times (b - b') + (a - a') \times \mathbf{c}(b') \\ &= \mathbf{c}(a') \times (b - b') + (a - a') \times \mathbf{c}(b). \end{aligned} \tag{9}$$

That the two expressions on the right hand side of (9) agree is a consequence of the identity

$$\mathbf{c}(g) \times h = g \times \mathbf{c}(h) \quad \text{in } (G_{k+1})_r(X \times Y) \tag{10}$$

if $g \in (G_k)_r(X)$, $h \in (G_1)_r(Y)$. To prove (10), we choose representatives $\gamma \in E_1^{k-1,-k}(X)$, $\eta \in E_1^{l-1,-l}(Y)$ for g and h. Then a representative for $\mathbf{c}(g) \times h - g \times \mathbf{c}(h)$ is

$$d_1(\gamma) \times \eta - \gamma \times d_1(\eta) = d_1(\gamma \times \eta)$$

by (3). This proves (10) and completes the explanation of (8).

We define $\mathbf{c}$: $(A \boxtimes B)_r(X \times Y) \to Z^{k+1}(X \times Y)$ by

$$\mathbf{c}((g, a, b) \bmod \sim) = \mathbf{c}(g) + \mathbf{c}(a) \times \mathbf{c}(b). \tag{11}$$

By (9), (5), and (7), this is independent of the choice of the representative. Finally we define $(A \boxtimes B)(U)$, $U \subset (X \times Y)_{(k+1)}$ open, by formula 3(4):

$$(A \boxtimes B)(U) = \{a \in (A \boxtimes B)_r(U) | \mathbf{c}(a)|_U = 0\}. \tag{12}$$

For $a \in A_r(X)$, $b \in B_r(Y)$ we denote the equivalence class of $(0, a, b)$ in (8) by $a \times b$. Then $\mathbf{c}(a \times b) = \mathbf{c}(a) \times \mathbf{c}(b)$. If $U \subset X_{(k)}$, $V \subset Y_{(1)}$ are open and $a \in A(U)$, $b \in B(V)$, then $a \times b \in (A \boxtimes B)(X \times Y - (X - U) \times (Y - V))$.

5.2. *Natural transformations involving* $\boxtimes$

We give a mere list of them without going into detail.

5.2.1. Biadditivity. For A, $B \in \mathrm{Ob}(\mathbf{CH}^k(X))$, C, $D \in \mathrm{Ob}(\mathbf{CH}^1(Y))$, we have isomorphisms

$$A \boxtimes (C \oplus D) \to A \boxtimes C \oplus A \boxtimes D \tag{13}$$

$$(A \oplus B) \boxtimes C \to A \boxtimes C \oplus B \boxtimes C \tag{14}$$

which satisfy the additivity condition [DM, 1.8] in each of the two variables and make the diagram

$$\begin{array}{ccc} (A \oplus B) \boxtimes (C \oplus D) & \longrightarrow & A \boxtimes (C \oplus D) \oplus B \boxtimes (C \oplus D) \\ \downarrow & & \downarrow \\ (A \oplus B) \boxtimes C \oplus (A \oplus B) \boxtimes D & \longrightarrow & A \boxtimes C \oplus A \boxtimes D \oplus B \boxtimes C \oplus B \boxtimes D \end{array} \tag{15}$$

commutative. If a, b, c, d are rational sections of A, B, C, D, then the isomorphisms (13) and (14) send $a \times (c \oplus d)$ to $a \times c \oplus a \times d$ and $(a \oplus b) \times c$ to $a \times c \oplus b \times c$.

5.2.2. Symmetry. Let s: $X \times Y \to Y \times X$ be the permutation of the factors. Since s interchanges the two expressions on the right hand side of (9), there is a canonical isomorphism $\mathbf{s}^*(A \boxtimes B) \to B \boxtimes A$ sending $s^*(a \times b)$ to $b \times a$.

5.2.3. Compatibility with pull-back. If f: $Z \to X$ is flat, $A \in \mathrm{Ob}(\mathbf{CH}^k(X))$, $B \in \mathrm{Ob}(\mathbf{CH}^1(Y))$, then there is an isomorphism $(\mathbf{f}^*A) \boxtimes B \to (\mathbf{f} \times \mathbf{Id}_Y)^*(A \boxtimes B)$ which maps $f^*(a) \times b$ to $(f \times \mathrm{Id}_Y)^*(a \times b)$.

Similarly, if i: $Z \to X$ is a closed immersion and if X and Y are quasiprojective, there is an isomorphism $(\mathbf{i}^! A) \boxtimes B \to (\mathbf{i} \times \mathbf{Id}_Y)^!(A \boxtimes B)$ which maps $i^!(a) \times b$ to $(i \times \mathrm{Id}_Y)^!(a \times b)$ if $a \in A_r(X)$ is a rational section whose cycle meets Z properly (cf. 4.14).

Finally, if X, Y, and Z are quasi-projective regular manifolds, we may factor g: $Z \to X$ into $Z \xrightarrow{i} S \xrightarrow{f} Y$, with a closed immersion i and f smooth and quasi-projective. We get an isomorphism

$$(\mathbf{g}^! A) \boxtimes B \to (\mathbf{g} \times \mathbf{Id}_Y)^!(A \boxtimes B)$$

by composing

$$\begin{aligned} \mathbf{g}^! A \boxtimes B &\to \mathbf{i}^! \mathbf{f}^* A \boxtimes B \to (\mathbf{i} \times \mathbf{Id}_Y)^!(\mathbf{f}^* A \boxtimes B) \\ &\to (\mathbf{i} \times \mathbf{Id}_Y)^!(\mathbf{f} \times \mathbf{Id}_Y)^*(A \boxtimes B) \to (\mathbf{g} \times \mathbf{Id}_Y)^!(a \boxtimes B). \end{aligned}$$

It is easy to see that this isomorphism does not depend on the factorization of g.

5.2.4. Compatibility with push-forward. If f: $Z \to X$ is proper, $A \in \mathrm{Ob}(\mathbf{CH}^k(Z))$, $B \in \mathrm{Ob}(\mathbf{CH}^1(Y)$, there is an isomorphism

$$(\mathbf{f}_* A) \boxtimes B \to (\mathbf{f} \times \mathbf{Id}_Y)_*(A \boxtimes B)$$

which maps $f_*(a) \times b$ to $(f \times \mathrm{Id}_Y)_*(a \times b)$.

There are some obvious compatibilities between the isomorphisms 5.2.1–4. We do not list them explicitly because they are not essential for our approach to Deligne's program.

5.3. The intersection product. Let X be a regular manifold, $A \in \mathrm{Ob}(\mathbf{CH}^k(X))$, $B \in \mathrm{Ob}(\mathbf{CH}^1(X))$. By $\Delta: X \to X \times X$ we denote the diagonal. This is a regular closed immersion. We put

$$A \cup B = \mathbf{\Delta}^!(A \boxtimes B) \in \mathrm{Ob}(\mathbf{CH}^{k+1}(X)). \tag{16}$$

If a and b are rational sections of A and B, we put $a \cup b = \Delta^!(a \times b)$. This is a rational section of $A \cup B$ if the supports of $\mathbf{c}(a)$ and $\mathbf{c}(b)$ intersect properly. If $a \in A(U)$, $b \in B(V)$, then $a \cup b \in (A \cup B)(U \cup V)$.

The following isomorphisms are derived from 5.2:

5.3.1. Biadditivity. Isomorphisms $A \cup (C \oplus D) \to A \cup C \oplus A \cup D$ and $(A \oplus B) \cup C \to A \cup C \oplus B \cup C$. They are additive in each variable and satisfy the analogue of (15).

5.3.2. Symmetry. An isomorphism $s: A \cup B \oplus B \cup A$ defined by composing

$$A \cup B = \mathbf{\Delta}^!(A \boxtimes B) = (\mathbf{s\Delta})^!(A \boxtimes B) \to \mathbf{\Delta}^!\mathbf{s}^*(A \boxtimes B) \to \mathbf{\Delta}^!(B \boxtimes A) = B \cup A.$$

The symmetry $s: A \cup A \to A \cup A$ may be different from the identity!

5.3.3. An associativity law and compatibilities with pull-back and push-forward. We do not define them explicitly because we shall not need them later.

5.4. *Example*

Let $p: X \to S$ be a proper morphism of relative dimension one between regular manifolds over a field. To line bundles L, M on X, Deligne associates the line bundle $\langle L, M\rangle$ on S which is Zariski-locally on S generated by sections $\langle l, m\rangle$, where l and m are rational sections of L and M whose divisors do not intersect. The relations

$$\langle gl, m\rangle = g(\mathrm{div}(m))\langle l, m\rangle; \ \langle l, gm\rangle = g(\mathrm{div}(l))\langle l, m\rangle$$

are satisfied. Cf. [D, 6.1].

Since $\mathbf{CH}^1(X)$ is the category of line bundles on X, we can consider the intersection product $L \cup M \in \mathrm{Ob}(\mathbf{CH}^2(X))$. There is a canonical isomorphism

$$\langle L, M\rangle \to \mathbf{p}_*(L \cup M)$$

$$\langle l, m\rangle \to p_*(l \cup m).$$

5.5. *Bloch's biextension*

By the identification $CH^p(X) = H^p(X, \mathscr{K}_p)$ and $C_q(X) = H^{q-1}(X, \mathscr{K}_q)$, there is a product $CH^p(X) \times G_p(X) \to G_{p+q}(X)$. Following the methods of [Gr], it is easy to show that this product coincides up to sign with the product defined by the biadditive functor $\cup$. More precisely, if $g \in G_q(X)$ is viewed as an automorphism of $B \in \mathrm{Ob}(\mathbf{CH}^q(X))$, then for any $A \in \mathrm{Ob}(\mathbf{CH}^p(x))$ the automorphism $\mathrm{Id}_A \cup g$ of $A \cup B$ is given by $[A] \cdot g \in G_{p+q}(X)$, where $[A] \in CH^p(X)$ is the class of A.

If $f: X \to S$ is a proper morphism of relative dimension n between regular manifolds over a field, we put

$$CH^p(X/S)^v = \{a \in CH^p(X) | \text{ For every Zariski-open } U \subset S \text{ and every } g \in G_{n+1-p}(f^{-1}(U)),\ f_*(a \cdot g) = 0 \text{ in } G_1(U) \text{ (i.e., } = 1 \text{ in } \mathcal{O}(U)^*)\}. \tag{17}$$

Let $p + q = n + 1$, $a \in CH^p(X/S)^v$, $b \in CH^q(X/S)^v$. We choose representatives $A \in \mathrm{Ob}(CH^p(X))$, $B \in \mathrm{Ob}(CH^q(X))$ for a and b. We have a line bundle

$$L_{A,B} = f_*(A \cup B) \tag{18}$$

on S. If $g \in G_p(X)$ is any automorphism of A, then $\mathbf{f}_*(g \cup \mathrm{Id}_B) = f_*([\mathrm{B}] \cdot g) = 1$. Consequently, if A' is another representative for A, then all isomorphisms $A \to A'$ define the same isomorphism $L_{A,B} \to L_{A',B}$. Since the same is true for representatives of b, the $L_{A,B}$ can be identified with one line bundle $L_{a,b}$.

By the biadditivity of our intersection product, we get isomorphisms

$$\begin{aligned} &L_{a,c} \otimes L_{b,c} \to L_{a+b,c} \\ &L_{a,c} \otimes L_{a,d} \to L_{a,c+d} \end{aligned} \tag{19}$$

satisfying the axioms of [SGA 7.I, Exp. VII, 2.1]. For instance, it follows from the analogue of (15) for the functor $\cup$ that the isomorphisms (19) satisfy [SGA 7.I, Exp. VII, (2.1.1)].

If $\mathscr{CH}^p(X/S)^v$ is the sheaf on S_{Zar} associated to the presheaf $U \to CH^p(f^{-1}(U)/U)^v$, we get a biextension of $\mathscr{CH}^p(X/S)^v \times \mathscr{CH}^{n+1-p}(X/S)^v$ by $\mathcal{O}_S^*$. This biextension is unique up to unique isomorphism of biextensions.

Using the crucial lemma 1 in Bloch's paper [B], it is easy to see that $CH^p(X ? S)^v$ is contained in the group of cycles which are homologically equivalent to zero on each geometric fibre (i.e. of cycles z on X such that for every geometric point s of S and every prime 1 prime to the characteristic of $k(s)$, the fundamental class of z in $H^{2p}(X_s, \mathbb{Z}_1(p))$ vanishes). If Bloch's biextension is also defined (i.e., if S and X are smooth and quasi-projective over a field), then there exists a canonical isomorphism between Bloch's biextension and the biextension constructed here.

References

[B] S. Bloch, Cyles and Biextensions, Contemporary Mathematics vol. 83, 1989, p. 19–30.

[D] P. Deligne, Le déterminant de la cohomology, *Contemporary Mathematics* 67 (1987), p. 93–177.

[DM] P. Deligne and J. Milne, *Tannakian Categories*, Lecture Notes in Mathematics 900, p. 101–228.

[F] W. Fulton, *Intersection Theory*, Springer-Verlag, 1984.

[FL] W. Fulton and S. Lang, *Riemann-Roch Algebra*, Springer-Verlag, 1985.

[Fr1] J. Franke, *Chow categories*, Proceedings of the 28. Mathematische Arbeitstagung, Bonn (MPI) 1988.

[Fr2] J. Franke, Chow categories, Manuscript, June 1988.

[G] H. Gillet, Riemann-Roch theorems for higher algebraic K-theory, *Advances in Math.* 40 (1981), 203–289.

[GN] C. Giffen and A. Neeman, *K-theory of triangulated categories*, Preprint, University of Virginia.

[Gr] D. Grayson, Products in K-theory and intersection algebraic cycles, *Inv. Math.* 17 (1978) 71–83.

[H] R. Hartshorne, *Algebraic geometry*, Springer-Verlag, 1977.

[M] J. Milne, *Etale cohomology*, Princeton University Press, 1980.

[Q] D. Quillen, *Higher algebraic K-theory I*, Lecture Notes in Math. No. 341, p. 85–147.

[S] C. Soulé, Opérations en K-théorie algébrique, *Can. J. Math.*, 37: 3 (1985), 488–550.

[SGA 6] P. Berthelot; A. Grothendieck and L. Illusie, *Théorie des Intersections et Théorème de Riemann-Roch*, Lecture Notes in Mathematics 225.

[SGA 7I] A. Grothendieck, *Groupes de Monodromie en Geometrie Algebrique*, Lecture Notes in Mathematics 288.

[W] F. Waldhausen, *Algebraic K-theory of generalized free products*, Part 1, Annals of Mathematics.

Compositio Mathematica **76**: 163–196, 1990.

Complex analytic compactifications of $\mathbb{C}^3$

MIKIO FURUSHIMA

Kumamoto National College of Technology, Nishi-goshi-machi, Kiguchi-gun, Kumamoto, 861-11, Japan ; Current address: Department of Mathematics, Ryukyu University, Nishihara-cho, Okinawa, 903-11, Japan)

Received 5 September 1988; accepted in revised form 28 August 1989

Introduction

The purpose of this paper is to give a proof of the announcement [7].

Let X be an n-dimensional compact connected complex manifold and Y an analytic subset of X. We call the pair (X, Y) a complex analytic compactification of $\mathbb{C}^n$ if $X - Y$ is biholomorphic to $\mathbb{C}^n$. By a theorem of Hartogs, Y is a divisor on X.

In this paper, we will consider only the case of $n = 3$. Let (X, Y) be a complex analytic compactification of $\mathbb{C}^3$. Assume that Y has at most isolated singularities. Then Y is normal. Thus, by Peternell-Schneider [18] (cf. Brenton [2]), X is projective. In particular, X is a Fano 3-fold of index $r(1 \leqslant r \leqslant 4)$ with $b_2(X) = 1$. In the case of $r \geqslant 2$, such a (X, Y) is completely determined (cf. [3], [5], [6], [18]). In the case of $r = 1$, by a detailed analysis of the singularities of the boundary Y, we can prove that such a compactification (X, Y) does not exist. Thus, we have:

THEOREM. *Let (X, Y) be a complex analytic compactification of $\mathbb{C}^3$. Assume that Y has at most isolated singularities. Then X is a Fano 3-fold of index $r(2 \leqslant r \leqslant 4)$ with $b_2(X) = 1$, and*

(1) $r = 4 \Rightarrow$ $(X, Y) \cong (\mathbb{P}^3, \mathbb{P}^2)$,

(2) $r = 3 \Rightarrow$ $(X, Y) \cong (\mathbb{Q}^3, \mathbb{Q}_0^2)$, *where $\mathbb{Q}^3$ is a smooth quadric hypersurface in $\mathbb{P}^4$ and $\mathbb{Q}_0^2$ is a quadric cone in $\mathbb{P}^3$,*

(3) $r = 2 \Rightarrow$ $(X, Y) \cong (V_5, H_5)$, *where V_5 is a complete intersection of three hyperplanes in the Grassmannian $G(2, 5) \hookrightarrow \mathbb{P}^9$, and H_5 is a normal hyperplane section of V_5 with exactly one rational double point of A_4-type.*

This paper consists of five sections. In Section 1, we will prove that $(X, Y) \cong (V_{22}, H_{22})$ if such a (X, Y) exists in the case of $r = 1$, where $V_{22} \hookrightarrow \mathbb{P}H^0(V_{22}, \mathcal{O}(-K_{V_{22}})) \cong \mathbb{P}^{13}$ is a Fano 3-fold of degree 22 in $\mathbb{P}^{13}$ (index 1, genus 12) and H_{22} is a normal hyperplane section which is rational (Proposition 1.13). In Section 2, we will determine the singularities of $Y = H_{22}$ (Proposition 2.5). In Sections 3 and 4, we will prove that such a $(X, Y) = (V_{22}, H_{22})$ does not exist. In Section 5, we will refer to a recent work of Peternell-Schneider [18] (c.f. Peternell [19]) on a projective compactification (X, Y) of $\mathbb{C}^3$ with $b_2(X) = 1$ (especially, the case where the boundary Y is non-normal), and prove that there is a compactification (X, Y) of $\mathbb{C}^3$ with a non-normal boundary Y in the case of the index $r = 1$.

Notations

- K_M: a canonical divisor on a projective manifold M.
- $b_i(M)$: the ith Betti number of M.
- $N_{C|M}$: the normal bundle of C in M.
- $c_1(\mathscr{F}_M)$: the first Chern class of a locally free sheaf $\mathscr{F}_M$ on M.
- $m(\mathcal{O}_{Y,x})$: the multiplicity of the local ring $\mathcal{O}_{Y,x}$ at x.

1. The structure in the case of $r = 1$

1. Let (X, Y) be an analytic compactification of $\mathbb{C}^3$ such that Y has at most isolated singularities. Assume that the index $r = 1$. Then X is a Fano 3-fold of index 1 with $\operatorname{Pic} X \cong \mathbb{Z}c_1\mathcal{O}_X(Y)$ ([3], [9], [18]). Then, by Proposition 1, Proposition 2 and Proposition 3 in [3], we have:

LEMMA 1.1. (1) $K_Y = 0$,
(2) $H^1(Y, \mathcal{O}_Y) = 0, H^2(Y, \mathcal{O}_Y) \cong \mathbb{C}$,
(3) $H^1(Y; \mathbb{Z}) = 0, H^2(Y; \mathbb{Z}) \cong \mathbb{Z}\, c_1(N_{Y|X})$.

Let Sing Y be the singular locus of Y and put $S := \{ y \in \operatorname{Sing} Y;\ y$ is not a rational singularity$\}$. Let $\pi\colon \tilde{Y} \to Y$ be the minimal resolution of singularities of Y and Z be the fundamental cycle of S associated with the resolution $(\tilde{Y}, \pi)$. We put $E := \pi^{-1}(\operatorname{Sing} Y), C := \pi^{-1}(S) = \bigcup_{i=1}^{s} C_i$ (C_i's are irreducible).

LEMMA 1.2. $S \neq \phi$

Proof. Let us consider the following exact sequence (see [2]):

$$\to H^1(Y; \mathbb{R}) \to H^1(\tilde{Y}; \mathbb{R}) \to H^1(E; \mathbb{R}) \to H^2(Y; \mathbb{R}) \to$$
$$\to H^2(\tilde{Y}; \mathbb{R}) \to H^2(E; \mathbb{R}) \to 0. \tag{1.1}$$

Assume that $S = \phi$. Then we have $K_{\tilde{Y}} = \mathcal{O}_{\tilde{Y}}$ and $H^1(E; \mathbb{R}) = 0$. By Lemma 1.1(3), $H^1(\tilde{Y}; \mathbb{R}) = 0$. Thus $\tilde{Y}$ is a K-3 surface. Since $b_2(Y) = 1$ and Y is projective, we have $b^+(Y) = 1$. On the other hand, by Brenton [2], $b^+(Y) = b^+(\tilde{Y})$. Thus we have $b^+(\tilde{Y}) = 1$. This is a contradiction. Therefore $S \neq \phi$. Q.E.D.

Thus, by Umezu [22], we have:

LEMMA *U*.

(1) $K_{\tilde{Y}} = -\Sigma_{i=1}^{s} n_i C_i (n_i > 0,\ n_i \in \mathbb{Z})$, *and thus* $\tilde{Y}$ *is a ruled surface over a non-singular compact algebraic curve R of genus* $q = \dim H^1(\tilde{Y}; \mathcal{O}_{\tilde{Y}})$ *(namely,* $\tilde{Y}$ *is bi-rationally equivalent to a* $\mathbb{P}^1$*-bundle over R),*

(2) *if* $q \neq 1$, *then S consists of one point with* $p_g := \dim(R^1\pi_* \mathcal{O}_{\tilde{Y}})_s = q + 1$,

(3) *if* $q = 1$, *then S consists of either one point with* $p_g = 2$ *or two points with* $p_g = 1$. *Moreover, in the second case of* (3), *both of the two points are simply elliptic.*

LEMMA 1.3. *S consists of one point with* $p_g = q + 1$ *and* $b_2(\tilde{Y}) = b_2(E) + 1$.

Proof. Assume that S consists of two points. By Lemma U(3), these two points are simply elliptic, and $C = \pi^{-1}(S) = C_1 \cup C_2$, where C_1, C_2 are distinct sections of $\tilde{Y}$. Since $b_2(Y) = 1$, by (1.1), we have $b_1(\tilde{Y}) = b_1(E)$. Since

$$2 = b_1(\tilde{Y}) = b_1(E) \geqslant b_1(C) = b_1(C_1) + b_1(C_2) = 4,$$

we have a contradiction. Therefore S consists of one point with $p_g = q + 1$. Since $b_1(\tilde{Y}) = b_1(E)$ and $b_2(Y) = 1$, we have $b_2(\tilde{Y}) = b_2(Y) + b_2(E) = 1 + b_2(E)$.

Q.E.D.

2. Let U be a strongly pseudoconvex neighborhood of C in $\tilde{Y}$. A cycle D on U is an integral combination of the C_i, $D = \Sigma d_i C_i$ $(1 \leqslant i \leqslant s)$, with $d_i \in \mathbb{Z}$. We denote the support of D by $|D| = \bigcup C_i, d_i \neq 0$. We put $\mathcal{O}_D = \mathcal{O}_U/\mathcal{O}_U(-D)$. Let K_U be a canonical divisor on U. We put $\chi(D) := \dim H^0(U, \mathcal{O}_D) - \dim H^1(U, \mathcal{O}_D)$. Then, by the Riemann-Roch theorem [21],

$$\chi(D) = -\tfrac{1}{2}\{(D \cdot D) + (D \cdot K_U)\}. \tag{1.2}$$

For two cycles A, B, we have, by (1.2),

$$\chi(A + B) = \chi(A) + \chi(B) - (A \cdot B). \tag{1.3}$$

LEMMA 1.4.

(1) $q = 0 \Rightarrow \tilde{Y}$ *is a rational surface, and* $K_{\tilde{Y}} = K_U = -Z$.

(2) $q \neq 0 \Rightarrow$ *there is an irreducible component* C_{i_1} *of C such that* C_{i_1} *is a section of* $\tilde{Y}$ *and the rest* $\overline{C - C_{i_1}} = \bigcup_{i \neq i_1} C_i (\neq \phi)$ *is contained in the singular fibers of* $\tilde{Y}$, *and* $-K_{\tilde{Y}} = Z + C_{i_1}$.

Proof. (1) Since $q = 0$, we have $p_g = 1$. Thus S consists of a minimally elliptic singularity. By Laufer [11], we have $K_{\tilde{Y}} = K_U = -Z$.

(2) Since $K_{\tilde{Y}} = -\Sigma_{i=1}^{s} n_i C_i$ $(n_i > 0, n_i \in \mathbb{Z})$, for a general fiber f of $\tilde{Y}$, we have

$$2 = (-K_{\tilde{Y}} \cdot f) = \sum_{i=1}^{s} n_i (C_i \cdot f).$$

Thus we have the following:

(i) there is an irreducible component C_{i_1} of C such that $n_{i_1} = 2$, $(C_{i_1} \cdot f) = 2$ and $(C_i \cdot f) = 0$ $(i \neq i_1)$,

(ii) there are two irreducible components C_1, C_2 of C such that $n_1 = n_2 = 1$, $(C_i \cdot f) = 1$ $(i = 1, 2)$, $(C_i \cdot f) = 0$ $(i \geqslant 3)$, and

(iii) there is an irreducible component C_1 of C such that $n_1 = 1$, $(C_1 \cdot f) = 2$, $(C_i \cdot f) = 0$ $(i \neq 1)$.

Claim 1. The case (ii) can not occur.

Indeed, by the adjunction formula, the curve C_i $(i = 1, 2)$ is a non-singular elliptic curve with $(C_1 \cdot C_2) = 0$ and there is no other irreducible component of C which intersects C_i $(i = 1, 2)$. Thus $C = C_1 \cup C_2$ $(C_1 \cap C_2 = \phi)$, namely, S consists of two points. This contradicts Corollary 1.4.

Claim 2. The case (iii) can not occur.

Indeed, by the adjunction formula, C_1 is a non-singular elliptic curve and there is no other irreducible component of C which intersects C_1. By Corollary 1.4, we have $C = C_1$, hence, $K_{\tilde{Y}} = -C_1$. This contradicts Lemma 1 and Lemma 2 in Umezu [22].

Thus we have the case (i). In particular, C_{i_1} is a section of the ruled surface $\tilde{Y}$ and C_i's $(i \neq i_1)$ are all contained singular fibers of $\tilde{Y}$. We also have $C - C_{i_1} \neq \phi$ by the same reason as above. Since $n_{i_1} = 2$, we have $-K_{\tilde{Y}} = 2C_{i_1} + \Sigma_{i \neq i_1} n_i C_i$ $(n_i > 0)$. We remark that the genus of C_{i_1} is equal to $q = h^1(\mathcal{O}_{\tilde{Y}}) \geqslant 1$.

Claim 3. $-K_{\tilde{Y}} = C_{i_1} + Z$.

Indeed, since $(-K_{\tilde{Y}} - C_{i_1}) \cdot C_i \leqslant 0$ $(1 \leqslant i \leqslant s)$, by definition of the fundamental cycle, $-K_{\tilde{Y}} - C_{i_1} \geqslant Z$. Now, assume that $-K_{\tilde{Y}} = C_{i_1} + Z + D$, where $D > 0$. For a general fiber f of $\tilde{Y}$, we have $2 = (-K_{\tilde{Y}} \cdot f) = (C_{i_1} \cdot f) + (Z \cdot f) + (D \cdot f)$. Since $C_{i_1} \subset |Z|$, we have $(C_{i_1} \cdot f) = (Z \cdot f) = 1$, and $(D \cdot f) = 0$. This means that the support $|D|$ is contained in the singular fibers of $\tilde{Y}$. Since

$$H^2(\tilde{Y}, \mathcal{O}_{\tilde{Y}}(-Z)) \cong H^0(\tilde{Y}; \mathcal{O}_{\tilde{Y}}(-C_{i_1} - D)) = 0 \quad \text{and } H^2(\tilde{Y}, \mathcal{O}_{\tilde{Y}}) = 0,$$

by the Riemann-Roch theorem, we have

$$0 \geqslant -\dim H^1(\tilde{Y}, \mathcal{O}_{\tilde{Y}}(-Z)) = \tfrac{1}{2}(Z \cdot Z + Z \cdot K_{\tilde{Y}}) + 1 - q.$$

Since $K_M = K_U$, by (1.2), we have $\chi(Z) \geqslant 1 - q$. Since $H^0(U; \mathcal{O}_Z) \cong \mathbb{C}$ (cf. [11,

p. 1260]), $\chi(Z) = 1 - \dim H^1(U, \mathcal{O}_Z) \leqslant 1$. Since S does not consists of a rational singularity, $\chi(Z) \neq 1$ by Artin [1]. Thus we have

$$1 - q \leqslant \chi(Z) \leqslant 0. \tag{1.4}$$

Since $1 - q = \chi(C_{i_1}) = \chi(-K_U - C_{i_1}) = \chi(Z + D) = \chi(Z) + \chi(D) - (D \cdot Z)$, we have

$$\chi(Z) = -\chi(D) + 1 - q + (D \cdot Z). \tag{1.5}$$

Since $(D \cdot Z) \leqslant 0$, by (1.4), (1.5), we have $\chi(D) \leqslant 0$. On the other hand, the support $|D|$ is contained in the singular fibers of $\tilde{Y}$. Thus, the contraction of $|D|$ in $\tilde{Y}$ yields rational singularities. Hence $\chi(D) \geqslant 1$. This is a contradiction. Therefore $D = 0$, namely, $-K_{\tilde{Y}} = Z + C_{i_1}$. Q.E.D.

COROLLARY 1.5. *Assume that* $q \neq 0$. *Then*
(1) $(C_{i_1} \cdot Z) = 2 - 2q$.
(2) $(Z \cdot Z) \leqslant (C_{i_1} \cdot C_{i_1})$.

Proof. Since $K_{\tilde{Y}} = -Z - C_{i_1}$, by the adjunction formula, we have (1) and (2). Q.E.D.

LEMMA 1.6.
(1) $q \neq 0 \Rightarrow b_2(\tilde{Y}) \leqslant 9 - 4q + \sqrt{9 + 8q}$.
(2) $q = 0 \Rightarrow 11 \leqslant b_2(\tilde{Y}) \leqslant 13$.

Proof. (1) By the Noether formula, we have

$$10 - 8q = (K_{\tilde{Y}} \cdot K_{\tilde{Y}}) + b_2(\tilde{Y}). \tag{1.6}$$

Since $K_{\tilde{Y}} = -Z - C_{i_1}$, we have

$$(K_{\tilde{Y}} \cdot K_{\tilde{Y}}) = (Z \cdot Z) + 2(Z \cdot C_{i_1}) + (C_{i_1} \cdot C_{i_1}). \tag{1.7}$$

By (1.6), (1.7) and Corollary 1.5, we have

$$b_2(\tilde{Y}) = 6 - 4q - (Z \cdot Z) - (C_{i_1} \cdot C_{i_1}). \tag{1.8}$$

$$\leqslant 6 - 4q - 2(Z \cdot Z). \tag{1.9}$$

Since $S = \{\text{one point}\}$ is a hypersurface singularity, we have

$$-(Z \cdot Z) \leqslant n := m(\mathcal{O}_{Y,S}) \quad \text{(Wagreich [23])}. \tag{1.10}$$

$$p_g \geqslant \tfrac{1}{2}(n-1)(n-2) \quad \text{(Yau [24])}. \tag{1.11}$$

Since $p_g = q + 1$, by (1.10), (1.11), we have

$$-(Z\cdot Z) \leqslant \tfrac{1}{2}(3 + \sqrt{9 + 8q}). \tag{1.12}$$

By (1.9), (1.12), we have the claim.

(2) By the Noether formula, we have

$$b_2(\tilde{Y}) = 10 - (K_{\tilde{Y}}\cdot K_{\tilde{Y}}). \tag{1.13}$$

Since $p_g = 1$ and S is a hypersurface singularity, by Laufer [11], we have $-3 \leqslant (Z\cdot Z) \leqslant -1$. By Lemma 1.4(1) and (1.13), we have the claim. Q.E.D.

COROLLARY 1.7. $0 \leqslant q \leqslant 3$.

Proof. Assume that $q \neq 0$. By Lemma 1.6(1), we have

$2 \leqslant b_2(\tilde{Y}) \leqslant 9 - 4q + \sqrt{9 + 8q}$. This implies $q \leqslant 3$. Q.E.D.

3. By the classification of Fano 3-Folds with the second Betti numbers one due to Iskovskih [9] (see also Mukai [15], [16]), we have:

(Table 1)

g	2	3	4	5	6	7	8	9	10	12
$\frac{1}{2}b_3(X)$	52	30	20	14	10	7	5	3	2	0

,

where $g := \frac{1}{2}(K_X^3) + 1 = \frac{1}{2}(Y^3) + 1$.

Since $2q = b_1(\tilde{Y}) = b_3(\tilde{Y}) = b_3(Y) = b_3(X)$ (cf. [2]), by Corollary 1.7, $0 \leqslant \frac{1}{2}b_3(X) \leqslant 3$. Thus, by the Table 1 above, we have $(g, q) = (9, 3), (10,2)$ or $(12, 0)$.

LEMMA 1.8. $q \neq 3$.

Proof. Assume that $q = 3$. By Lemma 1.6(1), we have $2 \leqslant b_2(\tilde{Y}) \leqslant -3 + \sqrt{33} < 3$, namely, $b_2(\tilde{Y}) = 2$. Hence, $\tilde{Y}$ is a $\mathbb{P}^1$-bundle over a smooth compact algebraic curve R of genus 3. Therefore, Y is a cone over R. This is a contradiction, by Table (I) in [3]. Q.E.D.

LEMMA 1.9. *Assume that $q \neq 0$. Then there is exactly one exceptional curve of the first kind in every singular fiber of the ruled surface $\tilde{Y}$, and the other irreducible components of the singular fiber are all contained in $E := \pi^{-1}(\text{Sing } Y)$.*

Proof. Since $q \neq 0$, the rest $\overline{E - C_{i_1}}$ must be contained in the singular fibers of $\tilde{Y}$. Let $F_1, \ldots, F_r$ be the singular fibers of $\tilde{Y}$, $1 + \alpha_i$ ($\alpha_i > 0$) the "number" of the irreducible components of F_i and δ_i the "number" of the irreducible components of F_i which are not contained in E. Then we have

$$\begin{cases} 1 + t = b_2(\tilde{Y}) = 2 + \sum_{i=1}^{r} \alpha_i. \\ \sum_{i=1}^{r} (1 + \alpha_i - \delta_i) + 1 = t. \end{cases}$$

Thus we have $\Sigma_{i=1}^{r} (1 - \delta_i) = 0$. Since each singular fiber F_i contains at least an exceptional curve of the first kind, we have $\delta_i \geqslant 1$ for $1 \leqslant i \leqslant r$, hence, $\delta_i = 1$ for $1 \leqslant i \leqslant r$. Q.E.D.

LEMMA 1.10. *Assume that* $q = 2$. *Then the dual graphs of all the exceptional curves in* $\tilde{Y}$ *look like the Figure* 1 *below.*

Proof. By Lemma 1.6(1), we have $2 \leqslant b_2(\tilde{Y}) \leqslant 6$. Since Y is not a cone (see Table 1 in [3], we have $b_2(\tilde{Y}) \neq 2$. If $b_2(\tilde{Y}) = 3$, then $\tilde{Y}$ contains two exceptional curves of the first kind in a singular fiber. This contradicts Lemma 1.9. Hence we have $4 \leqslant b_2(\tilde{Y}) \leqslant 6$. Thus, by (1.9), (1.12), we have $-4 \leqslant (Z \cdot Z) \leqslant -3$. We put $n := (C_{i_1} \cdot C_{i_1}) < 0$. Then, by Lemma 1.3 and (1.8), we have

(i) $b_2(\tilde{Y}) = 6 \Rightarrow (n, t) = (-4, 5)$ and $(Z \cdot Z) = -4$.
(ii) $b_2(\tilde{Y}) = 5 \Rightarrow (n, t) = (-3, 4)$ and $(Z \cdot Z) = -4$.
(iii) $b_2(\tilde{Y}) = 4 \Rightarrow (n, t) = (-3, 3)$ and $(Z \cdot Z) = -3$,
$= (-2, 3)$ and $(Z \cdot Z) = -4$.

Thus, we have
(a) $(Z \cdot Z) = -4 \Rightarrow (n, t) = (-2, 3), (-3, 4), (-4, 5)$.
(b) $(Z \cdot Z) = -3 \Rightarrow (n, t) = (-3, 3)$.

Since Sing $Y - S$ consists of rational double points, by Lemma 1.9 and (a), (b), the configuration of the exceptional curves of $\tilde{Y}$ can be easily described. Thus we have the lemma. Q.E.D.

Notation. The vertex $\boxed{k}$ represents a non-singular compact algebraic curve of genus 2 with the self-intersection number $-k$, (which is corresponding to the section C_{i_1} of $\tilde{Y}$), ⓚ a non-singular rational curve with the self-intersection number $-k$. We denote ② simply by ○. Adjacent to the graph, we write a basis $\{e_i\}$ $(0 \leqslant i \leqslant t)$ of $H^2(\tilde{Y}:\mathbb{Z})$, where $t = \dim H^2(E; \mathbb{R})$.

LEMMA 1.11. *Assume that* $q = 2$. *Then there is a canonical curve* D *of genus* 10 *and* $\deg D = 18$ *such that*

(i) Sing $Y \cap D = \phi$,
(ii) $\mathcal{O}_Y(Y) = \mathcal{O}_Y(D)$.

Proof. Since $q = 2$, by Table 1, X is a Fano 3-fold of degree 18 in $\mathbb{P}^{11}$ and Y is a hyperplane section of X (see [9]). For a sufficiently general hyperplane section H, we put $D = H \cdot Y$, which is desired. Q.E.D.

4. We put $\tilde{D} := \pi^{-1}(D) \hookrightarrow \tilde{Y}$. Since $D \cap \operatorname{Sing} Y = \phi$ by Lemma 1.11, we have $\tilde{D} \cong D$ (isomorphism), $(\tilde{D} \cdot \tilde{D}) = 18$ and $(\tilde{D} \cdot E_j) = 0$ for each irreducible component E_j of $E = \pi^{-1}(\operatorname{Sing} Y)$.

Let $\{e_i\}$ $(0 \leqslant i \leqslant t)$ be a basis of $H^2(\tilde{Y}; \mathbb{Z}) \cong \mathbb{Z}^{t+1}$ (see Fig. 1). Then, we have

$$c_1(\mathcal{O}(\tilde{D})) = \sum_{i=0}^{t} \alpha_i e_i \quad (\alpha_i \in \mathbb{Z}), \tag{$*$}$$

where $c_1(\mathcal{O}(\tilde{D})) \in H^2(\tilde{Y} : \mathbb{Z})$ is the first Chern class of $\mathcal{O}(\tilde{D})$, and

(i) the intersection number $e_i \cdot e_j$ is determined by the graph in Fig. 1,
(ii) $c_1(\mathcal{O}(\tilde{D})) \cdot e_i = 0$ $(0 \leqslant i \leqslant t)$,
(iii) $c_1(\mathcal{O}(\tilde{D})) \cdot c_1(\mathcal{O}(\tilde{D})) = 18$,
(iv) $d_{i_0} := c_1(\mathcal{O}(\tilde{D})) \cdot e_{i_0} \neq 0$, where e_{i_0} is a class corresponding to the exceptional curve of the first kind.

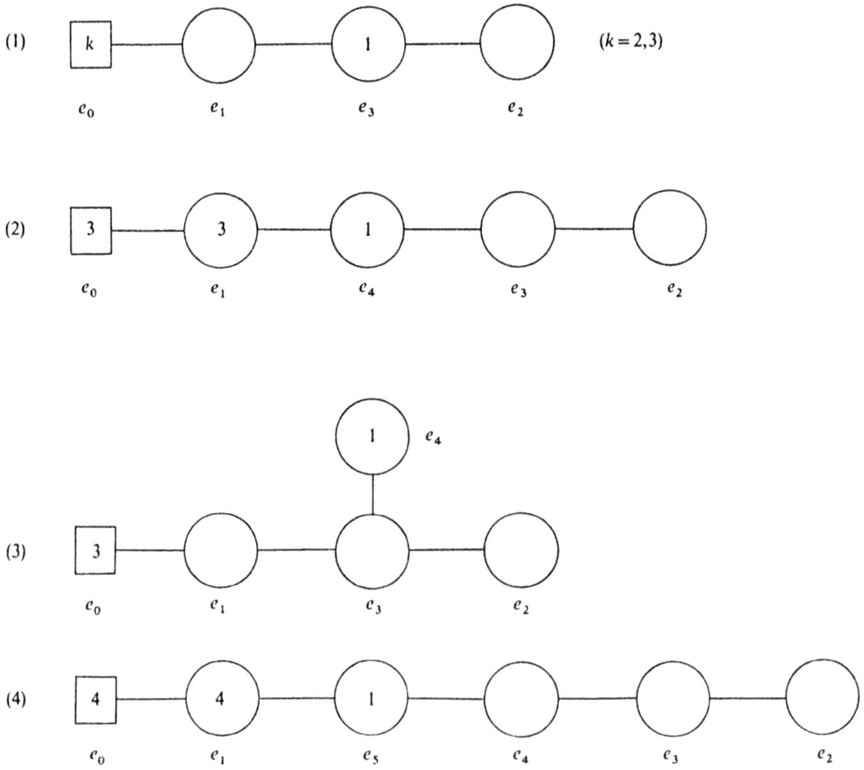

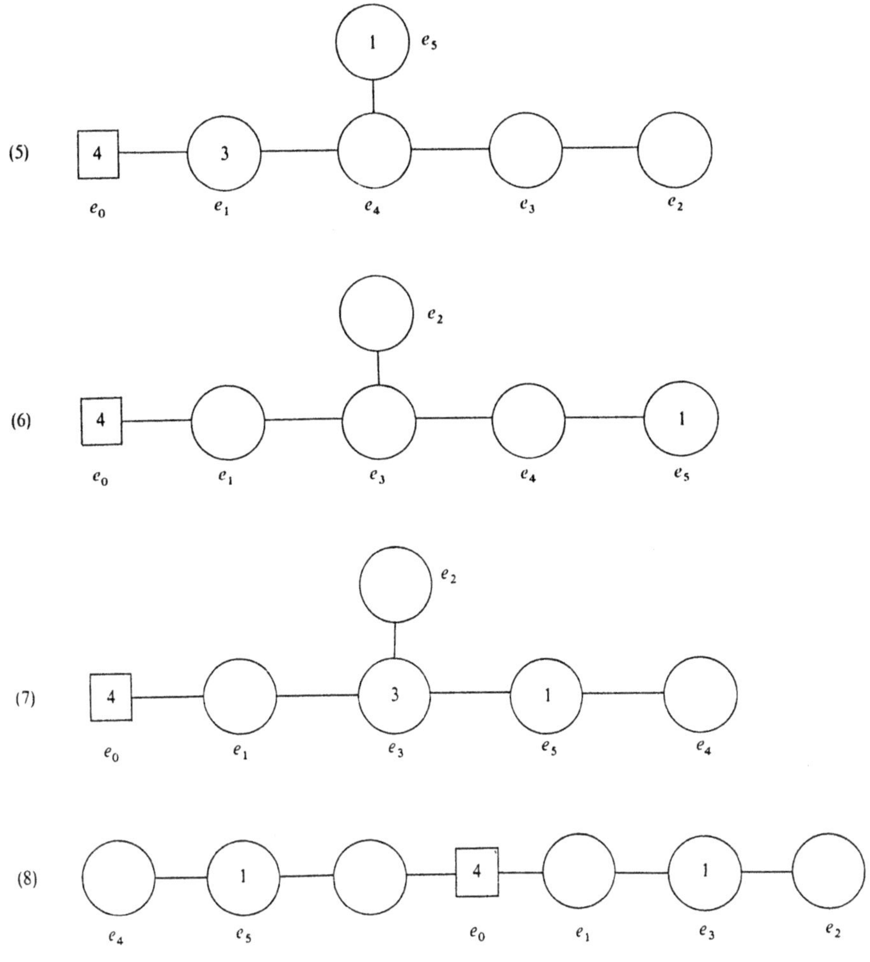

(Fig. 1).

By $(*)$ and (i)–(iv) above, for each graph in the Fig. 1, we have the equations of α_i $(0 \leqslant i \leqslant t)$ and d_{i_0} over $\mathbb{Z}$ below:

Case (1)

$$\begin{aligned} \alpha_1 - k\alpha_0 &= 0 \\ \alpha_3 - 2\alpha_1 + \alpha_0 &= 0 \\ \alpha_3 - 2\alpha_2 &= 0 \\ \alpha_2 - \alpha_3 + \alpha_1 &= d_3 \\ \alpha_3 \cdot d_3 &= 18 \end{aligned} \tag{C-1}$$

$\therefore \alpha_0^2 = 36/2k - 1$ $(k = 2, 3)$. Hence $\alpha_0 \notin \mathbb{Z}$.

Case (2).

$$\begin{aligned} \alpha_1 - 3\alpha_0 &= 0 \\ \alpha_4 - 3\alpha_1 + \alpha_0 &= 0 \\ \alpha_3 - 2\alpha_2 &= 0 \\ \alpha_4 - 2\alpha_3 + \alpha_2 &= 0 \\ \alpha_1 - \alpha_4 + \alpha_3 &= d_4 \\ \alpha_4 \cdot d_5 &= 18 \end{aligned} \tag{C-2}$$

$\therefore \alpha_0^2 = \frac{27}{4}$. Hence $\alpha_0 \notin \mathbb{Z}$.

Case (3).

$$\begin{aligned} \alpha_1 - 3\alpha_0 &= 0 \\ \alpha_3 - 2\alpha_1 + \alpha_0 &= 0 \\ \alpha_3 - 2\alpha_0 &= 0 \\ \alpha_4 + \alpha_1 - 2\alpha_3 + \alpha_2 &= 0 \\ \alpha_3 - \alpha_4 &= d_4 \\ \alpha_5 \cdot d_5 &= 18 \end{aligned} \tag{C-3}$$

$\therefore \alpha_0^2 = 8$. Hence $\alpha_0 \notin \mathbb{Z}$.

Case (4).

$$\begin{aligned} \alpha_1 - 4\alpha_0 &= 0 \\ \alpha_5 - 4\alpha_1 + \alpha_0 &= 0 \\ \alpha_3 - 2\alpha_2 &= 0 \\ \alpha_4 - 2\alpha_3 + \alpha_2 &= 0 \\ \alpha_5 - 2\alpha_4 + \alpha_3 &= 0 \\ \alpha_1 - \alpha_5 + \alpha_4 &= d_5 \\ \alpha_5 \cdot d_5 &= 18 \end{aligned} \tag{C-4}$$

$\therefore \alpha_0^2 = \frac{24}{5}$. Hence $\alpha_0 \notin \mathbb{Z}$.

Case (5).

$$\begin{aligned} \alpha_1 - 4\alpha_0 &= 0 \\ \alpha_4 - 3\alpha_1 + \alpha_0 &= 0 \\ \alpha_3 - 2\alpha_2 &= 0 \\ \alpha_4 - 2\alpha_3 + \alpha_2 &= 0 \\ \alpha_5 + \alpha_1 - 2\alpha_4 + \alpha_3 &= 0 \\ \alpha_4 - \alpha_5 &= d_5 \\ \alpha_5 \cdot d_5 &= 18 \end{aligned} \tag{C-5}$$

$\therefore \alpha_0^2 = \frac{81}{16}$. Hence $\alpha_0 \notin \mathbb{Z}$.

Case (6).

$$\begin{aligned} &\alpha_1 - 4\alpha_0 &&= 0 \\ &\alpha_3 - 2\alpha_1 + \alpha_0 &&= 0 \\ &\alpha_3 - 2\alpha_2 &&= 0 \\ &\alpha_4 + \alpha_1 - 2\alpha_3 + \alpha_2 &&= 0 \\ &\alpha_5 - 2\alpha_4 + \alpha_3 &&= 0 \\ &\alpha_4 - \alpha_5 &&= d_5 \\ &a_5 \cdot d_5 &&= 18 \end{aligned} \tag{C-6}$$

$\therefore \alpha_0^2 = 6$. Hence $\alpha_0 \notin \mathbb{Z}$.

Case (7).

$$\begin{aligned} &\alpha_1 - 4\alpha_0 &&= 0 \\ &\alpha_0 - 2\alpha_1 + \alpha_3 &&= 0 \\ &\alpha_3 - 2\alpha_2 &&= 0 \\ &\alpha_5 + \alpha_1 - 3\alpha_3 + \alpha_2 &&= 0 \\ &\alpha_5 - 2\alpha_4 &&= 0 \\ &\alpha_4 - \alpha_5 + \alpha_3 &&= d_5 \\ &\alpha_5 \cdot d_5 &&= 18 \end{aligned} \tag{C-7}$$

$\therefore \alpha_0^2 = \frac{16}{3}$. Hence $\alpha_0 \notin \mathbb{Z}$.

Case (8).

$$\begin{aligned} &\alpha_1 - 4\alpha_0 &&= 0 \\ &\alpha_3 - 2\alpha_1 + \alpha_0 &&= 0 \\ &\alpha_3 - 2\alpha_2 &&= 0 \\ &\alpha_1 - \alpha_3 + \alpha_2 &&= d_3 \\ &\alpha_5 - 2\alpha_4 &&= 0 \\ &\alpha_4 - \alpha_5 &&= d_5 \\ &\alpha_0 + \alpha_5 &&= 0 \\ &\alpha_5 d_5 + \alpha_3 d_3 &&= 18 \end{aligned} \tag{C-8}$$

$\therefore \alpha_0^2 = 6$. Hence $\alpha_0 \notin \mathbb{Z}$.

By the computations (C-1)–(C-8), we find that these equations have no integral solutions. Thus, we have:

LEMMA 1.12. $q \neq 2$.

By Lemma 1.8, Lemma 1.12, and Table 1, we have the following

PROPOSITION 1.13. (cf. [18], [19]). *Assume that the index* $r = 1$. *Then,*

$(X, Y) \cong (V_{22}, H_{22})$, where V_{22} is a Fano 3-fold of degree 22 in $\mathbb{P}^{13}$(index 1, genus 12) and H_{22} is a hyperplane section of V_{22} which is rational.

REMARK 1.14. Among Fano 3-folds of degree 22 in $\mathbb{P}^{13}$(index 1, genus 12), there is a special one, $V'_{22} \hookrightarrow \mathbb{P}^{12}$, which has been overlooked by Iskovskih [8] (see Mukai-Umemura [14]).

Recently, Mukai has succeeded in classifying Fano 3-folds of index 1 with $b_2(X) = 1$, applying the theory of vector bundles on K-3 surfaces (see [15], [16]).

2. Determination of the boundary

1. Let $(X, Y) = (V_{22}, H_{22})$ be as in Proposition 1.13. Since $q = 0$, by Lemma 1.3, S consists of one point x with $p_g = 1$, namely, x is a minimally elliptic singularity. We put $\operatorname{Sing} Y - \{x\} =: \{y_1, \ldots, y_k\}$ $(k \geqslant 0)$, and $B := \pi^{-1}(\{y_1, \ldots, y_k\})$. Then y_j's are all rational double points.

By Lemma 1.6(2), we have:

$$b_2(B) + b_2(C) = 10 \quad \text{if} \quad (Z \cdot Z) = -1, \tag{2.1}$$

$$b_2(B) + b_2(C) = 11 \quad \text{if} \quad (Z \cdot Z) = -2, \tag{2.2}$$

$$b_2(B) + b_2(C) = 12 \quad \text{if} \quad (Z \cdot Z) = -3. \tag{2.3}$$

2. Let T_0 (resp. T_i) be a contractible neighborhood of x (resp. y_i) in Y. We may assume that T_0, T_i $(1 \leqslant i \leqslant k)$ are disjoint. We put $T := \bigcup_{i=0}^{k} T_i$ and $\partial T := \bigcup_{i=0}^{k} \partial T_i$, where ∂T_i is the boundary of T_i. We put $T^* := T - \operatorname{Sing} Y$ and $Y^* := Y - \operatorname{Sing} Y$. Since $T^* \approx \partial T$ (deformation retract), by the Mayer-Vietoris exact sequence, we have

$$\begin{aligned} &\to H_i(\partial T; \mathbb{Z}) \to H_i(Y^*; \mathbb{Z}) \oplus H_i(T; \mathbb{Z}) \to \\ &\to H_i(Y; \mathbb{Z}) \to H_{i-1}(\partial T; \mathbb{Z}) \to . \end{aligned} \tag{2.4}$$

Since Sing Y is isolated in Y, we have $H_2(Y^*; \mathbb{Z}) \cong H^2(\tilde{Y}, E; \mathbb{Z}) \cong H^2(Y, \operatorname{Sing} Y; \mathbb{Z}) \cong H^2(Y; \mathbb{Z}) \cong \mathbb{Z}$. On the other hand, since $X = V_{22}$ is a Fano 3-fold of index 1 and the genus $g = 12$, we have $H^3(X; \mathbb{Z}) = 0$ (cf. [8], [15], [16]). Thus we have $H_3(Y; \mathbb{Z}) \cong H_3(X; \mathbb{Z}) \cong H^3(X; \mathbb{Z}) \cong 0$, and $H_1(Y^*; \mathbb{Z}) \cong H^3(Y; \mathbb{Z}) \cong H^3(X; \mathbb{Z}) \cong 0$ (cf. [2], [3]). Therefore we have finally the Poincaré's exact sequence:

$$0 \to H^2(Y; \mathbb{Z}) \xrightarrow{P_2} H_2(Y; \mathbb{Z}) \to H_1(\partial T; \mathbb{Z}) \to 0 \quad \text{(cf. [18])}. \tag{2.5}$$

By Lemma 2.5 in [18], we have

LEMMA 2.1 (cf. Peternell-Schneider [18]). *$H_1(\partial T; \mathbb{Z}) = H_1(\partial T_0; \mathbb{Z}) \oplus \mathscr{H} \cong \mathbb{Z}_{22}$, where $\mathscr{H} = \oplus_{i=1}^{k} H_1(\partial T_i; \mathbb{Z})$, namely, we have:*

(*Table* 2)

$H_1(\partial T_0; \mathbb{Z})$	$\mathbb{Z}_{22}$	$\mathbb{Z}_{11}$	$\mathbb{Z}_2$	0
$\mathscr{H}$	0	$\mathbb{Z}_2$	$\mathbb{Z}_{11}$	$\mathbb{Z}_{22}$

LEMMA 2.2. *For the rational double point $y_j \in \operatorname{Sing} Y - \{x\}$ $(1 \leqslant j \leqslant k)$, we have:*

(*Table* 3)

Type of y_i	A_n	D_{2n}(resp. D_{2n+1})	E_6	E_7	E_8
$H_1(\partial T_i; \mathbb{Z})$	$\mathbb{Z}_{n+1}$	$\mathbb{Z}_2 \oplus \mathbb{Z}_2$(resp. $\mathbb{Z}_4$)	$\mathbb{Z}_3$	$\mathbb{Z}_2$	0

Proof. Apply Lemma M below.

LEMMA M (Mumford [17]).

Let S be a smooth complex surface and consider a divisor $C = \bigcup_{i=1}^{n} C_i$ (C_i: a smooth rational curve) with normal crossings. Let ∂T be the boundary of a tubular neighborhood T of C in S. Then, $H_1(\partial T; \mathbb{Z})$ is generated by $\gamma_1, \ldots, \gamma_n$ with the fundamental relations:

$$\sum_{j=1}^{n} (C_i \cdot C_j) \cdot \gamma_j \quad (j = 1, 2, \ldots, n), \tag{\#}$$

where γ_j is a loop in ∂T which goes around C_j with positive orientation.

REMARK. By Lemma M, one can easily compute the homology group $H_1(\partial T_0 \colon \mathbb{Z})$ for each exceptional divisor C in Table L-1–Table L-9 below.

By Lemma 2.1, Lemma 2.2, and (2.1), (2.2), we have easily the following

LEMMA 2.3.
(1) $(Z \cdot Z) = -1 \Rightarrow H_1(\partial T_0 \colon \mathbb{Z}) \not\cong 0, \mathbb{Z}_2$
(2) $(Z \cdot Z) = -2 \Rightarrow H_1(\partial T_0; \mathbb{Z}) \not\cong 0.$

LEMMA 2.4.
(1) *The case of $(Z \cdot Z) = -1$. We have:*
  (i) $H_1(\partial T_0 \colon \mathbb{Z}) \cong \mathbb{Z}_{11} \Rightarrow b_2(C) = 1, 3, 9$ (*Table* L-1)
  (ii) $H_1(\partial T_0 \colon \mathbb{Z}) \cong \mathbb{Z}_{22} \Rightarrow b_2(C) = 2, 10$ (*Table* L-2)
(2) *The case of $(Z \cdot Z) = -2$. We have:*
  (i) $H_1(\partial T_0 \colon \mathbb{Z}) \cong \mathbb{Z}_2 \Rightarrow b_2(C) = 1$ (*Table* L-3)
  (ii) $H_1(\partial T_0; \mathbb{Z}) \cong \mathbb{Z}_{11} \Rightarrow b_2(C) = 2, 4, 10$ (*Table* L-4)
  (iii) $H_1(\partial T_0; \mathbb{Z}) \cong \mathbb{Z}_{22} \Rightarrow b_2(C) = 3, 11$ (*Table* L-5)

(3) *The case of* $(Z \cdot Z) = -3$. *We have:*

(i) $H_1(\partial T_0 \colon \mathbb{Z}) \cong 0 \Rightarrow b_2(C) = 1$ (*Table* L-6)

(ii) $H_1(\partial T_0; \mathbb{Z}) \cong \mathbb{Z}_2 \Rightarrow b_2(C) = 2$ (*Table* L-7)

(iii) $H_1(\partial T_0; \mathbb{Z}) \cong \mathbb{Z}_{11} \Rightarrow b_2(C) = 3, 5, 11$ (*Table* L-8)

(iv) $H_1(\partial T_0; \mathbb{Z}) \cong \mathbb{Z}_{22} \Rightarrow b_2(C) = 4, 12$ (*Table* L-9)

Proof. We will prove for the case (3)(iii). The proof for the other cases are similar. Since $(Z \cdot Z) = -3$ and $H_1(\partial T_0; \mathbb{Z}) \cong \mathbb{Z}_{11}$, we have $b_2(B) + b_2(C) = 12$ by (2.3), and $\mathscr{H} := \oplus_{i=1}^{r} H_1(\partial T_i; \mathbb{Z}) \cong \mathbb{Z}_2$. By the Table 3 Sing $Y - \{x\} = \{A_1\text{-type}\}, \{E_7\text{-type}\}$ or $\{A_1\text{-type} + E_8\text{-type}\}$, hence, $b_2(B) = 1, 7, 9$, respectively. By (2.3), we have $b_2(C) = 3, 5, 11$. Pick out the possible types of the dual graphs with $b_2(C) = 3, 5, 11$, from the Table 3 in Laufer [11], we have finally the Table L-8. We remark that there is no dual graph with $b_2(C) = 4$. Q.E.D.

From Lemma 2.4 and the Table L-1–Table L-9, we have directly the following

(*Table L-1*)

	Dual graph	$A_* \cdot A_*$	$H_1(\partial T_0 \colon \mathbb{Z})$
1	*Cu*	-1	0
2	*Tr*	$-2, -2, -3$	$\mathbb{Z}_3$
3	$A_{5,****}$	$-2, -2, -2, -3$	$\mathbb{Z}_2 \oplus \mathbb{Z}_2$
4	$E_{8,*}$	-3	0

(*Table L-2*)

	Type of x	$A_* \cdot A_*$	$H_1(\partial T_0; \mathbb{Z})$
5	*Ta*	$-2, -3$	$\mathbb{Z}_2$
6	$A_{6,****}$	$-2, -2, -2, -3$	$\mathbb{Z}_4$

(*Table L-3*)

	Type of x	$A_* \cdot A_*$	$H_1(\partial T_0; \mathbb{Z})$
7	*Cu*	-2	"$\mathbb{Z}_2$"

(*Table L-4*)

	Dual graph	$A_* \cdot A_*$	$H_1(\partial T_0; \mathbb{Z})$
8	*Ta*	$-2, -4$	$\mathbb{Z}_2^{\oplus 2}$
9	*Ta*	$-3, -3$	$\mathbb{Z}_5$

10	$A_{6,****}$	$-2, -2, -2, -4$	$\mathbb{Z}_2 \oplus \mathbb{Z}_4$
11	$A_{6,****}$	$-2, -2, -3, -3$	$\mathbb{Z}_2 \oplus \mathbb{Z}_6$
12	$A_{6,****}$	$-2, -3, -3, -2$	$\mathbb{Z}_{17}$
13	$A_{*,o} + A_{*,o} + A_{*,o}$ $+ A_{4,**o}$	$-2, -2, -2$ $-2, -2$	$\mathbb{Z}_2^{\oplus 4}$
14	$A_{*,o} + A_{*,o} + E_{7,o}$	$-2, -2$	$\mathbb{Z}_2^{\oplus 2}$
15	$A_{*,o} + A_{n,**o}$ $+ A_{m,**o}$ $(m + n = 4)$	$-2, -2, -2,$ $-2, -2$	$\mathbb{Z}_4^{\oplus 2}$ if $(m, n) = (1, 3)$ $\mathbb{Z}_2^{\oplus 4}$ if $(m, n) = (2, 2)$
16	$A_{*,o} + D_{7,*,o}$	$-2, -2$	$\mathbb{Z}_2^{\oplus 4}$
17	$A'_{7,**,o}$	$-2, -2$	$\mathbb{Z}_5$
18	$A_{2,**,o} + A'_{3,**o}$	$-2, -2, -2, -2$	$\mathbb{Z}_2 \oplus \mathbb{Z}_6$
19	$A_{1,**,o} + D_{5,*,o}$	$-2, -2, -2$	$\mathbb{Z}_2 \oplus \mathbb{Z}_4$
20	$A_{0,**o} + E_{7,o}$	$-2, -2$	$\mathbb{Z}_2 \oplus \mathbb{Z}_2$

(*Table L-5*)

	Dual graph	$A_* \cdot A_*$	$H_1(\partial T_0; \mathbb{Z})$
21	*Tr*	$-2, -2, -4$	$\mathbb{Z}_6$
22	*Tr*	$-2, -3, -3$	$\mathbb{Z}_8$
23	$A_{7,****}$	$-2, -2, -2, -4$	$\mathbb{Z}_2^{\oplus 3}$
24	$A_{7,****}$	$-2, -2, -3, -3$	$\mathbb{Z}_4 \oplus \mathbb{Z}_6$
25	$A_{7,****}$	$-2, -3, -3, -2$	$\mathbb{Z}_6 \oplus \mathbb{Z}_9$
26	$A_{*,o} + A_{*,o} +$ $+ A_{*,o} + A_{5,**,o}$	$-2, -2$ $-2, -2, -2$	$\mathbb{Z}_2^{\oplus 2} \oplus \mathbb{Z}_4$
27	$A_{*,o} + A_{n,**,o} +$ $+ A_{m,**,o}$ $(m + n = 5)$	$-2, -2, -2$ $-2, -2$	$\mathbb{Z}_2^{\oplus 2} \oplus \mathbb{Z}_4$
28	$A_{3,**,o} + A'_{3,**,o}$	$-2, -2, -2, -2$	$\mathbb{Z}_4 \oplus \mathbb{Z}_6$
29	$A_{2,**,o} + D_{5,*,o}$	$-2, -2, -2$	$\mathbb{Z}_2^{\oplus 3}$
30	$A_{1,**,o} + E_{7,o}$	$-2, -2$	$\mathbb{Z}_4$
31	$D_{9,*,o}$	-2	$\mathbb{Z}_2$

(*Table L-6*)

	Dual graph	$A_*\cdot A_*$	$H_1(\partial T_0; \mathbb{Z})$
32	*Cu*	-3	$\mathbb{Z}_3$

(*Table L-7*)

	Dual graph	$A_*\cdot A_*$	$H_1(\partial T_0; \mathbb{Z})$
33	*Ta*	$-2, -5$	$\mathbb{Z}_6$
34	*Ta*	$-3, -4$	$\mathbb{Z}_8$

(*Table L-8*)

	Dual graph	$A_*\cdot A_*$	$H_1(\partial T_0; \mathbb{Z})$
35	*Tr*	$-2, -2, -5$	$\mathbb{Z}_3^{\oplus 2}$
36	*Tr*	$-2, -3, -4$	$\mathbb{Z}_{13}$
37	*Tr*	$-3, -3, -3$	$\mathbb{Z}_4^{\oplus 2}$
38	$A_{1,****}$	$-2, -2, -2, -5$	$\mathbb{Z}_6^{\oplus 2}$
39	$A_{1,****}$	$-2, -2, -3, -4$	$\mathbb{Z}_{10}^{\oplus 2}$
40	$A_{1,****}$	$-2, -3, -3, -3$	$\mathbb{Z}_9^{\oplus 2}$
41	$A_{7,****}$	$-2, -2, -2, -5$	$\mathbb{Z}_6^{\oplus 2}$
42	$A_{7,****}$	$-2, -2, -3, -4$	$\mathbb{Z}_{10}^{\oplus 2}$
43	$A_{7,****}$	$-2, -3, -4, -2$	$\mathbb{Z}_{16}^{\oplus 2}$
44	$A_{7,****}$	$-2, -3, -3, -3$	$\mathbb{Z}_{15}^{\oplus 2}$
45	$A_{*,o} + A_{*,o} + A_{*,o} +$ $+ A_{5,**,o}$	$-3, -2, -2,$ $-2, -2$	$\mathbb{Z}_8^{\oplus 2}$
46	$A_{*,o} + A_{*,o} + A_{*,o} +$ $+ A_{5,**,o}$	$-2, -2, -2$ $-2, -3$	$\mathbb{Z}_{26}^{\oplus 2}$
47	$A_{*,o} + A_{n,**o} +$ $+ A_{m,**,o} (m + n = 5)$	$-3, -2, -2$ $-2, -2,$	$\mathbb{Z}_8^{\oplus 2}$
48	$A_{*,o} + A_{n,**,o} +$ $+ A_{m,**,o}\ (m + n = 5)$	$-2, -2, -2$ $-2, -3$	$\mathbb{Z}_{18}^{\oplus 2}$ if $(m, n) = (1, 4)$ $\mathbb{Z}_{10} \oplus \mathbb{Z}_{10}$ if $(m, n) = (2, 3)$ $\mathbb{Z}_{22}^{\oplus 2}$ if $(m, n) = (3, 2)$ $\mathbb{Z}_4 \oplus \mathbb{Z}_6 \oplus \mathbb{Z}_{12}$ if $(m, n) = (4, 1)$
49	$D_{9,*,o}$	-3	$\mathbb{Z}_5$
50	$A_{3,**,o} + A'_{3,**,o}$	$-3, -2, -2, -2$	$\mathbb{Z}_{33}$

51	$A_{3,**,o} + A'_{3,**,o}$	$-2, -2, -2, -3$	$\mathbb{Z}_{14}^{\oplus 2}$
52	$A_{2,**,o} + D_{5,*,o}$	$-3, -2, -2$	$\mathbb{Z}_{10}^{\oplus 2}$
53	$A_{2,**,o} + D_{5,*,o}$	$-2, -2, -3$	$\mathbb{Z}_{30}^{\oplus 2}$
54	$A_{1,**,o} + E_{7,o}$	$-3, -2$	$\mathbb{Z}_9$
55	$A_{1,*,o} + A_{1,*,o} +$ $+ E_{6,o}$	$-2, -2$	$\mathbb{Z}_3^{\oplus 2}$
56	$A_{1,*,o} + A_{7,*,o}$	$-2, -2$	$\mathbb{Z}_3 \oplus \mathbb{Z}_9$
57	$A_{4,*,o} + A_{4,*,o}$	$-2, -2$	$\mathbb{Z}_6^{\oplus 2}$

(*Table L-9*)

	Dual graph	$A_* \cdot A_*$	$H_1(\partial T_0; \mathbb{Z})$
58	$A_{8,****}$	$-2, -2, -2, -5$	$\mathbb{Z}_{12}$
59	$A_{8,****}$	$-2, -2, -3, -4$	$\mathbb{Z}_{20}$
60	$A_{8,****}$	$-2, -3, -4, -2$	$\mathbb{Z}_{34}$
61	$A_{8,****}$	$-2, -3, -3, -3$	$\mathbb{Z}_{48}$
62	$A_{*,o} + A_{*,o} + A_{*,o} +$ $+ A_{6,**,o}$	$-3, -2, -2$ $-2, -2$	$\mathbb{Z}_8^{\oplus 3}$
63	$A_{*,o} + A_{*,o} + A_{*,o} +$ $+ A_{6,**,o}$	$-2, -2, -2$ $-2, -3$	$\mathbb{Z}_{14}^{\oplus 3}$
64	$A_{*,o} + A_{n,**o} +$ $+ A_{m,**,o} (m + n = 6)$	$-3, -2, -2$ $-2, -2$	$\mathbb{Z}_8^{\oplus 3}$
65	$A_{*,o} + A_{n,**,o} +$ $+ A_{m,**,o} (m + n = 6)$	$-2, -2, -2$ $-2, -3$	$\mathbb{Z}_{36}$ if $(m, n) = (1, 5)$ $\mathbb{Z}_{10}^{\oplus 3}$ if $(m, n) = (2, 4)$ $\mathbb{Z}_{44}$ if $(m, n) = (3, 3)$ $\mathbb{Z}_4 \oplus \mathbb{Z}^{\oplus 2}$ if $(m, n) = (4, 2)$ $\mathbb{Z}_{13} \oplus \mathbb{Z}_{52}$ if $(m, n) = (5, 1)$
66	$A_{4,**,o} + A'_{3,**,o}$	$-3, -2, -2, -2$	$\mathbb{Z}_6 \oplus \mathbb{Z}_{12}$
67	$A_{4,**,o} + A'_{3,**,o}$	$-2, -2, -2, -3$	$\mathbb{Z}_{28}$
68	$A_{3,**,o} + D_{5,*,o}$	$-3, -2, -2$	"$\mathbb{Z}_{22}$"
69	$A_{3,**,o} + D_{5,*,o}$	$-2, -2, -3$	$\mathbb{Z}_{20}$
70	$A_{2,**,o} + E_{7,o}$	$-3, -2$	$\mathbb{Z}_{10}$
71	$A_{10,**,o}$	-2	$\mathbb{Z}_4$
72	$A_{4,*,o} + E_{6,o}$	-2	$\mathbb{Z}_6$

PROPOSITION 2.5. *Let* $(X, Y) = (V_{22}, H_{22})$ *be a compactification of* $\mathbb{C}^3$ *as in Proposition* 1.13. *Then,*

(a) Sing $Y = \{x\}$, *where* x *is a minimally elliptic singularity of* $A_{3,**o} + D_{5,*,o}$ – *Type* (*Table* L-9, (68)), *or*

(b) Sing $Y = \{x, y\}$, *where* x *is a minimally elliptic singularity of* Cu-*type* (*Table* L-3, (7)) *and* y *is a rational double point of* A_{10}-*type.*

In the Table L-1–Table L-9, we use the same terminology as that of the Table 1–Table 3 in Laufer [11, p. 1287–1294].

3. Non-existence of the case (a)

Assume that there is a compactification $(X, Y) = (V_{22}, H_{22})$ of the case (a) in Proposition 2.5. Let $\pi: \tilde{Y} \to Y$ be the minimal resolution of the singularity $x := \text{Sing}\, Y$ and Z the fundamental cycle of x associated with the resolution $(\tilde{Y}, \pi)$. By assumption, we have $K_{\tilde{Y}} = -Z$ and $(Z \cdot Z) = -3$. The dual graph of $\pi^{-1}(x)$ looks like the Fig. 2, where we denote by ○ (resp. ⊝3) a smooth rational curve with the self-intersection number -2 (resp. -3). We can represent $\tilde{Y}$ as a ruled surface $\nu: \tilde{Y} \to \mathbb{P}^1$ over $\mathbb{P}^1$ (see Fig. 3), where

$$\begin{cases} \nu^{-1}(0) =: \tilde{C} \cup f_1 \cup f_2 \cup f_3 \cup f_4 \cup l_1 \text{ and } \nu^{-1}(\infty) =: \bigcup_{i=5}^{10} f_i \cup \tilde{B} \text{ are} \\ \text{singular fibers.} \end{cases} \tag{3.1}$$

$$l_2 \text{ is a section} \tag{3.2}$$

$$\begin{cases} (l_1 \cdot l_1)_{\tilde{Y}} = (l_2 \cdot l_2)_{\tilde{Y}} = -3, \\ (f_i \cdot f_i)_{\tilde{Y}} = -2 (1 \leqslant i \leqslant 10), \\ (\tilde{C} \cdot \tilde{C})_{\tilde{Y}} = (\tilde{B} \cdot \tilde{B})_{\tilde{Y}} = -1. \end{cases} \tag{3.3}$$

$$\begin{cases} (\tilde{D} \cdot \tilde{C})_{\tilde{Y}} = 2, (\tilde{D} \cdot \tilde{B})_{\tilde{Y}} = 3, \text{ where } D = \pi(\tilde{D}) \text{ is a canonical} \\ \text{hyperplane section such that Pic } Y \cong \mathbb{Z} \cdot \mathcal{O}_Y(D), \text{ in particular,} \\ \deg D = (D \cdot D)_Y = 22. \end{cases} \tag{3.4}$$

$$\pi^{-1}(x) = \bigcup_{i=1}^{10} f_i \cup l_1 \cup l_2, \tag{3.5}$$

$$Z = f_4 + 2f_3 + 2f_2 + 2f_1 + l_1 + 2l_2 + \\ + 3f_5 + 4f_6 + 2f_7 + 3f_8 + 2f_9 + f_{10}. \tag{3.6}$$

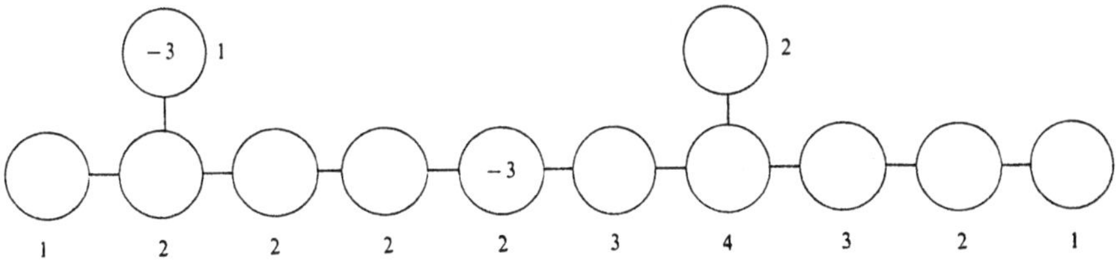

Fig. 2.

LEMMA 3.1.

(1) *there is no line in* X *through the point* $x = \operatorname{Sing} Y \in X$.

(2) $C_0 := \pi(C) \subset Y \subset X$ *is a unique conic on* X *through the point* x.

Proof. Since the multiplicity $m(\mathcal{O}_{Y,x})$ is equal to 3 by Laufer [11] and $\operatorname{Pic} X \cong \mathbb{Z} \cdot \mathcal{O}_X(Y)$, any line or any conic through the point x must be contained in Y. Now, since $(\tilde{C} \cdot \tilde{D})_Y = (C_0 \cdot D)_Y = 2$ and D is a hyperplane section, C_0 is a conic on X. Let F be a line or a conic on X through the point x. Then, we have $F \subset Y$.

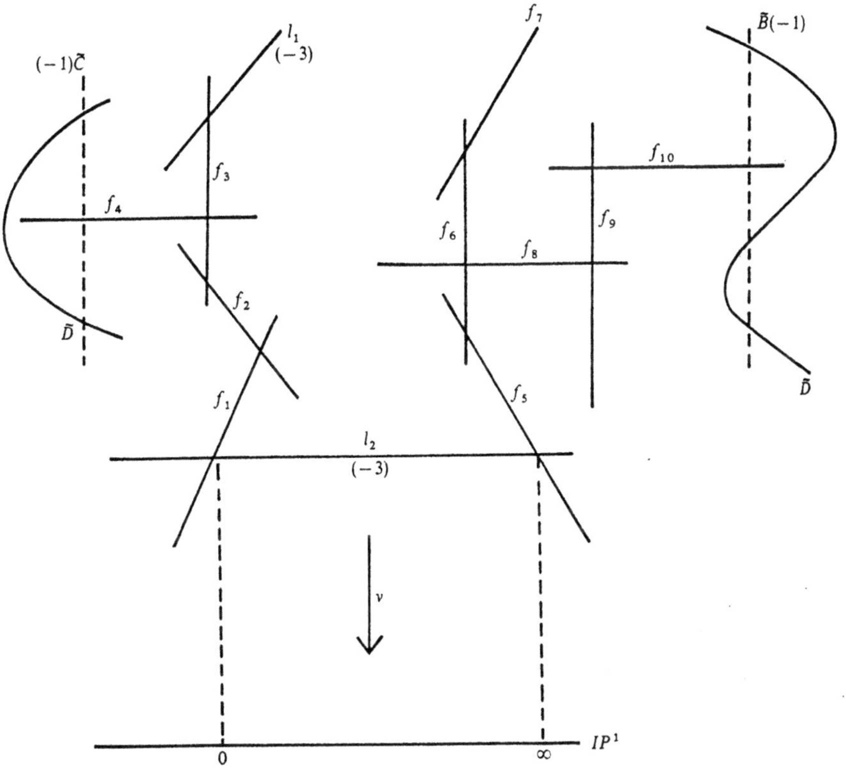

Fig. 3.

Let $\tilde{F}$ be the proper transform of F in $\tilde{Y}$. Since $\tilde{D}$ can be written as follows:

$$\tilde{D} = 2\tilde{C} + 4f_4 + 6f_3 + 2l_1 + 6f_2 + 6f_1 + \\ + 6l_2 + 12f_5 + 18f_6 + 9f_7 + 15f_8 + 12f_9 + 9f_{10} + 6\tilde{B}, \tag{3.7}$$

we have $(\tilde{D}\cdot\tilde{F}) \neq 1$, and also have $(\tilde{D}\cdot\tilde{F}) = 2$ if and only if $\tilde{F} = \tilde{C}$. This proves (1) and (2). Q.E.D.

2. Let us consider the triple projection of $X = V_{22}$ from the singularity $x = \operatorname{Sing} Y \in X$. For this purpose, we will consider the linear system $|H - 3x| := |\mathcal{O}_X(H) \oplus m_{x,X}^3|$, where H is a hyperplane section of X and $m_{x,X}$ is the maximal ideal of the local ring $\mathcal{O}_{X,x}$. Since the multiplicity $m(O_{Y,x})$ is equal to 3, we have $Y \in |H - 3x|$ (c.f. [16a]).

Let $\sigma_1 : X_1 \to X$ be the blowing up of X at the point x, and put $E_1 := \sigma^{-1}(x) \cong \mathbb{P}^2$. Let Y_1 be the proper transform of Y in X_1. Since $-K_X = H$ and $Y \in |H - 3x|$, we have

$$-K_{X_1} = \sigma_1^* H - 2E_1 \tag{3.8}$$

$$Y_1 = \sigma_1^* H - 3E_1. \tag{3.9}$$

By the adjunction formula, we have

$$-K_{Y_1} = E_1|_{Y_1}. \tag{3.10}$$

LEMMA 3.2. *$H^0(X_1, \mathcal{O}_{X_1}(\sigma_1^* H - 3E_1)) \cong \mathbb{C}^4$, and $H^i(X_1, \mathcal{O}_{X_1}(\sigma_1^* H - 3E_1)) = 0$ for $i > 0$.*

Proof. Let us consider the exact sequence

$$0 \to \mathcal{O}_{X_1} \to \mathcal{O}_{X_1}(Y_1) \to \mathcal{O}_{Y_1}(Y_1) \to 0. \tag{3.11}$$

Since $Y_1 = \sigma_1^* H - 3E_1$ and $H^i(X_1, \mathcal{O}_{X_1}) = 0$ for $i > 0$, we have only to prove $H^i(Y_1, \mathcal{O}_{Y_1}(Y_1)) = 0$ for $i > 0$ and $H^0(Y_1, \mathcal{O}_{Y_1}(Y_1)) \cong \mathbb{C}^3$.

By (3.10), we have

$$\mathcal{O}_{Y_1}(Y_1) = \mathcal{O}_{Y_1}(\sigma_1^* H - 3E_1) \\ = \mathcal{O}_{Y_1}(D_1 + 3K_{Y_1}), \tag{3.12}$$

where $D_1 := \sigma_1^* H|_{Y_1}$ is linearly equivalent to the proper transform of D in Y_1.

Claim. $\mathcal{O}_{Y_1}(D + 2K_{Y_1})$ is nef and big on Y_1. Indeed, there exists a birational morphism $\mu_1: \tilde{Y} \to Y_1$ such that $\pi = (\sigma_1|_{Y_1}) \circ \mu_1$. Then, we have $\mu_1^*(D_1 + 2K_{Y_1}) = \tilde{D} - 2Z$. It is easy to see that $\tilde{D} - 2Z$ is nef and big on $\tilde{Y}$ (see (3.6), (3.7)). Thus $\mathcal{O}_{Y_1}(D_1 + 2K_{Y_1})$ is nef and big on Y_1.

By the Kawamata–Vieweg vanishing theorem, we have $H^i(Y_1, \mathcal{O}_{Y_1}(D_1 + 3K_{Y_1})) = 0$ for $i > 0$, namely, $H^i(Y_1, \mathcal{O}_{Y_1}) = 0$ for $i > 0$. On the other hand, since $H^i(\tilde{Y}, \mathcal{O}_{\tilde{Y}}(D - 3Z)) = 0$ for $i > 0$, by the Riemann–Roch theorem, we have $H^0(\tilde{Y}, \mathcal{O}_{\tilde{Y}}(D - 3Z)) \cong \mathbb{C}^3$. Q.E.D.

By Lemma 3.2, the linear system $|H - 3x|$ defines a rational map $\Phi := \Phi_{|H-3x|}: X \dashrightarrow \mathbb{P}^3$, called a triple projection.

Let $\{g_1, g_2, g_3\}$ be a basis of $H^0(\tilde{Y}, \mathcal{O}_{\tilde{Y}}(\tilde{D} - 3Z))$ such that

$$\begin{cases} (g_1) = 11\tilde{C} + 10f_4 + 9f_3 + 2l_1 + 6f_2 + 3f_1 \\ (g_2) = 5\tilde{C} + 4f_4 + 3f_3 + 2f_2 + f_1 + 2f_5 + \\ \qquad + 4f_6 + 2f_7 + 4f_8 + 4f_9 + 4f_{10} + 4\tilde{B} \\ (g_3) = 8\tilde{C} + 7f_4 + 6f_3 + l_1 + 4f_2 + 2f_1 + \\ \qquad + f_5 + 2f_6 + f_7 + 2f_8 + 2f_9 + 2f_{10} + 2\tilde{B} \end{cases} \tag{3.13}$$

Since $2(g_3) = (g_1) + (g_2)$, $g := (g_1 : g_2 : g_3)$ defines a rational map $g: \tilde{Y} \dashrightarrow Q$ of $\tilde{Y}$ onto a conic $Q := \{w_2^2 = w_0 w_1\} \hookrightarrow \mathbb{P}^2(w_0 : w_1 : w_2)$. This implies that $\Phi(Y) = Q \cong \mathbb{P}^1$ and $W = \Phi(X)$ is a quadratic hypersurface in $\mathbb{P}^3$. Thus we have the following

LEMMA 3.3. *Let $\Phi := \Phi_{|H-3x|}: X \dashrightarrow \mathbb{P}^3$ be the triple projection from the point x. Then the image $W := \Phi(X)$ is an irreducible quadric hypersurface in $\mathbb{P}^3$, and $Q = \Phi(Y)$ is a smooth hyperplane section, namely, a conic in $\mathbb{P}^2$.*

3. Next, we will study the resolution of the indeterminancy of the rational map $\Phi: X \dashrightarrow \mathbb{P}^3$.

Let $\Phi^{(1)}_{|\sigma_1^* H - 3E_1|}: X_1 \dashrightarrow \mathbb{P}^3$ be a rational map defined by the linear system $|\sigma_1^* H - 3E_1|$. Then we have the diagram:

$$\begin{array}{ccc} X_1 & & \\ {\scriptstyle \sigma_1}\downarrow & \searrow^{\Phi^{(1)}} & \\ X & \underset{\Phi}{\dashrightarrow} & \mathbb{P}^3. \end{array}$$

Let $\Delta \subset X$ be a small neighborhood of x in X with a coordinate system

(Z_1, Z_2, Z_3). By Laufer [11], we may assume that

$$\begin{cases} \Delta \cap Y = \{Z_2 \cdot Z_3^2 = Z_1^3 \cdot Z_2 + Z_1 Z_3^3 + Z_1 Z_2^4\}, \\ x = (0, 0, 0) \in \Delta. \end{cases} \tag{3.14}$$

By an easy computation, we find that Y_1 has two rational double points q_1 of A_4-type and q_0 of D_6-type (cf. [11, Theorem 3.15]). Let $\mu_1\colon \tilde{Y} \to Y_1$ be the birational map as above. Then we have $\mu_1^{-1}(q_1) = f_1 \cup f_2 \cup f_3 \cup f_4, \mu_1^{-1}(q_0) = \bigcup_{j=5}^{10} f_j$.

We put $l_i^{(1)} := \mu_1(l_i)$, and $C_1 := \mu_1(\tilde{C}) \hookrightarrow Y_1 \subset X_1$. Then C_1 is the proper transform of C_0 in X_1, in particular, C_1 is a smooth rational curve in $Y_1 \subset X_1$ with $q_1 \in C_1, q_0 \notin C_1$. Moreover, we have $Y_1 \cdot E_1 = l_1^{(1)} + 2l_2^{(1)}$, where $l_1^{(1)}, l_2^{(1)}$ are two distinct lines in $E_1 \cong \mathbb{P}^2$, and $l_1^{(1)} \cap l_2^{(1)} = q_1 \in X_1$.

By (3.13), the base locus $Bs|\mathcal{O}_{Y_1}(Y_1)| = C_1 \ni q_1$. Since $H^1(X_1, \mathcal{O}_{X_1}) = 0$ by (3.12), we have the base locus $Bs|\sigma_1^* H - 3E_1| \ni C_1 \ni q_1$.

Since Pic $X \cong \mathbb{Z}\mathcal{O}_X(Y)$, the linear system $|H - 3x|$ has no fixed component. Thus, we have the following

LEMMA 3.4. *The linear system $|\sigma_1^* H - 3E_1|$ on X_1 has no fixed component, but has the base locus $Bs|\sigma_1^* H - 3E_1| = C_1 \ni q_1$.*

We need the following

LEMMA Mo (Morrison [13]). *Let S be a surface with only one singularity x of A_n-type in a smooth projective 3-fold X. Let $E \subset S \subset X$ be a smooth rational curve in X. Let $\mu\colon \tilde{S} \to S$ be the minimal resolution of the singularity of S and put*

$$\mu^{-1}(x) = \bigcup_{j=1}^{n} C_j,$$

where C_j's $(1 \leqslant j \leqslant n)$ are smooth rational curves with

$$\begin{aligned} &(C_j \cdot C_j)_S = -2 && (1 \leqslant j \leqslant n), \\ &(C_j \cdot C_{j+1})_S = 1 && (1 \leqslant j \leqslant n-1), \\ &(C_i \cdot C_j)_S = 0 && \text{if } |i-j| \geqslant 2. \end{aligned}$$

Let $\tilde{E}$ be the proper transform of E in $\tilde{S}$. Assume that

(i) $N_{\tilde{E}|\tilde{S}} \cong \mathcal{O}_{\tilde{E}}(-1)$,

(ii) $\deg N_{E|X} = -2$,

where $N_{\tilde{E}|\tilde{S}}$ (resp. $N_{E|X}$) *is the normal bundle of* $\tilde{E}$ (resp. E) *in* $\tilde{S}$ (resp. X). *Then we have*

(1) $N_{E|X} \cong \mathcal{O}_E \oplus \mathcal{O}_E(-2)$ *if* $x \in E$ *and* $(C_j \cdot \tilde{E}) = 1$ *for* $j = 1$ *or* n,

(2) $N_{E|X} \cong \mathcal{O}_E(-1) \oplus \mathcal{O}_E(-1)$ *if* $x \notin E$.

Proof. In the proof of Theorem 3.2 in Morrison [13], we have only to replace the conormal bundle $N^*_{\tilde{E}|\tilde{S}} = \mathcal{O}_{\tilde{E}}(2)$ by $N_{\tilde{E}|\tilde{S}} = \mathcal{O}_{\tilde{E}}(1)$. Q.E.D.

The indeterminancy of the rational map $\Phi^{(1)}: X_1 \dashrightarrow \mathbb{P}^3$ can be resolved by the following way:

Let us consider the following sequence of blowing ups:

$$X_6 \xrightarrow{\sigma_6} X_5 \xrightarrow{\sigma_5} X_4 \xrightarrow{\sigma_4} X_3 \xrightarrow{\sigma_3} X_2 \xrightarrow{\sigma_2} X_1,$$

where

(i) $\sigma_{j+1}: X_{j+1} \to X_j$ is the blowing up of X_j along $C_j \cong \mathbb{P}^1$ $(1 \leqslant j \leqslant 5)$,

(ii) C_{j+1} is the negative section of the $\mathbb{P}^1$-bundle $C'_j = \mathbb{P}(N^*_{C_j|X_j}) \cong \mathbb{F}_2 (1 \leqslant j \leqslant 4)$,

(iii) C_6 is a section of $C'_5 \cong \mathbb{P}^1 \times \mathbb{P}^1$ with $(C_6 \cdot C_6) = 0$.

Then we have the morphism $\bar{\Phi}: X_6 \to \mathbb{P}^3$ and a diagram:

$$\begin{array}{ccc} X_1 & \xleftarrow{\sigma} & X_6 \\ {\scriptstyle \sigma_1}\downarrow & \Phi^{(1)} & \downarrow{\scriptstyle \bar{\Phi}} \\ X & \underset{\Phi}{\dashrightarrow} & \mathbb{P}^3, \end{array} \tag{D-1}$$

where $\sigma := \sigma_2 \circ \sigma_3 \circ \sigma_4 \circ \sigma_5 \circ \sigma_6$.

This is a desired resolution of the indeterminancy of the rational map

$$\Phi^{(1)}: X_1 \dashrightarrow \mathbb{P}^3 \text{ (or } \Phi: X \dashrightarrow \mathbb{P}^3).$$

4. We will prove the facts above.

Notations:

- Y_{j+1}: the proper transform of Y_j in X_{j+1}.
- E_{j+1}: the proper transform of E_j in X_{j+1}.
- C_{j+1}: a section of $C'_j = \mathbb{P}(N_{C_j|X_j}{}^*)$.
- q_j: the singularity of Y_j of A_{5-j}-type (A_0-type means the smoothness).
- $\tilde{Y}_0$: the contraction of the exceptional set $\bigcup_{i=5}^{10} f_i$ in $\tilde{Y}$.
- $f^{(j+1)}$: a fiber of the $\mathbb{P}^1$-bundle $C'_j \hookrightarrow X_{j+1}$.
- $l_i^{(j+1)}$: the proper transform of $l_i^{(1)}$ in X_{j+1} $(i = 1, 2)$.

- $\mu_j\colon \tilde{Y}_0 \to Y_j$: a birational morphism with μ_j

$$\tilde{Y}_0 - \bigcup_{i=1}^4 f_i \xrightarrow{\sim} Y_j - \mu_j\left(\bigcup_{i=1}^4 f_i\right).$$

Step 1. Let $\sigma_2\colon X_2 \to X_1$ be the blowing up of X_1 along $C_1 \cong \mathbb{P}^1$. Since $(K_{X_1} \cdot C_1) = (\sigma_1^* H - 2E_1 \cdot C_1) = 0$, we have $\deg N_{C_1|X_1} = -2$. Since $q_1 \in C_1 \subset Y_1$ is the singularity of Y_1 of A_4-type and $(\tilde{C} \cdot f_1)_{\tilde{Y}} = 1$, by Lemma Mo, we have

$$N_{C_1|X_1} \cong \mathcal{O}_{C_1} \oplus \mathcal{O}_{C_1}(-2). \tag{3.15}$$

Thus we have $C_1' \cong \mathbb{F}_2$. It is easy to see that Y_2 has two rational double points q_2 of A_3-type and q_0 of D_6-type with $q_2 \in C_2 \subset Y_2, q_0 \notin C_2$. Since $(K_{X_2} \cdot C_2) = 0$, by Lemma Mo, we have

$$N_{C_2|X_1} \cong \mathcal{O}_{C_2} \oplus \mathcal{O}_{C_2}(-2). \tag{3.16}$$

In particular, we have

$$\begin{aligned} &\mu_2^{-1}(q_2) = f_2 \cup f_3 \cup f_4,\\ &\tilde{Y}_0 - (f_2 \cup f_3 \cup f_4) \cong Y_2 - \{q_2\},\\ &\mu_2(\tilde{C}) = C_2,\\ &\mu_2(f_1) = f^{(2)},\\ &\mu_2(l_1) = l_i^{(2)} \quad (i = 1, 2). \end{aligned} \tag{3.17}$$

(*Step* $k, 2 \leqslant k \leqslant 5$). Let σ_k: $X_k \to X_{k-1}$ be the blowing up of X_{k-1} along $C_{k-1} \cong \mathbb{P}^1$. Then Y_k has two rational double points q_k of A_{5-k}-type and q_0 of D_6-type with $q_k \in C_k \subset Y_k, q_0 \notin C_k$ $(k \leqslant 5)$. Since $(K_{X_k} \cdot C_k) = 0$, we have $\deg N_{C_k|X_k} = -2$. By Lemma Mo, we have

$$\begin{aligned} &N_{C_k|X_k} \cong \mathcal{O}_{C_k} \oplus \mathcal{O}_{C_k}(-2) \quad (2 \leqslant k \leqslant 4)\\ &N_{C_5|X_5} \cong \mathcal{O}_{C_5}(-1) \oplus \mathcal{O}_{C_5}(-1) \quad (k = 5). \end{aligned} \tag{3.18}$$

In particular,

$$\begin{aligned} &\mu_k^{-1}(q_k) = f_k \cup \cdots \cup f_4\\ &\tilde{Y}_0 - (f_k \cup \cdots \cup f_4) \cong Y_k - \{q_k\}\\ &\mu_k(\tilde{C}) = C_k,\\ &\mu_k(f_{k-1}) = f^{(k)}. \end{aligned} \tag{3.19}$$

Step 6. Let $\sigma_6: X_6 \to X_5$ be the blowing up of X_5 along $C_5 \cong \mathbb{P}^1$. By (3.18), we have $C'_5 = \sigma_5^{-1}(C_5) \cong \mathbb{P}^1 \times P^1$. Then we have an isomorphism $\mu_6: \tilde{Y}_0 \cong Y_6$. We identify $\tilde{Y}_0$ with Y_6 (see Fig. 6). Thus we put

$$\begin{aligned} &\mu_6(\tilde{C}) =: \tilde{C}, \quad \mu_6(\tilde{D}) =: \tilde{D} \\ &\mu_6(f_i) =: f_i, \quad \mu_6(l_i) =: l_i. \end{aligned} \tag{3.20}$$

Then $\tilde{C} = Y_6 \cdot C'_5$ gives another ruling on C'_5. Let $\bar{C}'_j$ $(1 \leqslant j \leqslant 4)$ be the proper transform of C'_j in X_6. Then we have Figure 4 (see also Pagoda (5.8) in Reid [20]).

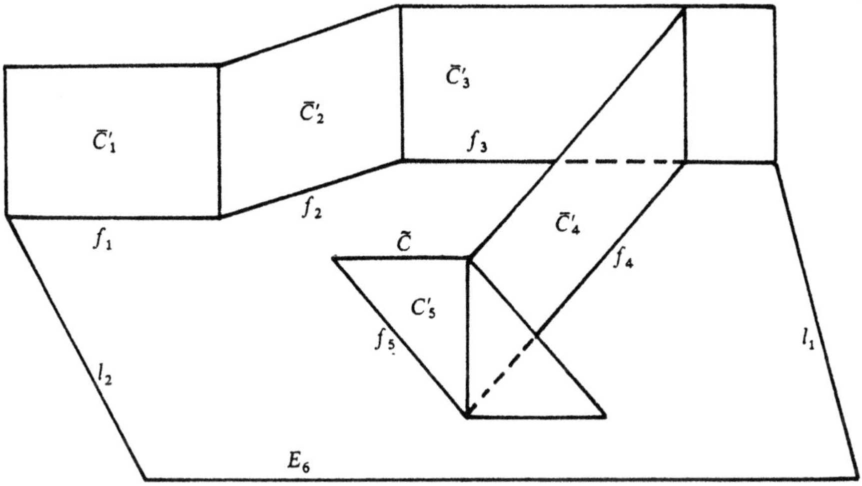

Figure 4

Now, since

$$\begin{aligned} Y_6 = {} & \sigma_6^*\sigma_5^*\sigma_4^*\sigma_3^*\sigma_2^*\sigma_1^* H - 3\sigma_6^*\sigma_5^*\sigma_4^*\sigma_3^*\sigma_2^* E_1 \\ & -5C'_5 - 4\bar{C}'_4 - 3\bar{C}'_3 - 2\bar{C}'_2 - \bar{C}'_1, \end{aligned} \tag{3.21}$$

we have

$$\begin{aligned} \mathcal{O}_{Y_6}(Y_6) &= \mathcal{O}_{Y_6}(\tilde{D} + 3K_{Y_6} - 5\tilde{C} - 4f_4 - 3f_3 - 2f_2 - f_1) \\ &\cong \mathcal{O}_{\tilde{Y}_0}(\tilde{D} - 3Z - 5\tilde{C} - 4f_4 - 3f_3 - 2f_2 - f_1) \\ &\cong \mathcal{O}_{\tilde{Y}_0}(2f), \end{aligned} \tag{3.22}$$

where f is a general fiber of $v: \tilde{Y} \to \mathbb{P}^1$(see Fig. 4). This shows that the linear

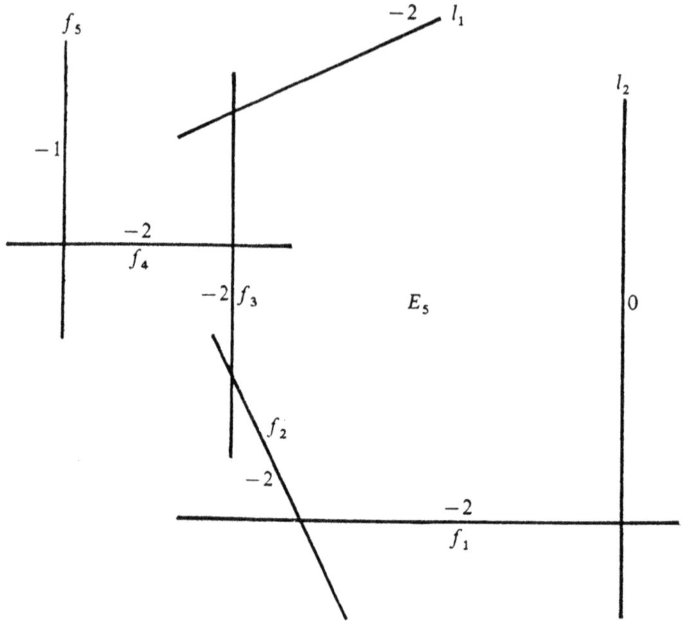

Figure 5

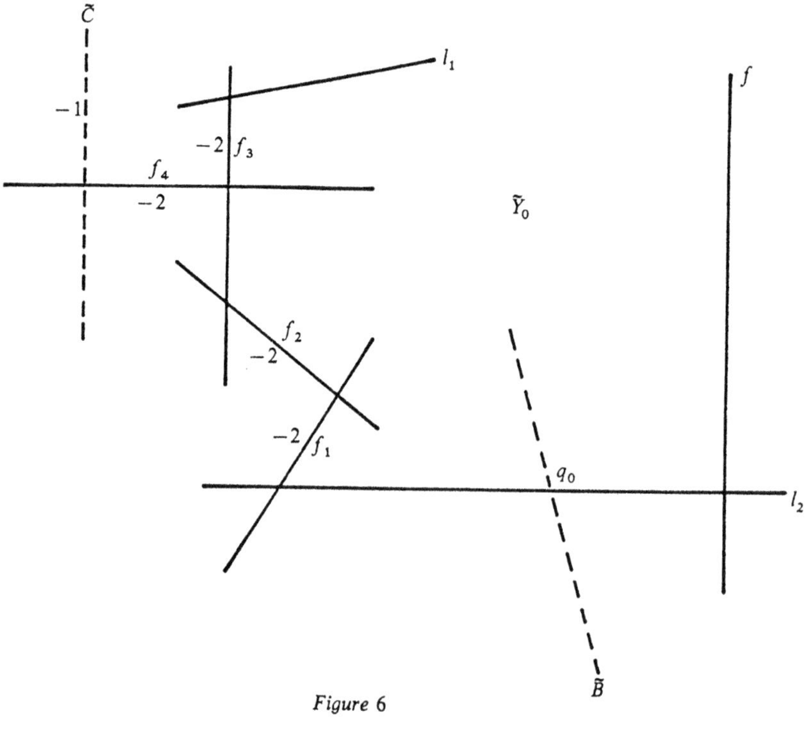

Figure 6

system $|\mathcal{O}_{Y_6}(Y_6)|$ has no fixed component and no base point. Therefore it defines a morphism $v_0: Y_6 = \tilde{Y}_0 \to Q$ of Y_6 onto a smooth conic $Q \cong \mathbb{P}^1$ in $\mathbb{P}^2$. Since $H^1(X_6, \mathcal{O}_{X_6}) = 0$ and Pic $X \cong \mathbb{Z}$, the linear system $|Y_6| = |\mathcal{O}_{X_6}(Y_6)|$ has no base locus. Therefore we have a morphism $\bar{\Phi} := \bar{\Phi}_{|Y_6|}: X_6 \to W \hookrightarrow \mathbb{P}^3$ defined by the linear system $|Y_6|$, and have the diagram (D-1), which is desired. It is easy to see that

$$\begin{aligned} &\Phi(Y_6) = \Phi(Y) = v_0(Y_6) = Q \cong l_2 \\ &\bar{\Phi}\left(\bigcup_{j=1}^{4} \bar{C}'_j \cup C'_5\right) = \bar{\Phi}(f_5) \quad \text{(a line in } \mathbb{P}^3). \end{aligned} \tag{3.23}$$

5. Since $N_{C_5|X_5} \cong \mathcal{O}_{C_5}(-1) \oplus \mathcal{O}_{C_5}(-1)$, by Reid [20], C'_5 can be blown down along C, and then the blowing downs can be done step-by-step. Finally, we have a smooth projective 3-fold V with $b_2(V) = 2$, morphisms $\bar{\Phi}_1: X_6 \to V$, $\bar{\Phi}_2: V \to W$, and a birational map $\rho: X_1 \to V$, called a flip, such that
(i) $\bar{\Phi} = \bar{\Phi}_2 \circ \bar{\Phi}_1$,
(ii) $X_1 - C_1 \overset{\rho}{\cong} (V - \bar{f}_3$, where $\bar{f}_3 := \Phi_1(f_3)$, (see (D-2)).

(D-2)

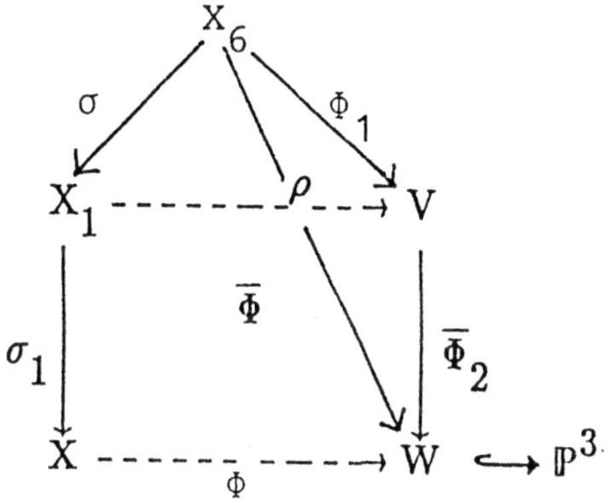

Since $-K_{X_1} = Y_1 + E_1$, by (ii) above, we have $-K_V = A + \Sigma$, where $A := \bar{\Phi}(Y_6)$ and $\Sigma := \bar{\Phi}_1(E_6)$. For a general fiber F of $\bar{\Phi}_2: V \to W$, since $\deg(K_F) = (K_V \cdot F) = -(\Sigma \cdot F) \leqslant -1$, we have $F \cong \mathbb{P}^1$ and $(\Sigma \cdot F) = 2$. Since $\bar{\Phi}_1(l_1)$ is a smooth rational curve contained in Σ, and since $\bar{\Phi}_2 \circ \bar{\Phi}_1(l_1) = \bar{\Phi}(l_1)$ is a point, Σ is a meromorphic double section of $\bar{\Phi}_2: V \to W$.

Let G be a scheme-theoric fiber. Then we have $(G \cdot \Sigma) = 2$. Since $V - (\Sigma \cup A) \cong X_1 - (Y_1 \cup E_1) \cong \mathbb{C}^3$ by assumption, it contains no compact analytic curve. Thus $\bar{\Phi}_2: V \to W$ is a conic bundle over W, and $\bar{\Phi}_2$ is the contraction of an extremal ray on V. Thus, W is smooth by Mori [12]. Since $\deg W = 2$, $W \cong \mathbb{P}^1 \times \mathbb{P}^1$, hence, $b_2(V) = 3$. This is a contradiction, since $b_2(V) = b_2(X_1) = 2$. Therefore we have:

Conclusion

The case (a) of Proposition 2.5 can not occur.

4. Non-existence of the case (b)

1. Assume that there is a compactification $(X, Y) = (V_{22}, H_{22})$ of the case (b) in Proposition 2.5. Then we have Sing $Y = \{x, y\}$, where x is a minimally elliptic singularity of *Cu*-type (Table L-3, (7)), and y is a rational double point of A_{10}-type. Let π: $\tilde{Y} \to Y$ be the minimal resolution of the singularities of Y and put $\tilde{C} := \pi^{-1}(x), \pi^{-1}(y) =: \bigcup_{j=1}^{10} B_j$. Then $\tilde{C}$ is an irreducible rational curve with a cusp, and $K_{\tilde{Y}} = -\tilde{C}, (\tilde{C} \cdot \tilde{C})_{\tilde{Y}} = -2$. We can easily see that $\tilde{Y}$ can be obtained from $\mathbb{P}^2$ by succession of 11 blowing ups at a smooth point p on a cubic curve $C_0 \hookrightarrow \mathbb{P}^2$ with a cusp (infinitely near points allowed). Let μ: $\tilde{Y} \to \mathbb{P}^2$ be the projection. Then $\tilde{C}$ is the proper transform of C_0 in $\tilde{Y}$ and $\mu^{-1}(p) = \bigcup_{j=1}^{11} B_j$, where B_{11} is the exceptional curve of the first kind (see Fig. 7).

We take sufficiently general hyperplane section H such that $D := H \cdot Y$ does not pass through the points x and y. Then D is a canonical curve of the genus $g = 12$ with $\deg D = 22$ in Y. Let $\tilde{D}$ be the proper transform of D in $\tilde{Y}$. Then we have

$$(\tilde{D} \cdot \tilde{D})_{\tilde{Y}} = \deg D = 22 \tag{4.1}$$

$$\tilde{D} = 3\tilde{C} + 2\tilde{G}, \tag{4.2}$$

where $\tilde{G}$ is the proper transform of a line $G \hookrightarrow \mathbb{P}^2$ with $p \notin G$ in $\tilde{Y}$. In particular, $(\tilde{G} \cdot \tilde{G})_{\tilde{Y}} = 1$.

LEMMA 4.1. *There is no line in X through the point x.*

Proof. Since the multiplicity $m(\mathcal{O}_{Y,x})$ is equal to two, any line through the point x is contained in Y. Let g be such a line in X, and $\tilde{g}$ be the proper transform of Y in $g \hookrightarrow \tilde{Y}$. Since $(D \cdot \tilde{g})_{\tilde{Y}} = (D \cdot g)_Y = (H \cdot g)_X = 1$, by (4.2), we have

$$3(\tilde{C} \cdot \tilde{g}) + 2(\tilde{G} \cdot \tilde{g}) = 1. \tag{4.3}$$

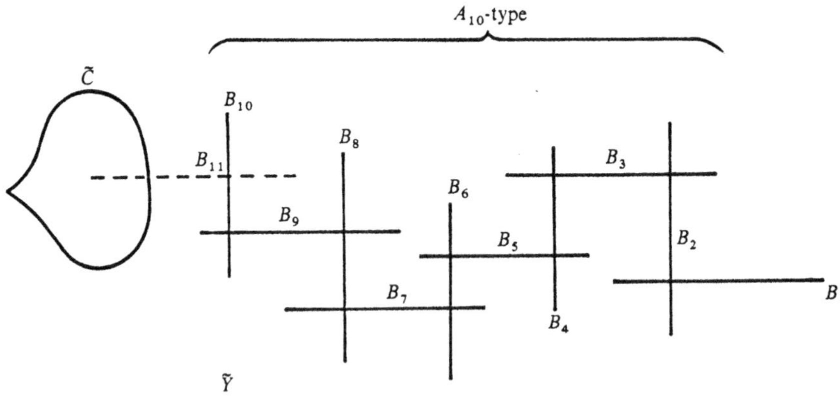

Figure 7

This is a contradiction. Q.E.D.

2. Let $\sigma: X_1 \to X$ be the blowing up of X at the point x, and put $E := \sigma^{-1}(x) \cong \mathbb{P}^2$. Let $Y_1 = \sigma^* H - 2E$ be the proper transform of Y in X. Then we have:

(i) $Y_1 \cdot E = 2l$, where l is a line in $E \cong \mathbb{P}^2$,
(ii) $\operatorname{Sing} Y_1 = l$,
(iii) $N_{l|X} \cong \mathcal{O}_l(-1) \oplus \mathcal{O}_l(1)$.

Let $\tau: X_2 \to X_1$ be the blowing up of X_1 along $l \cong \mathbb{P}^1$. By (iii) above, we have $L' := \tau^{-1}(l) \cong \mathbb{F}_2$. Let us denote the negative section (resp. a fiber) by s (resp. f). Let Y_2 be the proper transform of Y_1 in X_2. Then we have $Y_2 = \tau^* Y_1 - 2L'$. Let $\tilde{Y}_0$ be the contraction of the exceptional curve $\bigcup_{j=1}^{10} B_j$ in $\tilde{Y}$. Then $\tilde{Y}_0$ has a rational double point of A_{10}-type. By an easy computation, we have an isomorphism $\tilde{Y}_0 \overset{\nu}{\to} Y_2$. We identify Y_2 with $\tilde{Y}_0$ *via* ν. For simplicity, we put $\tilde{C} := \nu(\tilde{C}), \tilde{D} := \nu(\tilde{D}), \tilde{G} := \nu(\tilde{G})$. Then we have

$$L' \cdot Y_2 = \tilde{C}, \tag{4.4}$$

$$(\tilde{C} \cdot f)_{L'} = 2, (\tilde{C} \cdot s)_{L'} = 0, \tag{4.5}$$

(see Fig. 8).

3. We will study the linear system $|\mathscr{L}| := |\tau^* Y_1 - L'|$ on X_2.
Let us consider the exact sequence

$$0 \to \mathcal{O}_{X_2}(L') \to \mathcal{O}_{X_2}(\mathscr{L}) \to \mathcal{O}_{Y_2}(\mathscr{L}) \to 0. \tag{4.6}$$

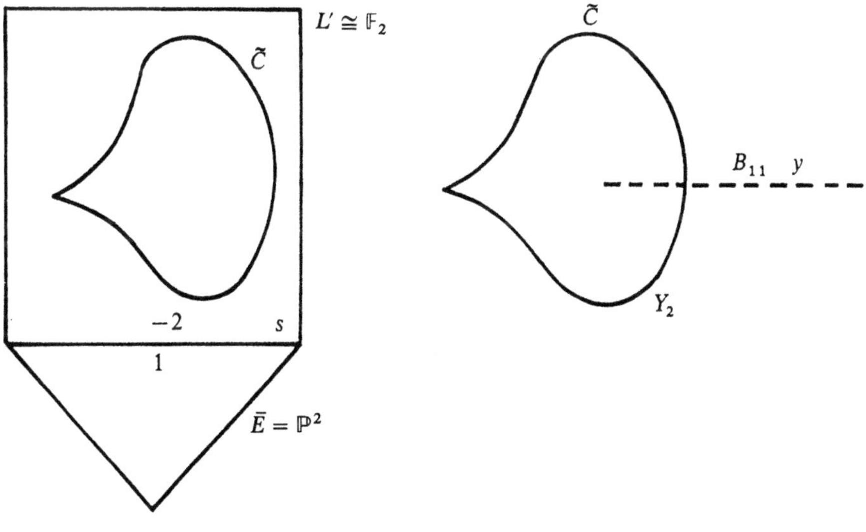

Figure 8

Since $\tau^* Y_1 - L' = \tau^*\sigma^* H - 2\bar{E} - 3L'$, we have

$$\mathcal{O}_{Y_2}(\mathcal{L}) = \mathcal{O}_{Y_2}(\tau^* Y_1 - L') = \mathcal{O}_{Y_2}(\tilde{D} - 3\tilde{C}) = \mathcal{O}_{Y_2}(2\tilde{G})$$

by (4.2), where $\bar{E} \cong \mathbb{P}^2$ is the proper transform of $E \cong \mathbb{P}^2$ in X_2. Since $H^i(Y_2, \mathcal{O}_{Y_2}(2\tilde{G})) = 0$ for $i > 0$, by the Riemann–Roch theorem, we have $H^0(Y_2, \mathcal{O}_{Y_2}(\mathcal{L})) \cong \mathbb{C}^6$. Since $H^0(X_2, \mathcal{O}(X_2, \mathcal{O}_{Y_2}(L')) \cong \mathbb{C}$ and $H^1(X_2, \mathcal{O}_{X_2}(L')) = 0$, we have finally the following exact sequence:

$$\begin{array}{ccccccc} 0 \to & H^0(X_2, \mathcal{O}_{X_2}(L')) & \to & H^0(X_2, \mathcal{O}_{X_2}(\mathcal{L})) & \to & H^0(Y_2, \mathcal{O}_{Y_2}(\mathcal{L})) & \to 0 \\ & \wr\| & & \wr\| & & \wr\| & \\ & \mathbb{C} \cong H^0(X_2, \mathcal{O}_{X_2}) & & \mathbb{C}^7 & & \mathbb{C}^6 & \end{array} \tag{4.6}$$

Since $\dim|\mathcal{L}| = 6$, we have a rational map $\Phi := \Phi_{|\mathcal{L}|}\colon X_2 \to \mathbb{P}^6$ defined by the linear system $|\mathcal{L}|$.

Since the linear system $|\mathcal{O}_{Y_2}(2\tilde{G})|$ has no base locus on Y_2, neither does $|\mathcal{L}|$ by (4.6). Therefore $\Phi\colon X_2' \to \mathbb{P}^6$ is a morphism X_2 to $\mathbb{P}^6$ with

$$\Phi|_{Y_2} = \varphi := \varphi_{|\mathcal{O}_{Y_2}(2\tilde{G})|}\colon Y_2 \to \mathbb{P}^5,$$

where $\varphi_{|\mathcal{O}_{Y_2}(2\tilde{G})|}$ is a morphism defined by $|\mathcal{O}_{Y_2}(2\tilde{G})|$.

Thus we have the following:

LEMMA 4.2. *$\Phi\colon X_2 \to \Phi(X_2) \hookrightarrow \mathbb{P}^6$ is a morphism of X_2 onto a 3-fold $\Phi(X_2)$ of degree 4 in $\mathbb{P}^6$. Moreover, the restriction $\Phi|_{Y_2}\colon Y_2 \to \Phi(Y_2) = \varphi(Y_2) \hookrightarrow \mathbb{P}^5$ gives an birational morphism of Y_2 onto a surface $\varphi(Y_2)$ of degree 4 in $\mathbb{P}^5$.*

Proof. Since $(\tau^* Y_1 - L')^3 = 4$, we have $\deg \Phi(X_2) = 4$. Q.E.D.

$$\begin{array}{ccccc} 0 \to & \mathcal{O}_{Y_2}(\tau^* Y_1 - 2L')) & \to \mathcal{O}_{X_2}(\mathcal{L}) \to & \mathcal{O}_{L'}(\mathcal{L}) & \to 0 \\ & \wr\| & & \wr\| & \\ & \mathcal{O}_{X_2}(Y_2) & & \mathcal{O}_{L'}(s + 3f) & \end{array} \tag{4.7}$$

Since $H^1(Y_2, \mathcal{O}_{Y_2}, \mathcal{O}_{Y_2}(Y_2)) \cong H^1(Y_2, \mathcal{O}_{Y_2}(2\tilde{G} - \tilde{C})) = 0$, we have a surjection

$$H^0(X_2 \mathcal{O}_{X_2}(\mathcal{L})) \twoheadrightarrow H^0(L', \mathcal{O}_{L'}(s + 3f)) \cong \mathbb{C}^6. \tag{4.8}$$

Since $\mathcal{O}_{L'}(s + 3f)$ is very ample on L', the morphism $\rho := \rho_{|s+3f|}\colon L' \to \rho(L') \hookrightarrow \mathbb{P}^5$ is an isomorphism of L' onto a smooth surface of degree 4 in $\mathbb{P}^5$. Thus we have the following

LEMMA 4.3. *The restriction* $\Phi|_{L'}: L' \to \Phi(L') \hookrightarrow \mathbb{P}^5$ *is an isomorphism of* L' *onto a smooth surface* $\Phi(L')$ *of degree* 4 *in* $\mathbb{P}^5$.

Finally, let us consider the exact sequence:

$$\begin{array}{ccccccc} 0 \to & \mathcal{O}_{X_2}(\tau^*(Y_1 - E)) & \to & \mathcal{O}_{X_2}(\mathcal{L}) & \to & \mathcal{O}_{\bar{E}}(\mathcal{L}) & \to 0 \\ & \wr\| & & & & \wr\| & \\ & \mathcal{O}_{X_2}(\tau^*(\sigma^* H - 3E)) & & & & \mathcal{O}_{\mathbb{P}^2}(1) & \end{array} \tag{4.9}$$

Then we have $H^1(X_2, \mathcal{O}_{X_2}(\tau^*(\sigma^* H - 3E))) = 0$, namely, we have a surjection

$$H^0(X_2, \mathcal{O}_{X_2}(\mathcal{L})) \twoheadrightarrow H^0(\bar{E}, \mathcal{O}_{\bar{E}}(\mathcal{L})) \cong \mathbb{C}^3. \tag{4.10}$$

Thus we have the following

LEMMA 4.4. *The restriction* $\Phi|_{\bar{E}}: \bar{E} \to \mathbb{P}^2$ *gives an isomorphism of* $\bar{E}$ *onto* $\mathbb{P}^2$.

5. Let γ be an irreducible curve in X_2 such that $(\tau^* Y_1 - L' \cdot \gamma) = 0$. Since $\tau^* Y_1 - L' = Y_2 + L'$, we have $(Y_2 \cdot \gamma) + (L' \cdot \gamma) = 0$. By Lemma 4.2, Lemma 4.3, Lemma 4.4, $\gamma \not\subset Y_2 \cup L'$. Thus $(Y_2 \cdot \gamma) = (L' \cdot \gamma) = 0$, namely, $(Y_1 \cdot \tau(\gamma)) = 0$. Hence, we have $Y_1 \cap \tau(\gamma) = \phi$ and $E \cap \tau(\gamma) \neq \phi$. This shows that there is no irreducible surface T in X_2 such that $\dim \Phi(T) \leqslant 1$. There are a finite numbers of conics in X through the point x (see [8]). Let γ be the proper transform of a conic in X through x. Then $\dim \Phi(\gamma) = 0$. In particular, there are a finite number of irreducible curves γ' in X_2 such that $\dim \Phi(\gamma') = 0$. Therefore we have the following

LEMMA 4.5. $\Phi: X_2 \to W := \Phi(X_2) \hookrightarrow \mathbb{P}^6$ *is a birational morphism of* X_2 *onto a 3-fold* W *of degree* 4 *in* $\mathbb{P}^6$. *In particular,* $b_2(X_2) = b_2(W) = 3$.

6. Since $\deg W = 4$ in $\mathbb{P}^6$, we have an equality

$$\deg W = \operatorname{codim} W + 1. \tag{4.11}$$

Since there is a smooth rational curve γ in X_2 such that $\dim \Phi(\gamma) = 0$ and $b_2(X_2) = b_2(W)$, the 3-fold W has a finite number of isolated singularities. Thus, W is a cone over a rational scroll or a cone over the Veronese surface. Hence, $b_2(W) = 1$. This is a contradiction, since $b_2(W) = 3$ by Lemma 4.5. Thus, we have:

Conclusion

The case (b) of Proposition 2.5 can not occur.

We have proved in Section 3 that the case (a) of Proposition 2.5 can not occur.

Therefore, in the case of the index $r = 1$, such a compactification of $\mathbb{C}^3$ does not exist. Thus we have the Theorem (see the Introduction).

5. Remarks and an example

1. Let (X, Y) be an analytic compactification of $\mathbb{C}^3$. Then we have (cf. [2], [3]):

$$
\begin{array}{lcl}
Y \text{ has at most isolated singularities.} & & \\
\quad\Updownarrow & & \\
Y \text{ is normal.} & \overset{\Rightarrow}{\not\Leftarrow} & Y \text{ is projective.} \\
\quad\Downarrow\ \not\Uparrow & & \\
Y \text{ is irreducible.} & \Leftrightarrow b_2(X) = 1 &
\end{array}
$$

In the case where Y is normal, we have determined the complete structure of such a (X, Y) (see Theorem in the Introduction).

On the other hand, we know that there is a non-normal hyperplane section E_5 of the Fano 3-fold V_5 such that $V_5 - E_5 \cong \mathbb{C}^3$([3]). This gives an example of a compactification (X, Y) of $\mathbb{C}^3$ with a non-normal irreducible boundary Y.

Recently, Peternell–Schneider [18] and Peternell [19] proved the following

THEOREM 5.1. *Let (X, Y) be a projective compactification of $\mathbb{C}^3$ with $b_2(X) = 1$. Assume that Y is non-normal. Then, X is a Fano 3-fold of the index $r(1 \leqslant r \leqslant 2)$, and*

(i) $r = 2 \Rightarrow (X, Y) \cong (V_5, E_5)$ *(up to isomorphism).*

(ii) $r = 1 \Rightarrow X \cong V_{22} \hookrightarrow \mathbb{P}^{13}$(or $V'_{22} \hookrightarrow \mathbb{P}^{12}$)
(Mukai–Umemura [14]).

2. Finally, we will prove that there is a non-normal hyperplane section H'_{22} of V'_{22} such that $V'_{22} - H'_{22} \cong \mathbb{C}^3$. Let $(a_0 : a_1 : \ldots : a_{12})$ be a homogeneous coordinate of $\mathbb{P}^{12}$. Then $V'_{22} \hookrightarrow \mathbb{P}^{12}$ can be written as follow (see p. 506 in [14]):

$$a_0 a_4 - 4a_1 a_3 + 3a_2^2 = 0$$

$$a_0 a_5 - 3a_1 a_4 + 2a_2 a_3 = 0$$

$$7a_0 a_6 - 12a_1 a_5 - 15a_2 a_4 + 20a_3^2 = 0$$

$$a_0 a_7 - 6a_2 a_5 + 5a_3 a_4 = 0$$

$$5a_0 a_8 + 12a_1 a_7 - 42a_2 a_6 - 20a_3 a_5 + 45a_4^2 = 0$$

$$a_0 a_9 - 6a_1 a_8 - 6a_2 a_7 - 28a_3 a_6 + 28a_4 a_5 = 0$$

$$a_0 a_{10} + 12a_1 a_9 + 12a_2 a_8 - 76a_3 a_7 - 21a_4 a_6 + 72a_5^2 = 0$$

$$a_0a_{11} + 24a_1a_{10} + 90a_2a_9 - 130a_3a_8 - 405a_4a_7 + 420a_5a_6 = 0$$
$$a_0a_{12} + 60a_1a_{11} + 534a_2a_{10} + 380a_3a_9 - 3195a_4a_8 - 720a_5a_7 + 2940a_6^2 = 0 \; (*)$$
$$a_1a_{12} + 24a_2a_{11} + 90a_3a_{10} - 130a_4a_9 - 405a_5a_8 + 420a_6a_7 = 0$$
$$a_2a_{12} + 12a_3a_{11} + 12a_4a_{10} - 76a_5a_9 - 21a_6a_8 + 72a_7^2 = 0$$
$$a_3a_{12} - 6a_4a_{11} - 6a_5a_{10} - 28a_6a_9 + 28a_7a_8 = 0$$
$$5a_4a_{12} + 12a_5a_{11} - 42a_6a_{10} - 20a_7a_9 + 45a_8^2 = 0$$
$$a_5a_{12} - 6a_7a_{10} + 5a_8a_9 = 0$$
$$7a_6a_{12} - 12a_7a_{11} - 15a_8a_{10} + 20a_9^2 = 0$$
$$a_7a_{12} - 3a_8a_{11} + 2a_9a_{10} = 0$$
$$a_8a_{12} - 4a_9a_{11} + 3a_{10}^2 = 0$$

In the affine part $\{a_0 = 1\} \cong \mathbb{C}^{12}(a_1, \ldots, a_{12})$, let us consider the following coordinate transformation:

$$\begin{aligned}
x_1 &= a_1 \\
x_2 &= a_2 \\
x_3 &= a_3 \\
x_4 &= a_4 - 4a_1a_3 + 3a_2^2 \\
x_5 &= a_5 - 3a_1a_4 + 2a_2a_3 \\
x_6 &= 7a_6 - 12a_1a_5 - 15a_2a_4 + 20a_3^2 \qquad (**) \\
x_7 &= a_7 - 6a_2a_5 + 5a_3a_4 \\
x_8 &= 5a_8 + 12a_1a_7 - 42a_2a_6 - 20a_3a_5 + 45a_4^2 \\
x_9 &= a_9 - 6a_1a_8 - 6a_2a_7 - 28a_3a_6 + 28a_4a_5 \\
x_{10} &= a_{10} + 12a_1a_9 + 12a_2a_8 - 76a_3a_7 - 21a_4a_6 + 72a_5^2 \\
x_{11} &= a_{11} + 24a_1a_{10} + 90a_2a_9 - 130a_3a_8 - 405a_4a_7 + 420a_5a_6 \\
x_{12} &= a_{12} + 60a_1a_{11} + 534a_2a_{10} + 380a_3a_9 - 3150a_4a_8 - 720a_5a_7 + 2940a_6^2
\end{aligned}$$

Then the Jacobian $|\partial(x_1, \ldots, x_{12})/\partial(a_1, \ldots, a_{12})| = 35 \neq 0$, and further we have

$$V'_{22} \cap \{a_0 = 1\} \cong \{(x_1, \ldots, x_{12});\, x_j = 0\ (4 \leqslant j \leqslant 12)\} = \mathbb{C}^3(x_1, x_2, x_3).$$

We put $H'_{22} := V_{22} \cap \{a_0 = 0\}$. Then H'_{22} is non-normal.

Therefore the pair (V'_{22}, H'_{22}) is a compactification of $\mathbb{C}^3$ with a non-normal boundary.

One can easily see that the singular locus of H'_{22} is a line in V'_{22}.

Question 1. Is there a non-normal hyperplane section E_{22} in V_{22} ($\neq V'_{22}$) such that $V_{22} - E_{22} \cong \mathbb{C}^3$?

Acknowledgements

The author would like to thank the Max-Planck-Institute für Mathematik in Bonn, especially Professor F. Hirzebruch for hospitality and encouragement, and would also like to thank Professor N. Nakayama for the stimulating conversations we had and for his valuable comments.

References

1. M. Artin, On isolated rational singularities of surfaces, *Amer. J. Math.* 88 (1966) 129–136.
2. L. Brenton, Some algebraicity criteria for singular surfaces, *Invent. Math.* 41 (1977) 129–147.
3. M. Furushima, Singular del Pezzo surfaces and complex analytic compactifications of the 3-dimensional complex affine space $\mathbb{C}^3$, *Nagoya Math. J.* 104 (1986) 1–28.
4. M. Furushima, Singular K3 surfaces with hypersurface singularities, Pacific J. Math. 125 (1986) 67–77.
5. M. Furushima and N. Nakayama, A new construction of a compactification of $\mathbb{C}^3$, *Tohoku Math. J.* 41 (1989), 543–560.
6. M. Furushima and N. Nakayama, The family of lines in the Fano 3-fold V_5, to appear in *Nagoya Math. J.* 116 (1989).
7. M. Furushima, Complex analytic compactifications of $\mathbb{C}^3$, *Proc. Japan Akad.* 64 Ser. A (1988) 25–26.
8. V.A. Iskovskih, Fano 3-fold I, *Math. U.S.S.R. Izvestija* 11 (1977) 485–527.
9. V.A. Iskovskih, Anticanonical models of three-dimensional algebraic varieties, *J. Soviet Math.* 13–14 (1980) 745–814.
10. Y. Kawamata, K. Matsuda and K. Matsuki, Introduction to the minimal model problem, in Algebraic Geometry, Sendai, *Advanced Studies in Pure Math.* 10, Kinokuniya, Tokyo and North Holland, Amsterdam (1987) 551–590.
11. H. Laufer, On minimally elliptic singularities, *Amer. J. Math.* 99 (1977) 1257–1295.
12. S. Mori, Threefolds whose canonical bundles are not numerical effective, *Ann. Math.* 116 (1982) 133–176.
13. D. Morrison, The birational geometry of surfaces with rational double points, *Math. Ann.* 271 (1985) 415–438.
14. S. Mukai and H. Umemura, Minimal rational threefolds, *Lecture Notes in Mathematics* 1016, Springer Verlag, Berlin, Heidelberg, New York (1983) 490–518.
15. S. Mukai, Curves, K3 surfaces and Fano 3-folds of genus $\leqslant 10$, Algebraic Geometry and Commutative Algebra in Honor of Masayoshi Nagata (1988), to appear.
16. S. Mukai, New classification of Fano 3-folds and Fano manifolds of coindex 3, Preprint (1980).
16a. S. Mukai: Private letter, May 1986.
17. D. Mumford, Topology of normal singularities of an algebraic surface and a criterion for simplicity, *Publ. Math. I.H.E.S.* 9 (1961) 5–22.
18. T. Peternell and M. Schneider, Compactifications of $\mathbb{C}^3$ (I), *Math. Ann.* 280 (1988) 129–146.
19. T. Peternell, Compactifications of $\mathbb{C}^3$ (II), *Math. Ann.* 283 (1989), 121–137.
20. M. Reid, Minimal models of canonical 3-folds, in Algebraic Varieties and Analytic varieties, *Advanced studies in Pure Math.* 1, Kinokuniya, Tokyo and North-Holland, Amsterdam (1983) 131–180.
21. J.P. Serre, Groupes algébriques et corps de classes, *Actualités Sci. Indust.* no. 1264. Hermann, Paris (1959).
22. Y. Umezu, On normal projective surfaces with trivial dualizing sheaf, *Tokyo J. Math.* 4 (1981) 343–354.
23. P. Wagreich, Elliptic singularities of surfaces, *Amer. J. Math.* 2 (1970) 419–454.
24. S.T. Yau, On maximally elliptic singularities, *Trans. A.M.S.* 257 (1980) 269–329.

Compositio Mathematica **76**: 197–201, 1990.

Analytic curves in power series rings

HERWIG HAUSER[1] & GERD MÜLLER[2]

[1]*Institut für Mathematik, Universität Innsbruck, 6020 Innsbruck, Austria;* [2]*Fachbereich Mathematik, Universität Mainz, 6500 Mainz, FRG*

Received 3 December 1988; accepted 20 July 1989

Let us state a standard result on algebraic group actions:

PROPOSITION. *For an analytic map germ $\gamma: S \to (V, v)$, S a reduced analytic space germ and (V, v) the germ in v of a finite dimensional complex vector space V, together with an algebraic subgroup G of* $\mathrm{GL}(V)$ *the following holds:*

(i) *The germ T of points t in S for which $\gamma(t)$ lies in the orbit $G \cdot v$ of G through v is analytic.*

(ii) *There is an analytic map germ $\phi: T \to (G, 1)$ such that $\gamma(t) = \phi(t) \cdot v$ for all t in T.*

Indeed, the orbit $G \cdot v$ is a locally closed submanifold of V, isomorphic to the homogeneous manifold G/G_v via the orbit map $G/G_v \to G \cdot v$, and the natural map $G \to G/G_v$ admits local sections.

The second part of the Proposition asserts that every analytic curve in a G-orbit is locally induced from an analytic curve in G. The object of the present article is to establish this statement in the case where $V = \mathcal{O}_n^p$ is a finite free module over the $\mathbb{C}$-algebra $\mathcal{O}_n$ of convergent power series in n variables and $G = \mathcal{K} = \mathrm{GL}_p(\mathcal{O}_n) \rtimes \mathrm{Aut}\, \mathcal{O}_n$ is the contact group acting naturally on $\mathcal{O}_n^p$. Recall that the orbits of $\mathcal{K}$ through f in $\mathcal{O}_n^p$ just correspond to the isomorphism classes of the analytic space germs in $(\mathbb{C}^n, 0)$ defined by f. Thus we shall obtain analytic trivializations of local families of space germs whose members lie in the same isomorphism class.

DEFINITION. Let S always denote a *reduced* analytic space germ. A map germ $\gamma: S \to E$ with values in some subset E of $\mathcal{O}_n^p$ is called analytic if there is an analytic map germ $G: (\mathbb{C}^n, 0) \times S \to \mathbb{C}^p$ such that $\gamma(s)(x) = G(x, s)$.

Choosing coordinates on $(\mathbb{C}^n, 0)$, the group $\operatorname{Aut}\mathcal{O}_n$ can be considered as a subset of $\mathcal{O}_n^n$. The analyticity of a map germ with values in $\operatorname{Aut}\mathcal{O}_n$ does not depend on this choice (cf. [M, sec. 6]) and thus analytic map germs with values in $\mathcal{K}$ are defined. We then have:

THEOREM 1. *Let $\gamma: S \to \mathcal{O}_n^p$ be an analytic map germ, S reduced.*

(i) *The germ T of points t in S for which $\gamma(t)$ lies in the orbit $\mathcal{K}\cdot\gamma(0)$ is analytic.*

(ii) *There is an analytic map germ $\phi: T \to \mathcal{K}$ with $\phi(0) = 1$ such that $\gamma(t) = \phi(t)\cdot\gamma(0)$ for all t in T.*

Proof. For $\mathfrak{m}_n \subset \mathcal{O}_n$ the maximal ideal and $k \in \mathbb{N}$ consider $A_k = \mathcal{O}_n/\mathfrak{m}_n^{k+1}$ and the algebraic group $\mathcal{K}_k = \mathrm{GL}_p(A_k) \rtimes \operatorname{Aut} A_k$ acting rationally on the finite dimensional vector space $V_k = A_k^p$. The composition $\gamma_k: S \to (V_k, \gamma_k(0))$ of γ with the natural map $\mathcal{O}_n^p \to V_k$ is analytic. By the Proposition the germ T_k of points t in S with $\gamma_k(t) \in \mathcal{K}_k \cdot \gamma_k(0)$ is analytic. Clearly $T_{k+1} \subset T_k$. As $\mathcal{O}_S$ is Noetherian the sequence becomes stationary, say $T_k = T^*$ for $k \gg 0$. The Proposition gives analytic $\phi_k: T^* \to (\mathcal{K}_k, 1)$ such that $\gamma_k(t) = \phi_k(t)\cdot\gamma_k(0)$ for $t \in T^*$. By Theorem 2 below there is an analytic $\phi: T^* \to \mathcal{K}$ with $\phi(0) = 1$ such that $\gamma(t) = \phi(t)\cdot\gamma(0)$ for all $t \in T^*$. This implies $T^* \subset T$. Obviously $T \subset T^*$ and Theorem 1 is proved.

THEOREM 2. *For two analytic map germs $\gamma, \eta: S \to \mathcal{O}_n^p$, S reduced, the following conditions are equivalent:*

(i) *There exists an analytic $\phi: S \to \mathcal{K}$ with $\phi(0) = 1$ such that $\gamma(s) = \phi(s)\cdot\eta(s)$ for all $s \in S$.*

(ii) *For any $k \in \mathbb{N}$ there exist analytic $\phi_k: S \to \mathcal{K}_k$* with *$\phi_k(0) = 1$ such that $\gamma_k(s) = \phi_k(s)\cdot\eta_k(s)$ for all $s \in S$.*

Proof. Embed S in $(\mathbb{C}^m, 0)$ and choose $G, H: (\mathbb{C}^{n+m}, 0) \to \mathbb{C}^p$ such that one has $\gamma(s)(x) = G(x, s)$ and $\eta(s)(x) = H(x, s)$. We have to find $u(x, s) \in \mathrm{GL}_p(\mathcal{O}_{n+m})$ and $y(x, s) \in \mathcal{O}_{n+m}^n$ such that:

$$u(x,0) = 1, \qquad y(x,0) = x, \qquad y(0,s) = 0,$$

and

$$H(y(x,s), s) \equiv u(x,s)\cdot G(x,s) \mod I(S)$$

where $I(S)$ is the ideal of $\mathcal{O}_m$ defining S in $(\mathbb{C}^m, 0)$. By condition (ii) this system of equations can be solved up to order k. A generalization of Artin's Approximation Theorem by Pfister and Popescu [P-P, Thm. 2.5] and Wavrik [W, Thm. 1] yields the solutions $u(x, s)$ and $y(x, s)$.

Let us now indicate some applications of Theorem 1. We first determine the tangent spaces to the orbits of the contact group $\mathcal{K}$ in $\mathcal{O}_n^p$:

DEFINITION. (a) The tangent vector in $\gamma(0)$ of an analytic curve $\gamma: (\mathbb{C}, 0) \to \mathcal{O}_n^p$

given by $G: (\mathbb{C}^n \times \mathbb{C}, 0) \to \mathbb{C}^p$ is defined as:

$$\frac{d\gamma}{ds}(0) = \frac{\partial G}{\partial s}\Big|_{s=0} \in \mathcal{O}_n^p.$$

(b) The tangent map at 0 of an analytic map germ $\gamma: (\mathbb{C}^m, 0) \to \mathcal{O}_n^p$ given by $G: (\mathbb{C}^n \times \mathbb{C}^m, 0) \to \mathbb{C}^p$ is defined as

$$T_0\gamma: \mathbb{C}^m \to \mathcal{O}_n^p, \quad v \to \sum_{i=1}^m v_i \cdot \frac{\partial \gamma}{\partial s_i}(0)$$

where

$$\frac{\partial \gamma}{\partial s_i}(0) = \frac{\partial G}{\partial s_i}\Big|_{s=0} \in \mathcal{O}_n^p.$$

(c) For a subset E of $\mathcal{O}_n^p$ and $g \in E$ let

$$T_g E = \left\{\frac{d\gamma}{ds}(0) \in \mathcal{O}_n^p, \quad \gamma: (\mathbb{C}, 0) \to E \text{ analytic with } \gamma(0) = g\right\}.$$

COROLLARY. *Let $g \in \mathcal{O}_n^p$ with $\mathscr{K}$-orbit $\mathscr{K} \cdot g$.*

(i) $T_g(\mathscr{K} \cdot g) = I(g) \cdot \mathcal{O}_n^p + m_n \cdot J(g)$
where $I(g)$ is the ideal of $\mathcal{O}_n$ generated by the components of g and $J(g)$ is the $\mathcal{O}_n$-submodule of $\mathcal{O}_n^p$ generated by the partial derivatives of g.

(ii) *Let $\gamma: (\mathbb{C}^m, 0) \to \mathcal{O}_n^p$ be analytic, $\gamma(0) = g$, restricting to $\gamma_{|S}: S \to \mathscr{K} \cdot g$ for some reduced $S \subset (\mathbb{C}^m, 0)$. Then $T_0\gamma(T_0 S) \subset T_g(\mathscr{K} \cdot g)$.*

Proof. (i) If $G(x, s) = u(x, s) \cdot g(y(x, s))$ with $u(x, 0) = 1, y(x, 0) = x$, then

$$\frac{\partial G}{\partial s}\Big|_{s=0} = \frac{\partial u}{\partial s}\Big|_{s=0} \cdot g + \frac{\partial g}{\partial x} \cdot \frac{\partial y}{\partial s}\Big|_{s=0}.$$

Theorem 1 therefore implies "$\subset$". The other inclusion is obvious.

(ii) Since any analytic $S \to \mathscr{K}$ can be extended to an analytic $(\mathbb{C}^m, 0) \to \mathscr{K}$, Theorem 1 gives an analytic map germ $\delta: (\mathbb{C}^m, 0) \to \mathcal{O}_n^p$, $\delta(0) = g$, such that $\delta(s) \in \mathscr{K} \cdot g$ for $s \in (\mathbb{C}^m, 0)$ and $\delta(s) = \gamma(s)$ for $s \in S$. Clearly $T_0\delta(\mathbb{C}^m) \subset T_g(\mathscr{K} \cdot g)$ and $T_0\delta(v) = T_0\gamma(v)$ for $v \in T_0 S$.

Next, let us interpret Theorem 1 geometrically.

For an analytic $\gamma: S \to \mathcal{O}_n^p$ given by $G: (\mathbb{C}^n, 0) \times S \to \mathbb{C}^p$ consider the space germ X defined in $(\mathbb{C}^n, 0) \times S$ by G. For fixed $s \in S$ the vector $\gamma(s) \in \mathcal{O}_n^p$ defines the germ in $\sigma(s) = (0, s)$ of the fiber of $\pi = pr_{|X}: X \to S$ over s. Conversely, a morphism $\pi: X \to S$ of space germs with section $\sigma: S \to X$ has an embedding $X \subset (\mathbb{C}^n, 0) \times S$ over S with $\sigma(S) = 0 \times S$, see [F, 0.35]. Moreover an analytic $\Phi: S \to \operatorname{Aut} \mathcal{O}_n$ given by $\Phi(s)(x) = y(x, s)$ induces an automorphism ϕ of $(\mathbb{C}^n, 0) \times S$ over S mapping $0 \times S$ onto itself: $\phi(x, s) = (y(x, s), s)$.

Combining these remarks we get:

THEOREM 1′. *For a morphism of analytic space germs $\pi: X \to S$, S reduced, with section $\sigma: S \to X$ denote by X_t, $t \in S$, the germ in $\sigma(t)$ of the fiber of π over t.*

(i) *The germ T of points t in S with $X_t \simeq X_0$ is analytic.*

(ii) *For any base change $\alpha: S' \to S$ with S' reduced the induced morphism $\pi': X' = X \times_S S' \to S'$ is trivial along the induced section $\sigma': S' \to X'$ if and only if α maps into T. (We say that π' is trivial along σ' if there is an isomorphism $X' \simeq X_0 \times S'$ over S' mapping $\sigma'(S')$ onto $0 \times S'$.)*

The universal property of (ii) applies in particular to the base change $T \subset S$ and then reads as follows: A local analytic family of analytic space germs with isomorphic members is trivial. This is a local analogon of a result of Fischer and Grauert [F-G] and Schuster [Sch, Satz 4.9]: A flat analytic family of compact analytic spaces with isomorphic members is locally trivial.

Theorem 1′ can be extended to the case where π does not come with a section σ:

THEOREM 3. *For a morphism of analytic space germs $\pi: X \to S$ with S reduced denote by $X(a)$, $a \in X$, the germ in a of the fiber of π through a.*

(i) *The germ Y of points a in X with $X(a) \simeq X(0)$ is analytic.*

(ii) *The restriction $\pi_Y: Y \to S$ is a mersion (i.e., has smooth special fiber $Y(0)$ and there is a germ $T \subset S$ such that π_Y maps into T and $Y \simeq Y(0) \times T$ over T).*

(iii) *For any base change $\alpha: S' \to S$ with S' reduced the induced morphism $\pi': X' = X \times_S S' \to S'$ is trivial if and only if α maps into T.*

(iv) *There is a germ Z with $X(0) \simeq Y(0) \times Z$.*

Proof. Choose embeddings $X \subset (\mathbb{C}^n, 0) \times S$ over S and $S \subset (\mathbb{C}^m, 0)$ and let $F: (\mathbb{C}^n \times \mathbb{C}^m, 0) \to \mathbb{C}^p$ define X. Let $\gamma: (\mathbb{C}^n \times \mathbb{C}^m, 0) \to \mathcal{O}_n^p$ be given by $\gamma(a)(x) = F(x + a_1, a_2)$. For fixed $a \in X$ the germ $\gamma(a) \in \mathcal{O}_n^p$ defines $X(a)$. Hence Theorem 1 yields the analyticity of Y.

Let $Y(a)$ be the germ in a of the fiber of π_Y through a. For $a \in Y$ fixed its reduction red $Y(a)$ is the germ of those points $b \in X(a)$ with $X(b) \simeq X(a)$. As $X(a)$ and $X(0)$ are isomorphic, red $Y(a)$ and red $Y(0)$ are isomorphic. In particular, $\dim Y(a) = \dim Y(0)$ for all $a \in Y$.

By the Corollary we have for $g = \gamma(0) \in \mathcal{O}_n^p$ and $v \in T_0 Y(0) = T_0 Y \cap (\mathbb{C}^n \times 0)$:

$$\sum_{i=1}^{n} v_i \cdot \frac{\partial g}{\partial x_i} \in I(g) \cdot \mathcal{O}_n^p + \mathfrak{m}_n \cdot J(g).$$

As g defines $X(0)$ in $(\mathbb{C}^n, 0)$ this signifies that there are $d = \dim T_0 Y(0)$ vectorfields $\xi_1, \ldots, \xi_d$ on $X(0)$ with $\xi_1(0), \ldots, \xi_d(0)$ linearly independent. A Theorem of Rossi [F, 2.12] implies $X(0) \simeq (\mathbb{C}^d, 0) \times Z$ for some Z. Hence, by definition, $Y(0)$ must have dimension at least d and therefore $Y(0) \simeq (\mathbb{C}^d, 0)$. This gives (iv). Moreover π_Y is a mersion by [F, 2.19, Cor. 2]. The universal property of (iii) is then a consequence of Theorem 1′.

We conclude by some remarks: In the absolute case $S = 0$ we have recovered a result of Ephraim [E, Thm. 0.2]. Another Corollary is Teissier's economy of the semi-universal deformation: In the semi-universal deformation of an isolated singularity $X(0)$ there are no fibers isomorphic to $X(0)$, [T, Thm. 4.8.4].

Finally, it is possible to provide the germs Y and T of Theorem 3 with canonical non-reduced analytic structures. The universal property then holds for arbitrary base changes. A detailed exposition of this non-reduced case is given in [H-M]. We also refer to results of Flenner and Kosarew [F-K] and Greuel and Karras [G-K]. Using deformation theory and Banach-analytic methods they treat the case of flat morphisms.

Acknowledgements

We would like to thank G.-M. Greuel and H. Flenner for stimulating discussions and suggestions.

References

[E] Ephraim, R.: Isosingular loci and the cartesian product structure of complex analytic singularities. *Trans. Am. Math. Soc.* 241 (1978) 357–371.

[F] Fischer, G.: Complex analytic geometry. *Springer Lect. Notes* 538, 1976.

[F-G] Fischer, W., Grauert, H.: Lokal-triviale Familien kompakter komplexer Mannigfaltigkeiten. *Nachr. Akad. Wiss. Göttingen, Math. Phys. Kl.* II 6 (1965) 89–94.

[F-K] Flenner, H., Kosarew, S.: On locally trivial deformations. *Publ. Res. Inst. Math. Sci.* 23 (1987) 627–665.

[G-K] Greuel, G.-M., Karras, U.: Families of varieties with prescribed singularities. *Compos. Math.* 69 (1989).

[H-M] Hauser, H., Müller, G.: The trivial locus of an analytic map germ. To appear.

[M] Müller, G.: Deformations of reductive group actions. *Math. Proc. Camb. Philos. Soc.* 106 (1989) 77–88.

[P-P] Pfister, G., Popescu, D.: Die strenge Approximationseigenschaft lokaler Ringe. *Invent. Math.* 30 (1975) 145–174.

[Sch] Schuster, H.W.: Zur Theorie der Deformationen kompakter komplexer Räume. *Invent. Math.* 9 (1970) 284–294.

[T] Teissier, B.: The hunting of invariants in the geometry of discriminants. In: Real and complex singularities, Oslo 1976, 565–677. Holm, P., (ed.) Sijthoff & Noordhoff 1977.

[W] Wavrik, J.J.: A theorem on solutions of analytic equations with applications to deformations of complex structures. *Math. Ann.* 216 (1975) 127–142.

Compositio Mathematica **76**: 203–242, 1990.

A Siegel Modular 3-fold that is a Picard Modular 3-fold*

BRUCE HUNT
Universität Kaiserslautern, FB Mathematik, Postfach 3049, 675 Kaiserlautern, FRG

Received 26 August 1988; accepted 4 October 1989

$-\infty$ Introduction

Let $\mathscr{D}$ be a bounded symmetric domain. $\Gamma \subset \mathrm{Aut}(\mathscr{D})$ be a discrete, properly discontinuous group. If Γ is cocompact and acts freely, it has been known for several decades (Kodaira: [K], Hirzebruch: [Hi]) that $\Gamma\backslash\mathscr{D}$ is then an algebraic variety, and in fact of general type. The Hirzebruch proportionality theorem then tells us the (ratios of) Chern numbers of $X = \Gamma\backslash\mathscr{D}$, which allows us to recover $\mathscr{D}$ from the Chern numbers of X if we know only that X is of the form $\Gamma\backslash\mathscr{D}$ for *some* $\mathscr{D}$. The group Γ is then of course just the fundamental group $\pi_1(X)$. So it can't happen, for example, that $X = \Gamma\backslash\mathscr{D} = \Gamma\backslash\mathscr{D}'$ for 2 non-isomorphic bounded symmetric domains $\mathscr{D}$ and $\mathscr{D}'$.

Arithmetically the condition Γ cocompact means that Γ has $\mathbb{Q}$-rank zero. Although occasionally such groups occur in algebraic geometry (Shimura curves, for example), in most cases Γ will only be of finite covolume, and the space $\Gamma\backslash\mathscr{D}$ occur as moduli spaces of some sort, usually as moduli spaces of varieties with special properties (e.g. classes of abelian varieties, K3-surfaces, curves, etc.) More generally, the *period domains* according to Griffiths [GS] (classifying spaces of Hodge structures) are all of the form $\Gamma\backslash G/H$, where G is a real simple non-compact Lie group and H is a compact subgroup, Γ a discrete subgroup. If one has a Torelli theorem for a corresponding class of varieties, then these spaces, locally homogenous complex manifolds are the moduli spaces for that class of varieties. These spaces are all fibre bundles over some symmetric spaces, for example

$$\mathrm{SO}(a,b)/\mathrm{U}(h^0) \times \cdots \times \mathrm{U}(h^{m-1}) \times \mathrm{SO}(h^m) \to \mathrm{SO}(a,b)/\mathrm{SO}(a) \times \mathrm{SO}(b),$$
$$(a = h^0 + h^2 + \cdots + h^{2m}, b = h^1 + h^3 + \cdots + h^{2m-1})$$
$$\mathrm{Sp}(2a,\mathbb{R})/\mathrm{U}(h^0) \times \cdots \times \mathrm{U}(h^m) \to \mathrm{Sp}(2a,\mathbb{R})/\mathrm{U}(a),$$
$$(a = h^0 + h^1 + \cdots + h^m).$$

*Research supported in part by NSF Grant 8500994A2

Returning to the situation in which $\Gamma\backslash\mathscr{D}$ is a moduli space, in general $\Gamma\backslash\mathscr{D}$ will not be compact, and the compactification is instrumented by adding moduli of the *degenerations* at the boundary. This leads one to consider (smooth) algebraic varieties X which contain $\Gamma\backslash\mathscr{D}$ as a Zariski-open subset. We call such X locally symmetric.

Locally symmetric varieties, together with a divisor D (usually assumed to be normal crossings) such that $X - D = \Gamma\backslash\mathscr{D}$, still contain a lot of "symmetry", that is, structure determined by the group structure of $\mathrm{Aut}(\mathscr{D})$, and so we can think of a correspondence of pairs (X, D) and pairs $(\mathscr{D}, \Gamma)$, where now $\Gamma = \pi_1(X - D)$. Thus it is now quite conceivable that given an algebraic variety X, there exist 2 different divisors D_1 and D_2 such that (X, D_1) corresponds to $(\mathscr{D}_1, \Gamma_1)$ and (X, D_2) corresponds to $(\mathscr{D}_2, \Gamma_2)$ in the above sense, but $\mathscr{D}_1$ and $\mathscr{D}_2$ are not isomorphic, nor even of the same $\mathbb{R}$-rank for that matter. This phenomena seemed worthy of study as soon as it occurred somewhere. For surfaces no such examples were, up to now, known.

Let X be the following modification (blow-up) of $\mathbb{P}^3$ (for notations and details see 2.5.1. below). Let $D_1, \ldots, D_4$ be the 4 coordinate planes (regular tetrahedron), $D_5, \ldots, D_{10}$ the 6 symmetry planes of this tetrahedron, and $D_{11}, \ldots, D_{15}$ be the 5 exceptional $\mathbb{P}^2$'s gotten by blowing up $\mathbb{P}^3$ at the corners and the center of the tetrahedron. Finally let $E_1, \ldots, E_{10}$ be the $\mathbb{P}^1 \times \mathbb{P}^1$'s gotten by blowing up along the 10 3-fold lines of the 10 planes (i.e. of the *arrangement* $D = D_1 \cup \cdots \cup D_{10}$, see 2.5). The proof and study of the following result is the object of this paper:

THEOREM. (i) (X, D) *corresponds to* $(\mathbb{S}_2, \Gamma(2))$ *in the above sense, where* $\mathbb{S}_2$ = *Siegel upper half space of degree* 2, $\Gamma(2)$ = *principal congruence subgroup of level* 2.

(ii) (X, E) *corresponds to* $(\mathbb{B}^3, \Gamma(1-\rho))$ *in the above sense,* $\mathbb{B}^3$ = *complex hyperbolic* 3-*ball,* $\Gamma(1-\rho)$ = *principal congruence subgroup of* $\mathrm{U}((3,1), \mathcal{O}_K)$, $K = \mathbb{Q}(\sqrt{-3})$, $\rho = e^{2\pi i/3}$. *Here* $D = \Sigma D_i, i = 1, \ldots, 15$, $E = \Sigma E_\lambda, \lambda = 1, \ldots, 10$.

There are several directions into which this result can be further investigated. On the group level, one can associate to the discrete group Γ a Tits building with scaffolding. Here we find that the corresponding Tits building with scaffoldings are *dual* to each other. It is not yet clear whether this is implied by the double structure as locally symmetric space or whether this is perhaps as additional coincidence of this particular example.

It is not difficult to determine the ring of modular forms of $\Gamma(2)$, using theta constants (Igusa [I1], [I2]). As for the ring corresponding to the group $\Gamma(1-\rho)$, the structure can be determined by utilizing results of Holzapfel [Ho1], together with results of Deligne-Mostow [DM]. Using in addition a result of Shimura [S] which characterises (X, E) as a moduli space of Picard curves, one can even describe the ring in terms of theta constants. Viewing things this way, the duality above turns into projective duality of varieties (the singular Baily-Borel embeddings), and this duality is in fact classical (cf. [B]).

It turns out that both (X, D) and (X, E) have moduli interpretations, and it is easthetically pleasing to see this duality in terms of degenerations. $X - D - E$ parametrises on the one hand hyperelliptic curves $y^2 = p_6(x)$, genus 2 curves, and on the other hand Picard curves $y^3 = p_6(x)$, genus 4 curves, the correspondence being given by the zeros of $p_6(x)$. But where zeroes of $p_6(x)$ coincide ($x \in D \cup E$), the types of degenerations seem to be dual to each other. For the hyperelliptic curves D corresponds to curves aquiring doublepoints, E to curves splitting into 2 components. For the Picard curves it is the other way around. This is summarised in the table 5.4.

For each such family of curves, one can consider the Picard-Fuchs differential equation corresponding to the dependency of the periods on the moduli. Both our families of curves are associated with the famous hypergeometric differential equation, but with different parameters. Here we use results of [DM] on the one hand, and of Sasaki and Yoshida [KSY] on the other.

The variety we are studying in this paper is probably one of the most thoroughly studied algebraic 3-folds, so we don't claim to be deriving new results on this variety. Rather, our object is to study in detail the 2 structures of locally symmetric spaces and their interrelations. The Siegel picture (§1) is very well known whence we only sketch the necessary statements and facts; we refer to [V] for general results and to [LW] for combinatorial and cohomological results. The Picard picture (§2) is, to the best of our knowledge, essentially new, so we give this side of the picture in much greater detail. However, this Picard picture is an almost straightforward 3-dimensional generalisation of the Picard picture of a surface for which extremely detailed results are available, namely Holzapfel's monograph [Hol]. Hence here there is also little which is original.

As this subject also has an interesting history, as well as being a model for our efforts in dimension 3, we now recall some of the background and results of the surface case. As long ago as 1769, L. Euler considered the following partial differential equation in connection with acoustics:

$$(E_a) \quad \frac{\partial^2 z}{\partial x \partial y} - \frac{a}{x-y}\frac{\partial z}{\partial x} + \frac{a}{x-y}\frac{\partial z}{\partial y} = 0.$$

About a century later Riemann constructed solutions of this equation by an inversion process, during which hypergeometric functions occurred. In 1880 Appell gave the generalisation to several variables of the hypergeometric function which also occurs in recent work of Mostow and Deligne. In 1881 E. Picard studied these and found the famous integral representation for Appell's hypergeometric series:

$$\int_0^1 u^\alpha(1-u)^\beta(1-xu)^\gamma(1-yu)^\delta\, du.$$

He in particular studied the integral

$$I_{g,h}(t_1,t_2) := \int_g^h \frac{dx}{\sqrt[3]{x(x-1)(x-t_1)(x-t_2)}}, \quad g,h \in \{0,1,t_1,t_2,\infty\}$$

and discovered that this function (of t_1 and t_2) is a special solution of the Euler equation $(E_{1/3})$! In fact, 3 of these integrals form a fundamental system of solutions for a system of 3 partial differential equations.

One recognises immediately the integrals above as forming a base of the $(1,0)$-differentials on the trigonal curve

$$y^3 = x(x-1)(x-t_1)(x-t_2),$$

which is a genus 3 curve with Galois action by $\mathbb{Z}/3\mathbb{Z}$. These are the so-called Picard curves, of which we will be studying a suitable generalisation (we also call our curves, genus 4 curves, Picard curves).

We now recall for the reader's benefit some of Holzapfel's results in the study of the family of Picard curves. Consider the following subgroups of $\mathrm{Aut}(\mathbb{B}^2) = \mathrm{PSU}(2,1)$:

Eisenstein lattice	$\tilde{\Gamma} := \mathrm{PU}(2,1;\mathcal{O}_K), K = \mathbb{Q}(\sqrt{-3})$
Special Eisenstein lattice	$\Gamma := \mathrm{PSU}(2,1;\mathcal{O}_K)$,

$$\mathcal{O}_K = \text{ring of integers in } K.$$

Letting $\rho = e^{2\pi i/3}$ be a primitive cube root of unity, $(1-\rho)$ is an ideal in $\mathcal{O}_K$ (note that $\mathbb{Q}(\sqrt{-3}) = \mathbb{Q}(\rho)$ since $\rho = \dfrac{-1+\sqrt{-3}}{2}$. One considers also the congruence subgroups $\Gamma' = \Gamma(1-\rho)$, and $\tilde{\Gamma}' = \tilde{\Gamma}(1-\rho)$ (here we are using Holzapfel's notation). The following results are proved in [Ho1]:

I. The monodromy group of the system of partial differential equations alluded to above is $\tilde{\Gamma}'$.

II. The rings of automorphic forms for $\tilde{\Gamma}'$ and Γ are:

$$\bigoplus_{m=0}^{\infty} [\tilde{\Gamma}', m]_\chi = \mathbb{C}[\xi_1, \xi_2, \xi_3], \quad \bigoplus_{m=0}^{\infty} [\tilde{\Gamma}, m]_\chi = \mathbb{C}[G_2, G_3, G_4],$$

where the ξ_i have weight 1 and the G_j have weight j.

III. $\tilde{\Gamma}'\backslash\mathbb{B}^2 \cong \mathbb{P}^2$-{4 points}.

IV. If F_1, F_2 and F_3 are 3 fundamental solutions of the system alluded to above, then the automorphic forms of weight 1, $(\xi_1:\xi_2:\xi_3)$ give (up to coordinate transformations) the inverse to the many-valued function $(F_1:F_2:F_3)$, in other words, $(\xi_1:\xi_2:\xi_3)\circ(F_1:F_2:F_3)$ is $1-1$, and

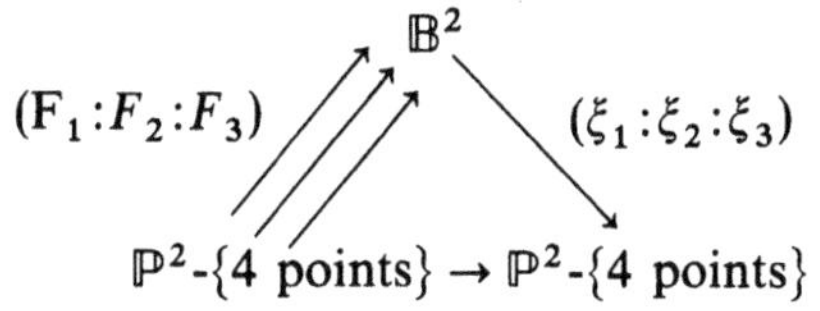

commutes.

V. There is also a commutative diagram

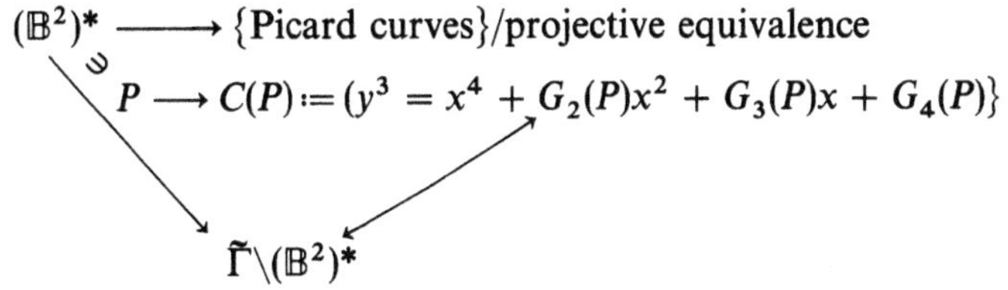

where $(\mathbb{B}^2)^* = \mathbb{B}^2 \cup \{\tilde{\Gamma}$-rational cusps}, $\tilde{\Gamma}\backslash(\mathbb{B}^2)^*$ is the Baily-Borel compactification.

VI. The (image of) $\tilde{\Gamma}$-fixed points of $\mathbb{B}^2$ are the set of Picard curves with automorphism group larger than $\mathbb{Z}/3\mathbb{Z}$.

In our work below we are able to give reasonable generalisations of I, II, III and IV to dimension 3. It would be challenging and extremely interesting to also get a nice equation in terms of modular forms (V) and to generalise VI also to dimension 3.

Finally we would like to remark that during the last year it has started to emerge that this example is the first in some finite list of examples with similar properties. An upcoming paper with Weintraub will give more details on the other known examples related to this one.

Acknowledgements

The author would like to thank the organisers of the conference at the Humboldt Universitat for the invitation and the opportunity to speak on this topic there. Also conversations with R.-P. Holzapfel and J.-M. Feustel were helpful. Finally

I am greatly indebted to S. Weintraub for many explanations and conversations on this and related topics, in particular for the technicalities of 2.1.

0. Notations and conventions

All varieties considered are over the complex number field $\mathbb{C}$. We use freely the standard notions of hemitian symmetric spaces. For a non-compact locally hermitian symmetric space X, X^* usually denotes the Baily-Borel compactification, $X^\wedge$ some desingularisation.

Throughout, we use the following notations:

$\mathbb{P}^n$	projective space (over $\mathbb{C}$)
$\mathbb{B}^n$	the complex n-ball
Σ_n	symmetric group on n letters
A_n	alternating group on n letters
Γ	some arithmetic group, particular cases of which are:
$\Gamma(2), \Gamma(4)$	principal congruence subgroups (Siegel case)
$\Gamma(1-\rho), S\Gamma(1-\rho), \Gamma(1-\rho)^2$	lattices in Picard groups
$\mathbb{F}_p$	field with p elements (p prime), which is sometimes also denoted $\mathbb{Z}/p\mathbb{Z}$
$\mathrm{Sp}(n, \mathbb{R}), \mathrm{U}(p, q), \mathrm{SU}(p, q)$	the usual classical groups
K	number field,
$\mathcal{O}_K$	ring of integers in K
$M_n(\mathbb{C})$	$n \times n$ matrices with $\mathbb{C}$-coefficients
$\pi_1(X)$	fundamental group.

1. A Siegel Modular 3-fold

1.1. Let $\mathbb{S}_2 = \mathrm{Sp}(2, \mathbb{R})/\mathrm{U}(2) = \{Z \in M_2(\mathbb{C}) \mid Z = {}^tZ.\ \mathrm{Im}(Z) > 0\}$ be the Siegel upper half-space of degree 2, a 3-dimensional, $\mathbb{R}$-rank 2 bounded symmetric domain. $\mathrm{Sp}(2, \mathbb{Z})$ is a lattice in $\mathrm{Sp}(2, \mathbb{R})$ which has $\mathbb{Q}$-rank 2. The action of $\Gamma = \mathrm{Sp}(2, \mathbb{Z})$ on $\mathbb{S}_2$ is $Z \mapsto (AZ + B)(CZ + D)^{-1}$. We are particularly interested in the principal congruence subgroup of level 2, defined by the following exact sequence:

$$1 \to \Gamma(2) \to \mathrm{Sp}(2, \mathbb{Z}) \to \mathrm{Sp}(2, \mathbb{Z}/2\mathbb{Z}) \to 1. \tag{1.1.1}$$

$\Gamma(2)$ is thus a normal subgroup, of index equal to $|\mathrm{Sp}(2, \mathbb{Z}/2\mathbb{Z})| = 720$, since

$Sp(2, \mathbb{Z}/2\mathbb{Z}) = \Sigma_6$, the symmetric group on 6 letters, as is well known. $\Gamma(2)$ does not act freely, but the quotient is smooth [I1], [I3], [C]. Let $X(2)$ denote the non-compact quotient $\Gamma(2)\backslash\mathbb{S}_2$.

1.2. A compactification of $X(2)$ is constructed in the standard way, i.e. Baily-Borel. Adjoin to $\mathbb{S}_2$ the rational (with respect to Γ) boundary components, which are copies of $\mathbb{S}_1$ in dimension 1 (rank 1), and points in dimension 0 (rank 0). The action of $\Gamma(2)$ extends to the rational boundary of $\mathbb{S}_2$, and the quotient $\Gamma(2)\backslash\mathbb{S}_2^*$ is a compact Hausdorff space. The action of $\Gamma(2)$ on one of the $\mathbb{S}_1$'s on the boundary is via the principal congruence subgroup of level 2 of $Sp(1, \mathbb{Z}) = Sl(2, \mathbb{Z})$, which has 3 inequivalent cusps. Since $\Gamma(2) \subset Sp(2, \mathbb{Z})$ has 15 inequivalent 1-dimensional cusps as well as 15 inequivalent 0-dimensional ones, one gets the following configuration on $X(2)^* = \Gamma(2)\backslash\mathbb{S}_2^*$:

15 curves $C_i = \Gamma(2)\backslash\mathbb{S}_1^*$
15 points $P_{ij} = C_i \cap C_j$ = cusp of C_i and C_j.
3 different C_i meet at each P_{ij}.
Each C_i contains 3 cusps P_{ij}

1.3. In order to describe the boundary components precisely it is convenient to work in $(\mathbb{Z}/2\mathbb{Z})^4$. Since any two cusps are equivalent under $Sp(2, \mathbb{Z})$ the exact sequence 1.1.1 implies the natural action of $Sp(2.\mathbb{Z}/2\mathbb{Z})$ on $(\mathbb{Z}/2\mathbb{Z})^4$ gives exactly the action of $\Gamma(2)$ on the boundary components (see also [LW]):

$$\text{1-dimensional cusps of } X(2)^* \leftrightarrow \begin{cases} l \in (\mathbb{Z}/2\mathbb{Z})^4, & 1 \neq 0 \\ l = (\varepsilon_1, \varepsilon_2, \varepsilon_3, \varepsilon_4), & \varepsilon_j = 0 \text{ or } 1. \end{cases}$$

$$\text{0-dimensional cusps of } X(2)^* \leftrightarrow \begin{cases} h = l_1 \wedge l_2, \text{ 2-dimensional} \\ \text{isotropic subspaces.} \end{cases}$$

In this scheme the curves C_i are numbered by 4-tuples $(\varepsilon_1, \ldots, \varepsilon_4)$, $\varepsilon_j = 0, 1$. Then $C_i \cap C_j$ is the 2-plane spanned by

$$\begin{pmatrix} \varepsilon_1^i & \varepsilon_2^i & \varepsilon_3^i & \varepsilon_4^i \\ \varepsilon_1^j & \varepsilon_2^j & \varepsilon_3^j & \varepsilon_4^j \end{pmatrix},$$

and the third curve C_k meeting $C_i \cap C_j$ is (ε_μ^k), $\varepsilon_\mu^k = \varepsilon_\mu^i + \varepsilon_\mu^j$.

1.4. A desingularisation of $X(2)^*$ was constructed by Igusa in [I3] by blowing up along the sheaf of ideals defining the boundary. In [V] van der Geer explains how to obtain the desingularisation directly by means of toroidal embeddings of $X(2)$. The result is the same, and is as follows: there are 15 divisors $D_1, \ldots, D_{15}$, each itself an algebraic fibre space $D_i \to C_i$, whose generic fibre is a Kummer curve $= \mathbb{P}^1$ (=elliptic curve/involution) and whose 3 special fibres are degenerate conics in $\mathbb{P}^2$, consisting of 2 copies of $\mathbb{P}^1$ meeting at a point

(1.4.1)

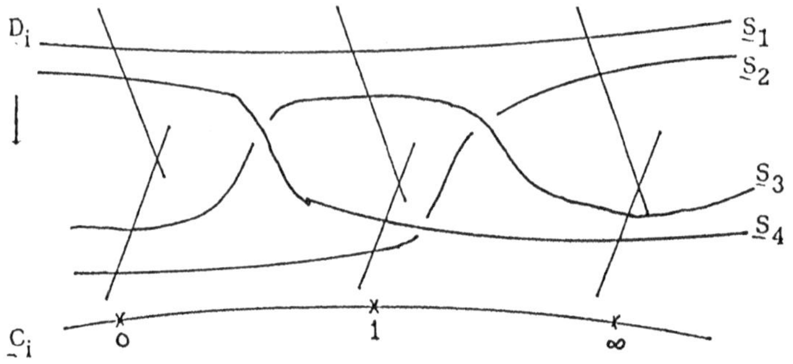

The fibre space $D_i \to C_i$ has 4 sections $S_1, \ldots, S_4$. Another way to view the D_i is as $\mathbb{P}^2$ blown up in the 4 3-fold points of the line arrangement in $\mathbb{P}^2$:

1.4.2

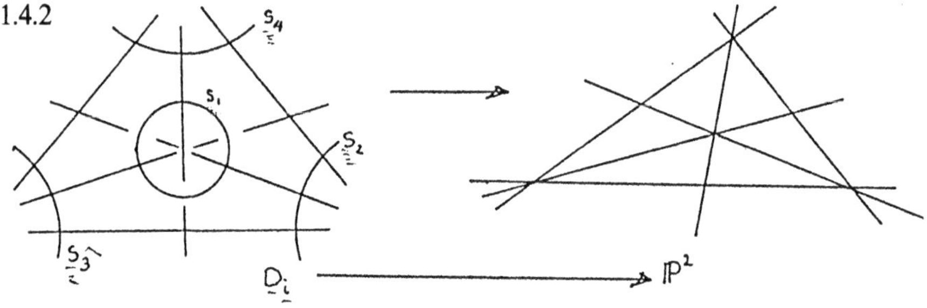

The fibering $D_i \to C_i$ is given by the pencil of conics passing through the 4 points, and the singular fibres are the 6 lines of the arrangement, 2 of them at a time forming a degenerate conic, and the sections S_i are the exceptional $\mathbb{P}^1$'s of the blow-up. Therefore on each D_i one can blow down the 4 sections, the result being $\mathbb{P}^2$. The D_i intersect 2 at a time along the singular fibres and 3 at a time at the double points of those fibres. This describes the structure of the normal crossings divisor $D = \Sigma D_i$. For more details on the intersection behavior see 2.5. We denote the Igusa desingularisation by $X(2)^\wedge$.

1.5. We now describe another important set of divisors, the Humbert surfaces of discriminant 1. For each natural number $\Delta \equiv 0$ or $1 \bmod(4)$ there is such a Humbert surface H_Δ ([V], §2), but we will only describe H_1 here. The diagonal $\mathbb{S}_1 \times \mathbb{S}_1 \subset \mathbb{S}_2$ has 10 inequivalent transforms under $\Gamma(2)$, and the action of $\Gamma(2)$ restricted to each copy is by $\Gamma_1(2) \times \Gamma_1(2) \subset \mathrm{Sl}(2, \mathbb{Z}) \times \mathrm{Sl}(2, \mathbb{Z})$. (Here $\Gamma_1(N)$ denotes the principle congruence subgroup of level N and degree 1, i.e. in $\mathrm{SL}(2, \mathbb{Z})$.) Let $E_1, \ldots, E_{10}$ be the images in $X(2)^*$ of these diagonals. Then each $E_j = \Gamma_1(2)\mathbb{S}_1^* \times \Gamma_1(2)\backslash\mathbb{S}_1^*$ is a copy of $\mathbb{P}^1 \times \mathbb{P}^1$. The E_j

are disjoint, but intersect the D_i in $X(2)^\wedge$ at the sections of each. Each E_λ intersects 6 D_i, 3 in each direction:

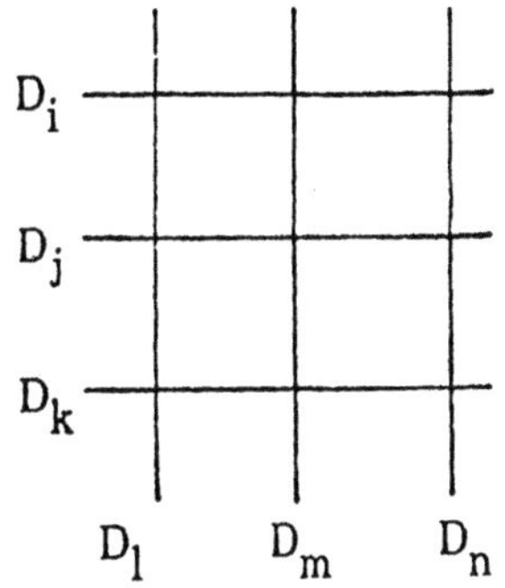

$$\begin{matrix} \varepsilon_1^i & \varepsilon_2^i & \varepsilon_3^i & \varepsilon_4^i \\ \varepsilon_1^j & \varepsilon_2^j & \varepsilon_3^j & \varepsilon_4^j \\ \varepsilon_1^k & \varepsilon_2^k & \varepsilon_3^k & \varepsilon_4^k \end{matrix}$$

1.6. To describe the incidences $E_\lambda \cap D_i \neq \varnothing$, we follow [LW, §2]. Let $\Delta = \{\delta, \delta^\perp\}$, an unordered pair of δ, a non-singular plane and $\delta^\perp$, its orthogonal complement. Such Δ are in $1-1$ correspondence with the E_λ, and $E_\lambda \cap D_i \neq \varnothing$ iff $(\varepsilon_\alpha^i) \in \delta$ or $\in \delta^\perp$. Hence the E_λ can be numbered by $\{(\varepsilon_\alpha^i) \wedge (\varepsilon_\alpha^j), (\varepsilon_\alpha^k) \wedge (\varepsilon_\alpha^l)\}$, for example. We just give one example of this. Say E_1 will be numbered by $(1,0,0,0) \wedge (0,0,1,0)$ and $(0,1,0,0) \wedge (0,0,0,1)$, and the other D_i's meeting E_1 are $(1,0,1,0)$ and $(0,1,0,1)$ which, plugging into the above scheme, describes all intersections quite explicitly. The action of $\mathrm{Sp}(2, \mathbb{Z}/2\mathbb{Z})$ induces an action of the E_λ, i.e. $gE_\lambda = E_\mu$, E_λ associated to Δ and E_μ associated to Δg, for $g \in \mathrm{Sp}(2, \mathbb{Z}/2\mathbb{Z})$.

1.7. Finite covers

We just state a result here which follows from Theorem 3.3.2. and the proof of Theorem 2.7.1, but whose statement belongs here. Let $\Gamma(4)$ denote the principle congruence subgroup of level 4. This is a normal subgroup of $\Gamma(2)$ and $P\Gamma(2)/P\Gamma(4) = (\mathbb{Z}/2\mathbb{Z})^9$ (it is the projective groups that are acting effectively). This is the Galois group of the Fermat cover of degree 2 branched along the arrangement H (see 2.7. and [Hu] for Fermat covers), and in fact

THEOREM 1.7.1. *The (smooth) Fermat cover $Y(2, H)$ branched along H,*

$$Y(2, H) \to \hat{\mathbb{P}}^3$$

is the (Igusa compactification of the) Siegel modular 3-fold of level 4.

2. A Picard Modular 3-fold

2.1. Let $\mathbb{B}^3 := SU(3,1)/S(U(3) \times U(1)) = \{z \in \mathbb{C}^3 \mid \sum |z_i|^2 < 1\}$ be the complex hyperbolic 3-ball, a 3-dimensional, $\mathbb{R}$-rank 1 bounded symmetric domain. The best-known lattices in SU(3, 1) are the Picard modular groups. For each square-free integer d let $K = \mathbb{Q}(\sqrt{-d})$ be the imaginary quadratic field associated with d and $\mathcal{O}_K$ = the ring of integers in K. Then $\mathcal{O}_K \subset \mathbb{C}$ is a lattice, and SU(3, 1; $\mathcal{O}_K$) is the *Picard Modular group of discriminant D* (D the discriminant of K) which is a lattice in SU(3,1) (note that whereas Sp(2, $\mathbb{R}$) is a group with real coefficients, so integer coefficients give a lattice, SU(3, 1) is a group of complex matrices, so we need coefficients in a lattice in $\mathbb{C}$). We shall be concerned in this paper with the field of Eisenstein numbers $K = \mathbb{Q}(\sqrt{-3})$, and the corresponding Picard modular group. Actually, the group more basic to our applications is the lattice U(3, 1; $\mathcal{O}_K$). These lattices are related as follows: SU $\lhd$ U, and U/(SU = $\mathbb{Z}/3\mathbb{Z}$, since an element $\in$ U(3, 1; $\mathcal{O}_K$) may have determinant $= \rho$ or ρ^2, as well as 1. We will refer to U as the Picard lattice, and to SU as the special Picard (or Eisenstein) lattice.

The action of Aut($\mathbb{B}_3$) = PU(3, 1) (PU since the center acts trivially) is by fractional linear transformations. For $\gamma \in \mathrm{Aut}(\mathbb{B}^3)$, $z = (z_1, z_2, z_3) \in \mathbb{B}^3$, this action is described as follows:

$$\gamma z = \frac{1}{\Sigma a_{i4} z_i} \begin{pmatrix} \Sigma a_{i1} z_i \\ \Sigma a_{i2} z_i \\ \Sigma a_{i3} z_i \end{pmatrix}, \quad \gamma = (a_{ij}), \quad 1 \leqslant i, j \leqslant 4. \tag{2.1.1}$$

The jacobian of the action is $(\Sigma a_{i4} z_i)^{-1}$ at $z = (z_i)$. When considering lattices Γ in U(3, 1) or SU(3, 1) we will, without mentioning it, take their images in PU(3, 1). However, as this is a potential source of confusion, we describe this in some detail. Since the result is quite different in dimensions 2 and 3 we describe both. Let Γ, $\tilde{\Gamma}, \Gamma', \tilde{\Gamma}'$ be as in the introduction (Holzapfel's notation) and let a P in front of one of the groups denote the projectivised group. Let $Z(G)$ denote the center of a group G. Then:

$$Z(\tilde{\Gamma}) = \{\pm 1, \pm\rho, \pm\rho^2\}, \qquad Z(\Gamma) = \{1, \rho, \rho^2\}$$

$$Z(\tilde{\Gamma}') = \{1, \rho, \rho^2\}, \qquad Z(\Gamma') = \{1, \rho, \rho^2\}.$$

Thus we have the diagrams:

$$\begin{array}{ccc} \Gamma & \xrightarrow{1-6} & \tilde{\Gamma} \\ \downarrow {\scriptstyle 3-1} & & \downarrow {\scriptstyle 6-1} \\ P\Gamma & \xrightarrow{1-3} & P\tilde{\Gamma} \end{array} \qquad\qquad \begin{array}{ccc} \Gamma' & \xrightarrow{1-3} & \tilde{\Gamma}' \\ \downarrow {\scriptstyle 3-1} & & \downarrow {\scriptstyle 3-1} \\ P\Gamma' & \xrightarrow{1-3} & P\tilde{\Gamma}' \end{array}$$

and then a diagram relating the congruence subgroups:

$$\begin{array}{ccccccccc}
1 & \longrightarrow & P\tilde{\Gamma}' & \longrightarrow & P\tilde{\Gamma} & \longrightarrow & PU(2,1;\mathbb{F}_3) = \Sigma_4 & \longrightarrow & 1 \\
 & & \uparrow & & \uparrow & & \wr\| & & \\
1 & \longrightarrow & P\Gamma' & \longrightarrow & P\Gamma & \longrightarrow & PSU(2,1;\mathbb{F}_3) = \Sigma_4 & \longrightarrow & 1
\end{array}$$

the bottom sequence of which Holzapfel proves is exact [Ho1, 1.3.1]. Both inclusions $P\Gamma' \subset P\tilde{\Gamma}'$ and $P\Gamma \subset P\tilde{\Gamma}$ are of index 3.

We now give the corresponding diagrams in dimension 3, and fix, for the rest of this paper, the following notations:

$$\Gamma := U(3,1;\mathcal{O}_K), \quad \Gamma(1-\rho) \text{ congruence subgroup,}$$
$$S\Gamma := SU(3,1;\mathcal{O}_K), \quad S\Gamma(1-\rho) \text{ congruence subgroup.}$$

Here we have

$$Z(\Gamma) = \{\pm 1, \pm\rho, \pm\rho^2\}, \quad Z(S\Gamma) = \{\pm 1\}$$
$$Z(\Gamma(1-\rho)) = \{1, \rho, \rho^2\}, \quad Z(S\Gamma(1-\rho)) = \{1\},$$

giving the following diagrams:

$$\begin{array}{ccc}
S\Gamma & \xrightarrow{1-6} & \Gamma \\
\downarrow{\scriptstyle 2-1} & & \downarrow{\scriptstyle 6-1} \\
PS\Gamma & \xrightarrow{1-2} & P\Gamma
\end{array}
\qquad\qquad
\begin{array}{ccc}
S\Gamma(1-\rho) & \xrightarrow{1-3} & \Gamma(1-\rho) \\
\downarrow{\scriptstyle 1-1} & & \downarrow{\scriptstyle 3-1} \\
PS\Gamma(1-\rho) & \xrightarrow{1-1} & P\Gamma(1-\rho).
\end{array}$$

Furthermore, from the Atlas of finite simple groups we have

$$U(3,1;\mathbb{F}_3) = \Sigma_6 \times 2$$
$$PU(3,1;\mathbb{F}_3) = \Sigma_6$$
$$SU(3,1;\mathbb{F}_3) = A_6 \times 2$$
$$PSU(3,1;\mathbb{F}_3) = A_6.$$

We have the following sequence

$$\begin{array}{ccccccccc}
1 & \longrightarrow & P\Gamma(1-\rho) & \longrightarrow & P\Gamma & \longrightarrow & PU(3,1;\mathbb{F}_3) \cong \Sigma_6 & \longrightarrow & 1 \\
 & & \wr\| & & \uparrow & & \uparrow & & \\
1 & \longrightarrow & PS\Gamma(1-\rho) & \longrightarrow & PS\Gamma & \longrightarrow & PSU(3,1;\mathbb{F}_3) \cong A_6 & \longrightarrow & 1
\end{array}$$

(where A_6 is the alternating group on 6 letters), and $PS\Gamma(1-\rho) \subset P\Gamma(1-\rho)$ is an isomorphism, whereas $PS\Gamma \subset P\Gamma$ is a subgroup of index 2.

Now just like $\Gamma(2)$ in Section 1, $\Gamma(1-\rho)$ does not act on $\mathbb{B}^3$ freely but it turns out that the quotient is nonetheless smooth. The subgroup $\Gamma(1-\rho)^2$ does act freely, and the singularities of $\Gamma(1-\rho)\backslash\mathbb{B}^3$ can be analysed by studying the cover corresponding to $\Gamma(1-\rho)^2 \subset \Gamma(1-\rho)$. We will describe this in 2.8 below. Let $Y(1-\rho)$ denote the non-compact quotient $\Gamma(1-\rho)\backslash\mathbb{B}^3$.

2.2. We now discuss the boundary components of $\mathbb{B}^3$ with respect to Γ. To do this, think of $\mathbb{B}^3 \subset V^4$, a 4-dimensional vector space over $\mathbb{C}$ with hermitian form

$$\Phi(x, y) = x_1\bar{y}_1 + x_2\bar{y}_2 + x_3\bar{y}_3 - x_4\bar{y}_4.$$

Let $\mathscr{B} \subset V$ be the positive cone, i.e. $\mathscr{B} = \{v \subset V\, \Phi(v, v) > 0\}$. This looks as follows:

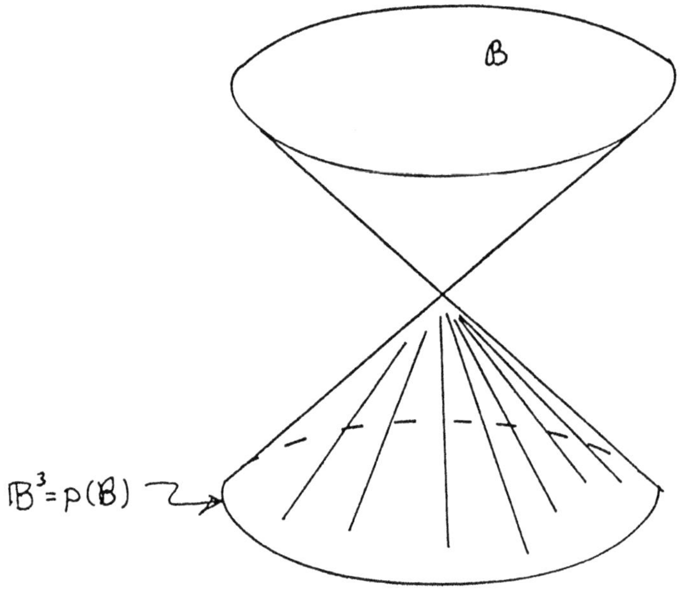

Then $\mathbb{B}^3 = p(\mathscr{B})$, $p\colon V \to \mathbb{P}(V)$ the natural projection. From this one sees that $\partial\mathbb{B}^3 = p(\partial\mathscr{B}) = p(\mathscr{E})$, $\mathscr{E} \subset V$ the set of isotropic vectors, $\mathscr{E} = \{v \in V \mid \Phi(v, v) = 0\}$. With this picture in mind it is obvious that the K-boundary components (or the boundary components with respect to Γ) are: (here we should fix an embedding $K \subset \mathbb{C}$),

$$\partial_K\mathbb{B}^3 = \{v \in K^4 \mid \Phi(v, v) = 0\}/K^* = p(\mathscr{E}_K), \mathscr{E}_K = \mathscr{E} \cap K^4;$$

these are the isotropic lines in $\mathbb{P}(K^4)$. Hence the number of Γ-cusps is the order of $\Gamma\backslash\partial_K\mathbb{B}^3$. This question is discussed in [Z, I] for any discriminant and corresponding Picard group. The answer is: Γ_d has $\mu(d)$ cusps, where $\mu(d) = \#\,|\text{equivalence classes of hermitian unimodular lattices in } K^4|$. This number has been calculated by Hashimoto [HK] and for $d = 3$ there is a unique equivalence class (1 cusp).

2.3. Hence, to determine the number of cusps of $\Gamma(1-\rho)$, just consider the exact sequence 2.1.2, and the ensuing action of $\Gamma/\Gamma(1-\rho) = \mathrm{PU}(3,1;\mathbb{Z}/3) = \Sigma_6$ on $(\mathbb{Z}/3)^4$. The number of cusps is just the number of totally isotropic lines $\{v \in \mathbb{Z}/3^4 \mid \Phi(v,v) = 0\}/\pm 1$ (note that $\mathbb{F}_3^* = \pm 1$), just reducing the form $\Phi \bmod 3$. This works as follows. The form on $\mathbb{C}^4$ is $z_1\bar{z}_1 + z_2\bar{z}_2 + z_3\bar{z}_3 - z_4\bar{z}_4$. The involution $z \to \bar{z}$ in $\mathcal{O}$ (where it is given by $\rho \to \rho^2$) descends to the trivial involution on $\mathbb{Z}/3$ (as it must, $\mathbb{Z}/3$ having no non-trivial automorphisms) which can be seen by the fact that $\rho \equiv 1 \bmod (1-\rho)$ so $\rho^2 \equiv 1^2 = 1 \equiv \rho \bmod (1-\rho)$. It is easily checked that there are, up to sign, 10 isotropic vectors: $(1,0,0,1)$, $(1,0,0,-1)$, $(0,1,0,1)$, $(0,1,0,-1)$, $(0,0,1,1)$, $(0,0,1,-1)$, $(1,1,1,0)$, $(1,1,-1,0)$, $(1,-1,1,0)$ and $(-1,1,1,0)$. I am indebted to Steve Weintraub for this nice exposition.

2.4. A desingularisation of $Y(1-\rho)^*$ is constructed by blowing up at the cusps. The resolving divisor is A/ι, $A = E \times E$ an abelian surface which is a product of 2 copies of the elliptic curve with complex multiplication by $\sqrt{-3}$ and ι is the involution $(z_1, z_2) \mapsto (-z_1, -z_2)$, in other words the resolving divisors are $\mathbb{P}^1 \times \mathbb{P}^1$'s. This can be extracted from the standard construction ([He], [Ho1]). This is of course the same as a desingularisation by means of toroidal embeddings which in the case of $\mathbb{Q}$-rank 1 yields isolated resolving divisors. Let $Y(1-\rho)^\wedge$ denote this resolution. We will see below that $Y(1-\rho)^\wedge$ is actually already smooth.

2.5. We now describe another important set of subvarieties, which we call modular subvarieties. Let $D = \mathbb{B}^2 \subset \mathbb{B}^3$ be a totally geodesic embedding such that Γ_D act properly discontinously, that is $\Gamma_D := \{\gamma \in \Gamma \mid \gamma D = D\} \subset \mathrm{Aut}(\mathbb{B}^2)$ is properly discontinous. That is to say the diagram

$$\begin{array}{ccc} D & \subset & \mathbb{B}^3 \\ {\scriptstyle \Gamma_D}\downarrow & & \downarrow{\scriptstyle \Gamma} \\ \Gamma_D\backslash\mathbb{B}^2 & \subset & \Gamma\backslash\mathbb{B}^3 \end{array}$$

commutes. This is what is usually called a modular embedding $(\Gamma_D, D) \subset (\Gamma, \mathbb{B}^3)$. We will be interested in the Γ_D such that $\Gamma_D = \mathrm{U}(2,1;\mathcal{O}_K(1-\rho))$ is the Picard (surface) group. We can then draw on the very detailed results of Holzapfel for these surfaces (i.e. the whole book [Hol]). As we will see below there are 15 modular subvarieties of this kind on $Y(1-\rho)$. These 15 meet in a union of special curves on each copy, the Γ-reflection discs of [Hol, 1.3.3]. To describe their intersections (which will be identified as such below) we use the figure alluded to in the introduction. Let $H_1, \ldots, H_{10}$ be the 10 planes consisting of the 4 facet planes and 6 symmetry planes of a regular tetrahedron in $\mathbb{P}^3$:

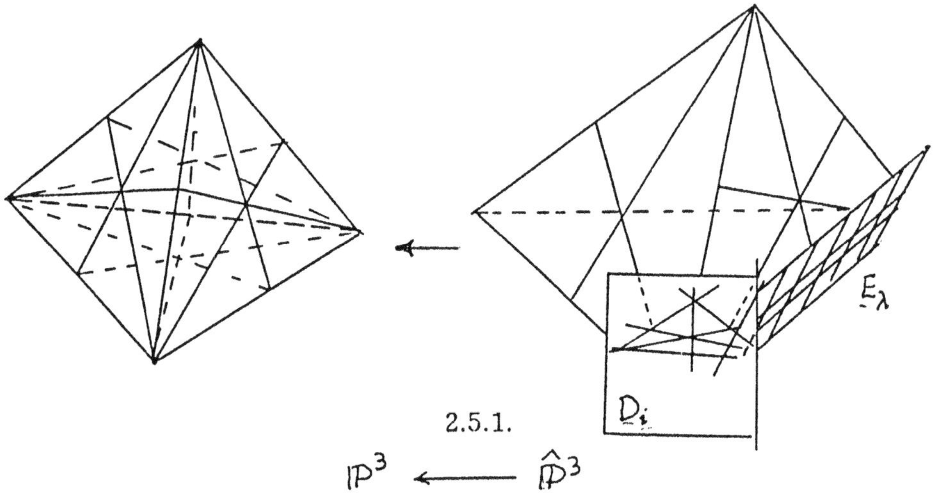

2.5.1.

This arrangement has the following singularities (i.e. is not a normal crossings divisor because of):

5	6-fold points	(4 corners and the center)
10	3-fold lines	(6 edges and 4 diagonals).

The divisor $H = \sum H_i$ is turned into a normal crossings divisor by blowing up $\mathbb{P}^3$ at the 5 points, then along the 10 lines just mentioned. Let $\hat{\mathbb{P}}^3$ denote this blow up, $[H_i] =: D_i$ the proper transforms, $D_{11}, \ldots, D_{15}$ the exceptional $\mathbb{P}^2$'s and $E_1, \ldots, E_{10}$ the exceptional $\mathbb{P}^1 \times \mathbb{P}^1$'s. Then under the isomorphism $Y(1-\rho)^\wedge = \hat{\mathbb{P}}^3$ (see 2.7) $D_1, \ldots, D_{15}$ are the modular varieties just introduced. Notice that after the blow-up each of the $D_i, i = 1, \ldots, 15$, is identical. Each is the blow up of $\mathbb{P}^2$ at 4 points in general position, the 3-fold points of the linear arrangement 1.4.2, hence in each D_i there are 10 $\mathbb{P}^1$'s, all of which have self-intersection (-1) in each D_i. These are of course the same surfaces occurring in 1.4.

Σ_5 acts in a natural way permuting the 10 $\mathbb{P}^1$'s; in fact $D_i - P$, ($P = \sum P_i$, P_i the exceptional divisors under the modification $D_i \to \mathbb{P}^2$) is a GIT-quotient, arising as follows: Let (x_i, y_i), $i = 0, \ldots, 4$ be a set of homogenous coordinates on

$(\mathbb{P}^1)^5$. Let $X \subset (\mathbb{P}^1)^5$ be the Zariski open subset consisting of those (x_i, y_i) such that no 3 of the 5 are identical. PGL(2, $\mathbb{C}$) acts freely on X, and the quotient can be compactified to $\mathbb{P}^2$ blown up in 4 points ([Y], p. 140). The action of Σ_5 is then just permutation of the factors on X.

On the other hand we have the natural action of Σ_4 on the D_i as described in [Ho1, 1.3.ff]. This comes about as soon as you have chosen a subset of 4 (disjoint) out of the 10 $\mathbb{P}^1$'s to be blown down, i.e. identified a set of cusps.

2.6. We now give the combinatorial description of the D_i in $\mathbb{Z}/3\mathbb{Z}^4$. Obviously, in K^4 each such $D_i(K)$ (K-valued points) is given by the intersection of the cone of 2.1 with a hyperplane, fixed by Γ as in 2.5. However, in $\mathbb{Z}/3\mathbb{Z}^4$ there is no distinction between signature (3, 0) and (2, 1), so we must find the hyperplanes $H \subset K^4$ such that Φ restricted to $H \cap \mathbb{B}$ has signature (2, 1), *then* take their images in $\mathbb{Z}/3\mathbb{Z}^4$. Note that we can find representatives of the cusps (cf. 2.3) in $\mathbb{Z}^4 \subset K^4$:

$$(1,0,0,1),(0,1,0,1),(0,0,1,1),(-1,0,0,1),(0,-1,0,1),$$
$$(0,0,-1,1),(1,-2,-2,3),(1,2,2,3),(2,1,2,3),(2,2,1,3).$$

Letting the cusps (now in $\mathbb{Z}^4$) be denoted by v_i ($i = 1, \ldots, 10$), there are $\binom{10}{3} = 120$ sets of 3 of them. For each such triple, say (v_i, v_j, v_k), we can find an orthogonal base of the 3-space they span:

$$w_1 = v_i + v_j$$
$$w_2 = v_i - v_j$$
$$w_3 = -(v_j, v_k)v_i - (v_i, v_k)v_j + (v_i, v_j)v_k.$$

(Here (,) denotes the form Φ for notational simplicity). Using this base we can calculate the signature: (note $(v_i, v_j) < 0$ for any i, j)

$$(w_1, w_1) = (v_i + v_j, v_i + v_j) = 2(v_i, v_j) < 0$$
$$(w_2, w_2) = (v_i - v_j, v_i - v_j) = -2(v_i, v_j) > 0$$
$$(w_3, w_3) = -2(v_i, v_j)(v_j, v_k)(v_i, v_k) > 0,$$

so on *any* such 3-space, the form Φ has signature (2, 1). Of these 120, there are exactly 15 which contain a 4th cusp, and the images of these 15 subspaces of $\mathbb{Z}^4$ in $\mathbb{Z}/3\mathbb{Z}^4$ give the combinatorial description of the modular subvarieties. This amounts then, a postiori, to a linear combination, *in* $\mathbb{Z}/3\mathbb{Z}$, of the cusps given in 2.3. For example, the 3-plane spanned by (v_1, v_2, v_8) also contains v_9: $-v_1 + v_2 + v_8 \equiv v_9 (\mathrm{mod}\ 3)$.

2.7. We now come to the proof of

THEOREM 2.7.1. $Y(1-\rho)^\wedge = \hat{\mathbb{P}}^3$, *the* D_i *are the modular subvarieties of* 2.5.

$i = 1, \ldots, 15$ and the E_λ are the compactification divisors of 2.4. $\lambda = 1, \ldots, 10$.

Proof. Let $Y(3, H)$ be the Fermat cover of degree 3 associated to the arrangement H (see [Hu] for details on this construction), i.e. the variety whose function field is

$$\mathbb{C}[{}^{x_1}/x_0, {}^{x_2}/x_0, {}^{x_3}/x_0][({}^{l_2}/l_1)^{1/3}, \ldots, ({}^{l_{10}}/l_1)^{1/3}]$$

where $\{l_i = 0\} = H_i$. In [Hu] I constructed a desingularisation and calculated the Chern numbers of $Y(3, H)$, as well as the logarithmic Chern classes of $(Y(3, H), \bar{E})$, where $\bar{E} = \pi^{-1}(\sum E_\lambda)$. It turned out that $\bar{c}_1{}^3(Y, \bar{E}) = 3\bar{c}_1\bar{c}_2(Y, \bar{E})$ (logarithmic Chern numbers) so by Kobayashi's generalisation of Yau's theorem quoted in [Hu], (also proved by Yau), $Y - \bar{E}$ is a smooth, non-compact ball quotient, Y its compactification. The desingularisation described in [Hu] is affected by blowing up $\mathbb{P}^3$ in exactly the same manner as above, so the smooth cover $Y \to \hat{\mathbb{P}}^3$ is a branched cover of $\hat{\mathbb{P}}^3$, or put differently, $\hat{\mathbb{P}}^3$ is a ball quotient; we just have to identify the group. Let Γ_Y be the group such that $\Gamma_Y \backslash \mathbb{B}^3 = Y - \bar{E}$. Then $\Gamma \supset \Gamma_Y$, $\Gamma = \pi_1(\hat{\mathbb{P}}^3 - E - D)$, as $\Gamma_Y \backslash \mathbb{B}^3$ is a cover of $\hat{\mathbb{P}}^3$ which is unramified over $\hat{\mathbb{P}}^3 - E - D$. The quotient $\Gamma/\Gamma_Y = (\mathbb{Z}/3\mathbb{Z})^9$ is the Galois group of $Y \to \hat{\mathbb{P}}^3$.

Utilising known results on the hypergeometric differential equation, Γ is the monodromy group of Appell's equation, number 1 in the [DM] list in dimension 3. Later we will identify this differential equation with the Picard Fuchs equation of the periods of Picard curves, whose monodromy group is easily identified with $\Gamma(1 - \rho)^\wedge = \mathrm{U}(3, 1; \mathcal{O}_K(1 - \rho))$ (see §5–§6). The identification of this particular group is thus by means of the scheme:

$$(\text{Fermat cover}) \xrightarrow{1} ([\text{DM}]\text{-}\#) \xrightarrow{2} (\text{Picard curves}) \xrightarrow{3} (\text{monodromy group}).$$

Step 1 was done in [Hu]. Step 2 will be done in Sections 5–6, Step 3 in Section 6. The statements about D_i and E_λ follow from [Ho1] (identification of the D_i) and direct calculations (showing E_λ is an abelian variety as in 2.4).

Let us just mention that much of this is more or less well-known.

2.8. Finite covers

In this section we clarify a few questions which were left untouched up till now. We consider the following coverings:

$$\begin{array}{ccccccc}
\Gamma(4)\backslash\mathbb{S}_2 & \subset & (\Gamma(4)\backslash\mathbb{S}_2)^\wedge & & (\Gamma(1-\rho)^2\backslash\mathbb{B}^3)^\wedge & \supset & \Gamma(1-\rho)^2\backslash\mathbb{B}^3 \\
& \searrow & (\mathbb{Z}/2\mathbb{Z})^9 \searrow & & \swarrow (\mathbb{Z}/3\mathbb{Z})^9 & \swarrow & \\
& & \Gamma(2)\backslash\mathbb{S}_2 & \subset \ \hat{\mathbb{P}}^3 \ \supset & \Gamma(1-\rho)\backslash\mathbb{B}^3 & &
\end{array}$$

We explain now the inclusion $Y(2) \subset \hat{\mathbb{P}}^3$, which will be given an easy proof in the next section. Delete $\cup D_i$ from $\hat{\mathbb{P}}^3$. Then the inclusion $Y(2) \subset \hat{\mathbb{P}}^3$ is such that the Humbert surfaces are the divisors $E_1, \ldots, E_{10}$. As mentioned above, each D_i is a Kummer modular surface (compactification divisor), and $\hat{\mathbb{P}}^3$ is the Igusa desingularisation of $Y(2)^*$.

Taking that for granted, one has two natural covers,

$$Y(4)^\wedge \to Y(2)^\wedge \quad \text{and}$$
$$Y(1-\rho)^{2\wedge} \to Y(1-\rho)^\wedge$$

(using obvious notations). We claim these are in fact both Fermat covers, of degrees 2 and 3, respectively. To see this, first note that both $P\Gamma(4)/P\Gamma(2)$ and $P\Gamma(1-\rho)/P\Gamma(1-\rho)^2$ are abelian. In fact, it is true that $\Gamma(\mathfrak{D})/\Gamma(\mathfrak{D}^2)$ is abelian for any ideal $\mathfrak{D}$. (Steve Weintraub pointed this out to me. Just calculate $(A+B)^2 \bmod(\mathfrak{D}^2)$.) The coefficients are in $\mathbb{Z}/4\mathbb{Z}/\mathbb{Z}/2\mathbb{Z} = \mathbb{Z}/2\mathbb{Z}$ and $\mathcal{O}_K(1-\rho)/\mathcal{O}_K(1-\rho)^2 = \mathbb{Z}/3\mathbb{Z}$, respectively, and a matrix in $\mathrm{PSp}(2, \mathbb{Z})$ and $\mathrm{PU}(3, 1; \mathcal{O}_K)$, respectively, will have 9 independent entries.

Once we know the groups are correct, we just have to note that the fixed point set under these Galois groups which consists of the union $\cup D_i \cup E_j$ of 2.5, are hermitian symmetric , and in fact identical in the Fermat covers as well as in $Y(4)^\wedge \to Y(2)^\wedge$ and $Y(1-\rho)^{2\wedge} \to Y(1-\rho)^\wedge$, respectively. This also allows us to count modular subvarieties and compactification divisors:

Siegel: $10 \cdot 2^4 = 160$ modular subvarieties, $15 \cdot 2^3 = 120$ compactification divisors,

Picard: $15 \cdot 3^3 = 135$ modular subvarieties, $10 \cdot 3^4 = 270$ compactification divisors.

We remark that this discussion finishes, modulo the proof of 3.3.3. below, the proof of 1.7.1. above.

3. Modular forms

In this paragraph we prove the theorem stated in the introduction, utilising for the proof modular forms. First we recall the structure of $\mathrm{R}(\Gamma(2))$, a result due to Igusa [I1]. We then deduce the structure of $\mathrm{R}(\Gamma(1-\rho))$. It turns out that these rings are dual to each other, i.e. the projective varieties $\mathrm{Proj}(\mathrm{R}(\Gamma(2)))$ and $\mathrm{Proj}(\mathrm{R}(\Gamma(1-\rho)))$ are dual. This implies they are birational, and our theorem follows.

3.1. Theta Constants

In this section we review the work of Igusa [I1]–[I2]. For later use we will need theta constants of genus 2 and 4, so in this section we give definitions and results for any g. Let $\tau \in \mathbb{S}_g$, $z \in \mathbb{C}^g$, and $m = (m', m'') \in \mathbb{Q}^{2g}$.

DEFINITION 3.1.1. The theta function of degree g and characteristic m is

$$\theta_m(\tau, z) = \sum_{n \in \mathbb{Z}^g} \exp(\tfrac{1}{2}{}^t(n + m')\tau(n + m') + {}^t(n + m')(z + m'')).$$

The corresponding theta constant is

$$\theta_m(\tau) := \theta_m(\tau, 0).$$

Igusa has studied these theta constants. Some of his results are:

LEMMA 3.1.2. $\theta_m(\tau) \equiv 0 \Leftrightarrow$ mmod(1) *satisfies* $\exp(4\pi i^t(m')(m'')) = -1$.

The Siegel modular group $\Gamma_g(1) := \mathrm{Sp}(g, \mathbb{Z})$ acts on the arguments (τ, z) by:

$$M(\tau, z) = ((A\tau + B)(C\tau + D)^{-1}, (C\tau + D)^{-1}z) \tag{3.1.2a}$$

and on the characteristic itself by

$$M\colon m = (m', m'') \mapsto \begin{pmatrix} D & -C \\ -B & A \end{pmatrix} m + \frac{1}{2}\begin{pmatrix} \mathrm{diag}\,(C^tD) \\ \mathrm{diag}\,(A^tB) \end{pmatrix}. \tag{3.1.2b}$$

The behavior of the thetas under M is given by

LEMMA 3.1.3. (Igusa's transformation law), [I1], p. 226

$$\theta_{Mm}(M(\tau, z)) = \kappa(M)\exp(2\pi i\phi_m(M))\det(C\tau + D)^{1/2} \times$$
$$\times \exp(\pi i^t z(C\tau + D)^{-1}Cz)\theta_m(\tau, z),$$

where $\kappa(M)$ *is some* 8*th root of unity,*

$$\phi_m(M) = -\tfrac{1}{2}{}^t m' BDm' + {}^t m''{}^t ACm'' - 2{}^t m'{}^t BCm'' - {}^t\mathrm{diag}(A^tB)(Dm' - Cm'').$$

In particular, for the theta constants the formula becomes

$$\theta_{Mm}(M\tau) = \kappa(M)\exp(2\pi i\phi_m(M))\det(C\tau + D)^{1/2}\theta_m(\tau).$$

3.2. The ring of modular forms for $\Gamma(2)$.

We review the results of Igusa describing $R(\Gamma(2))$, the ring of modular forms for $\Gamma(2)$. For the rest of this section we take the following particular case of 3.1: $g = 2$,

$m = (m', m'') \in \frac{1}{2}\mathbb{Z}^4$. There are 16 such characteristics, 6 of which yield odd theta functions ($\theta_m(\tau, z) = -\theta_m(\tau, -z)$) and hence zero theta constants (this is a special case of 3.1.2). There are 10 even characteristics, and among their fourth powers there are 5 linear relations (Riemann theta formula). In fact,

THEOREM 3.2. [I1, p. 397] *Let*

$$\begin{aligned} y_0 &= \theta^4_{(0110)}(\tau), \quad y_1 = \theta^4_{(0100)}(\tau), \quad y_2 = \theta^4_{(0000)}(\tau), \\ y_3 &= \theta^4_{(1000)}(\tau) - \theta^4_{(0000)}(\tau), \quad y_4 = -\theta^4_{(1100)}(\tau) - \theta^4_{(0000)}(\tau). \\ \chi_{10} &= \Pi\theta^2_m \end{aligned}$$

Then,

$$R(\Gamma(2)) = \mathbb{C}[y_0, \ldots, y_4, \chi]/_{\mathscr{E}}, \ \mathscr{E} \textit{ the ideal generated by}$$

$$\mathscr{E} = \begin{cases} R_1 = (y_0y_1 + y_0y_2 + y_1y_2 - y_3y_4)^2 - 4y_0y_1y_2(\Sigma y_i) \\ R_2 = \chi^2_{10} - \frac{1}{4}f(y_0, \ldots, y_4), \ f \textit{ homogenous of degree } 5. \end{cases}$$

This theorem implies in particular, that the (singular) quartic defined by R_1 is the Baily-Borel embedding of $X(2)$. The y_i are modular forms of weight 2 with respect to $\Gamma(2)$, as follows from the transformation law 3.1.3. A more symmetric description of the same variety is given by van der Geer in [V, §5]. This is done by taking all thetas $\theta^4_m(\tau)$, and the map into $\mathbb{P}^9$,

$$\phi: (\Gamma(2)\backslash\mathbb{S}_2)^* \to \mathbb{P}^4 \subset \mathbb{P}^9, \tag{3.2.1}$$

The image being onto the $\mathbb{P}^4$ which is cut out by the 5 linear relations. The equation for $X(2)^*$ is then ([V], 5.2)

$$\left(\Sigma\theta^8_m\right)^2 - 4\left(\Sigma\theta^{16}_m\right) = 0. \tag{3.2.2}$$

In this description the action of $\mathrm{Sp}(2, \mathbb{Z}/2\mathbb{Z})$ on $X(2)^*$ is the action on the characteristics 3.1.2b. It is known that each theta vanishes along exactly one of the Humbert surfaces, so we can identify this action with the action of Σ_6 on H_1 described in 1.6. This replaces the rather artificial action 3.1.2b by the much more natural one of Σ_6 on $(\mathbb{Z}/2\mathbb{Z})^4$ described in 1.6. This observation is due to van der Geer [V] and Lee and Weintraub [LW].

3.3. The ring of modular forms on $Y(1-\rho)^*$

We now proceed in the following manner: first we recall the (well-known) identification of $Y(1-\rho)^*$ with the Segre Cubic. Then, since we know the coordinate ring of the Segre cubic, we know the coordinate ring of $Y(1-\rho)^*$. Finally it is not difficult to see that the natural coordinates used are indeed modular forms.

The Segre Cubic

There is a unique cubic 3-fold S in $\mathbb{P}^4$ with 10 ordinary doublepoints. S is given most symmetrically by the following 2 equations in the homogenous variables $x_0, \ldots, x_5$ on $\mathbb{P}^5$:

$$\sum x_i = 0$$
$$\sum x_i^3 = 0.$$

The double points are $(1, 1, 1, -1, -1, -1)$ and its permutations under Σ_6, which acts naturally on S by permuting coordinates. There are 15 $\mathbb{P}^2$'s lying on S:

$$x_{\sigma(1)} + x_{\sigma(4)} = x_{\sigma(2)} + x_{\sigma(5)} = x_{\sigma(3)} + x_{\sigma(6)} = 0, \quad \sigma \in \Sigma_6.$$

The GIT-Quotient

It was already mentioned above that the following was proved in [LW], see also [KLW]:

LEMMA 3.3.1. $\tilde{S} = \hat{\mathbb{P}}^3 =$ GIT-*quotient* #1 *in* [DM]. ($\tilde{S}$ *denotes the (big) resolution of* S).

From this and 2.7.1 it follows first that $S = Y(1-\rho)^*$ and from this that the coordinate ring of $Y(1-\rho)^*$ coincides with that of S, i.e.

$$R(Y(1-\rho)^*) = \mathbb{C}[x_0, \ldots, x_5]/\mathscr{E} \tag{3.3.2}$$

$$\mathscr{E} = \begin{cases} R_1 = \sum x_i \\ R_2 = \sum x_i^3. \end{cases}$$

An important additional bit of information we get from the argument of 2.7. is the following: The D_i, being in the branch locus of the cover $Y \to \hat{\mathbb{P}}^3$, are pointwise fixed under the Galois group, hence subball quotients. (The only submanifolds in the universal cover $\mathbb{B}^3$ fixed under automorphisms are subballs.) This implies on the one hand the fact (that we already know) that the D_i are modular subvarieties, and on the other implies that the divisors D_i are the zero-loci of modular forms (since roots of their quotients give the field extension of the finite cover $Y \to \hat{\mathbb{P}}^3$ described in 2.5). Hence there are 15 modular forms δ_i, $i = 1, \ldots, 15$, with the property that each vanishes identically along precisely one of the D_i and does not vanish identically at any other. Taking all of these forms as coordinates we get a map

$$\begin{aligned} &\phi\colon Y(1-\rho) \to \mathbb{P}^5 \subset \mathbb{P}^{14} \\ &t \to \delta_i(t) \in S \end{aligned}$$

onto the Baily-Borel compactification (the x_i above will be linear combinations of the δ_i) much in the spirit of van der Geer's (3.2.1.–3.2.2. above).

THEOREM 3.3.3. *$\hat{\mathbb{P}}^3$ is the common resolution of singularities of $X(2)^*$ and $Y(1-\rho)^*$, i.e. we have a diagram*

$$\begin{array}{ccc} & \hat{\mathbb{P}}^3 \quad \text{GIT} & \\ \swarrow \sigma_1 & & \sigma_2 \searrow \\ \text{Siegel } X(2)^* & & Y(1-\rho)^* \quad \text{Picard.} \end{array}$$

Using the notation in the introduction, we have

$(\hat{\mathbb{P}}^3, D)$ *corresponds to* $(\mathbb{S}_2, \Gamma(2))$
$(\hat{\mathbb{P}}^3, E)$ *corresponds to* $(\mathbb{B}^3, \Gamma(1-\rho))$.

This follows from the above results as follows: we know $X(2)^*$ is the quartic described in 3.2, $Y(1-\rho)^*$ is the Segre cubic, and classically (1880's !) it is known that these varieties are dual to each other. This implies they are birational, and σ_1 and σ_2 have been explicitly described above. σ_1 blows down the D_i to $\mathbb{P}^1$'s, σ_2 blows down the E_j to ordinary double points.

3.4. Theta constants of degree 4

Assume for the moment we have a modular embedding $(\mathbb{B}^3, \Gamma(1-\rho)) \subset (\mathbb{S}_4, \Gamma(?))$, where $\Gamma(?)$ is a not further specified level subgroup. We will see later in Section 5 that this exists via Jacobians of Picard curves. Consider the diagram

$$\begin{array}{ccc} \mathbb{B}^3 & \subset & \mathbb{S}_4 \\ \downarrow & & \downarrow \\ \Gamma(1-\rho)\backslash\mathbb{B}^3 & \subset & \Gamma(?)\backslash\mathbb{S}_4 =: X_4(?) \end{array}$$

It follows from Igusa's results ([I2, Cor., p. 235]) that theta constants can be used to embed $X_4(?)$ in projective space (ala Satake-Baily-Borel). Restricting these thetas to $\Gamma(1-\rho)\backslash\mathbb{B}^3$ yields Picard modular forms. This is the idea behind Feustel's proof (in dimension 2) of

THEOREM 3.4.1 [F, II]. *The modular forms* ξ_i *mentioned below in* 6.1.8 *for* $U(2,1;\mathcal{O}_K(1-\rho))$ *can be written as follows in terms of theta constants*:

$$y_i := \theta\begin{bmatrix} 0 & 1/6 & 0 \\ i/3 & 1/6 & i/3 \end{bmatrix}, \quad i = 0, 1, 2.$$

$$\begin{aligned} \xi_1 &= \sum y_i^3 \\ \xi_2 &= -3y_1^3 + y_2^3 + y_3^3 \\ \xi_3 &= y_1^3 - 3y_2^3 + y_3^3 \\ \xi_4 &= y_1^3 + y_2^3 - 3y_3^3. \end{aligned}$$

This gives at least partial information on $\Gamma(?)$: it contains the group $\Gamma(6, 36)$. Since we have D_i given by the equations, say

$$x_1 = -x_4, \quad x_2 = -x_5, \quad x_3 = -x_6,$$

we can use x_1, x_2 and x_3 as coordinates on $\Gamma(1-\rho)\backslash\mathbb{B}^2$, so by 3.4.1, the x_i can be written in terms of genus 3 thetas when restricted to the D_i. Hence in principle at least, we can take the modular forms in 3.3.2 to be genus 4 theta constants which restrict on the modular subvarieties D_i to the theta constants in 3.4.1. It would be very interesting to get some explicit results in this direction. This is, however, a highly non-trivial task. One would first have to find an explicit embedding $(\mathbb{B}^3, \Gamma) \subset (\mathbb{S}_4, \Gamma(?))$ together with 15 compatible sub-modular embeddings

$$\begin{array}{ccc} (\mathbb{B}^3, \Gamma) & \subset & (\mathbb{S}_4, \Gamma(?)) \\ \cup & & \cup \\ (\mathbb{B}^2, \Gamma_{2,1}) & \subset & (\mathbb{S}_3, \Gamma_3(?)). \end{array}$$

The 15 subballs will give the intersection of $(\mathbb{B}^3, \Gamma)$ with the zero locus of the sought for theta constants. Let us briefly describe what this looks like in local coordinates. Say

$$\mathbb{B}^2 = \{(z_1, z_2) \in \mathbb{C} \mid |z_i|^2 < 1\},$$
$$\mathbb{B}^2 = \{z_3 = 0\} \subset \mathbb{B}^3 = \{(z_1, z_2, z_3) \in \mathbb{C}^3 \mid \sum |z_i|^2 < 1\},$$

and

$$\mathbb{S}_3 = \begin{bmatrix} \tau_0 & \tau_1 & \tau_2 \\ \tau_1 & \tau_3 & \tau_4 \\ \tau_2 & \tau_4 & \tau_5 \end{bmatrix} \subset \mathbb{S}_4 = \begin{bmatrix} \tau_0 & \tau_1 & \tau_2 & x_1 \\ \tau_1 & \tau_3 & \tau_4 & x_2 \\ \tau_2 & \tau_4 & \tau_5 & x_3 \\ x_1 & x_2 & x_3 & \tau_6 \end{bmatrix}.$$

The modular embeddings are:

$$\mathbb{B}^2 = \{z_1 = \tau_2, z_2 = \tau_4\} \quad \text{or} \quad \{z_1 = x_1, z_2 = x_2\}, \quad \mathbb{B}^3 = \{z_i = x_i\}.$$

These 4 spaces parameterise:

$\mathbb{S}_4$ = principally polarised abelian 4-folds
$\cup$
Ξ = Shottky divisor = locus of jacobians = genus 4 curves
$\cup$
$\mathbb{B}^3$ = Jacobians with complex multiplication = Picard curves
$\mathbb{S}_4 \supset \mathbb{S}_3$ = genus 3 curves (= degenerating abelian 4-folds)

$\mathbb{B}^2$ = Picard curves (Jacobians with complex multiplication)
$\|$
$\mathbb{B}^3 \cap \mathbb{S}_3$ = Degenerate Jacobians with complex multiplication.

Next one would have to find the right level groups acting on $\mathbb{S}_3$ and $\mathbb{S}_4$ and show the equivariance of the diagram above with respect to the groups involved. Finally, one would have to describe the genus 3 thetas as special values of genus 4 theta constants on $\mathbb{S}_4$. As the embeddings involved are rather complicated (see for example those given by Resnikoff-Tai), this would seem to be a formidable task.

4. Tits buildings with scaffoldings

4.1. Let $\Gamma \subset G_{\mathbb{Q}}$ be an arithmetic subgroup, and $P_1, \ldots, P_k$ a complete set of Γ-inequivalent maximal parabolics. One can construct a simplicial complex associated to Γ, the *Tits Building*, as follows:

vertices: $v_i \leftrightarrow P_i$
1-edges: $v_{ij} \leftrightarrow P_i \subset P_j$ inclusions
2-faces: $v_{ijk} \leftrightarrow P_i \subset P_j \subset P_k$ flags
… …

First, let $\Gamma = \mathrm{Sp}(2, \mathbb{Z})$. Then there are 2 maximal parabolics P_1, P_2, corresponding to the 0-dimensional and 1-dimensional boundary components, respectively. There is one inclusion $P_1 \subset P_2$. Hence the Tits building is:

$$\mathrm{Sp}(2, \mathbb{Z}): \quad v_1 \circ\!\!-\!\!\!-\!\!\circ\, v_2.$$

The building for $\Gamma_{\sqrt{-3}} = \mathrm{SU}(3, 1; \mathcal{O}_K)$ is even more boring, being just one point:

$$\mathrm{SU}(3, 1; \mathcal{O}_K) \quad v_1 \circ$$

Now we consider the congruence subgroups $\Gamma(2)$ and $\Gamma(1-\rho)$. The building of $\Gamma(1-\rho)$ is still not too interesting, consisting of 10 disjoint verticies:

$$\Gamma(1-\rho)\ v_1 \circ \cdots \circ v_{10}.$$

The interesting Tits building is that of $\Gamma(2)$. There are 15 verticies $v_1, \ldots, v_{15}$ corresponding to the 1-dimensional components, and 15 verticies $w_1, \ldots, w_{15}$ corresponding to the 0-dimensional components. Each w_i has 3 inclusions into v_j's, and v_j contains 3 w_i:

(4.1.1.)

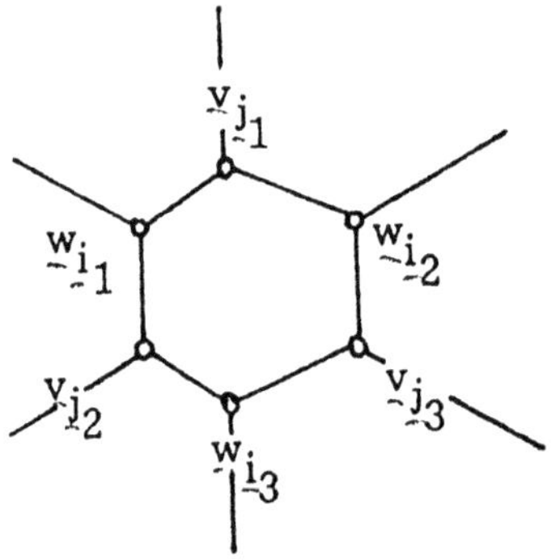

There are therefore $3(15+15).\frac{1}{2} = 45$ edges, see Figure 4.1.2.

We now "blow-up" this Tits building, i.e. replace each w_i by a 2-simplex:

and we note that the blown-up Tits building is the dual complex of the normal crossings divisor D. This means we have 1 vertex for each component, 1 edge for each double intersection and 1 2-simplex for each triple point:

$$v_i \leftrightarrow D_i \text{ divisor} \qquad (15)$$

$$v_i v_j \leftrightarrow D_i \cap D_j \qquad (45)$$

$$\Delta v_i v_j v_k = w_\lambda \leftrightarrow D_i \cap D_j \cap D_k \qquad (15)$$

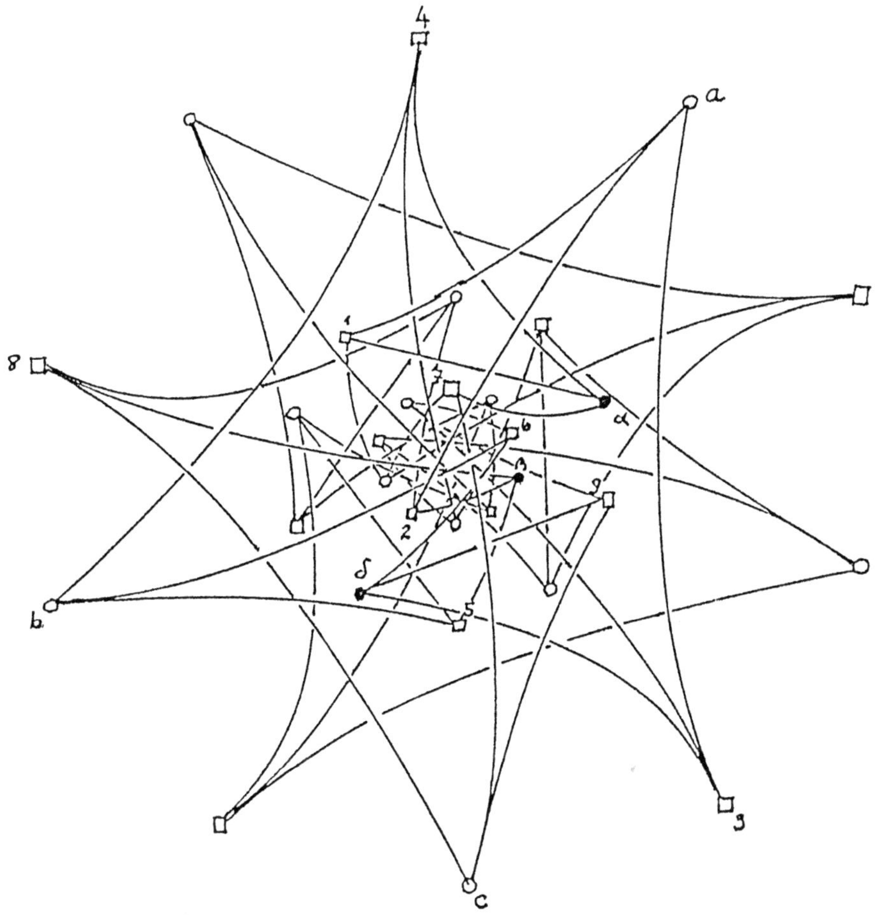

Figure 4.1.2.

4.2. Modular Scaffoldings

We consider a fixed locally symmetric space $(X, \Delta) = (\mathscr{D}, \Gamma)$ (we recall that this notational convenience explained in the introduction means $X - \Delta = \Gamma\backslash\mathscr{D}$), and modular subvarieties $(D_i, \Delta \cap D_i) = (\mathscr{D}_i, \Gamma)$ for $i \in$ Index set.

DEFINITION 4.2.1. A finite set $(\mathscr{D}_i, \Gamma_i) \subset (\mathscr{D}, \Gamma)$, $i = 1, \ldots, N$, is called a *Modular scaffolding of* $(\mathscr{D}, \Gamma)$, if:

(i) Each $(\mathscr{D}_i, \Gamma_i) \subset (\mathscr{D}, \Gamma)$ is a modular embedding.

(ii) Each intersection $(D_i \cap D_j, \Delta \cap (D_i \cap D_j)) = (\mathscr{D}_{ij}, \Gamma_{ij})$ is a modular embedding in both $(\mathscr{D}_i, \Gamma_i)$ and $(\mathscr{D}_j, \Gamma_j)$.

(iii) $\exists_{\lambda_i, \mu_j \in \mathbb{Q}}\, c_1(X) = \sum \lambda_i D_i + \sum \mu_j \Delta_j, \Delta = \sum \Delta_j$.

Now it is an easy matter to see that the modular subvarieties discussed in 1.5 and

2.5 form modular scaffoldings of $X(2)$ and $Y(1-\rho)$, respectively. The necessary Chern class calculations can be found in [V, §3], for example.

4.3. Scaffoldings on buildings

We now introduce a notion due to Lee-Weintraub ([LW1]), in a somewhat different fashion. According to 4.2 a modular scaffolding on (X, Δ) will be a normal crossings divisor.

DEFINITION 4.3.1. A *scaffolding* on the Tits building of (X, Δ) is the dual complex of a modular scaffolding.

This dual complex is a simplicial complex consisting of 1 vertex for each component, 1 edge for each simple intersection, and so on. Now just looking at the example above, we have:

PROPOSITION 4.3.2. *The scaffolding of* $Y(1-\rho)$ *is the blown-up Tits building of* $X(2)$. *The scaffolding of* $X(2)$ *is the Tits building of* $Y(1-\rho)$.

We now would like to combine the Tits building with the scaffolding of (X, Δ). The easy way to do this is to take the graph of the normal crossings divisor $\Delta + D$, Δ the compactification divisor and D the modular scaffolding as above. We call this complex the Tits building with scaffolding, Tbws.

PROPOSITION 4.3.3. *The Tits building with scaffolding of* $\Gamma(2)$ *and of* $\Gamma(1-\rho)$ *are the same, and the scaffolding of* $\Gamma(2)$ *is isomorphic to the Tits building of* $\Gamma(1-\rho)$ *and vice versa.*

This is what we mean by saying that $\Gamma(2)$ and $\Gamma(1-\rho)$ have dual Tits buildings with scaffolding. Now going back to the diagramm in 3.3.3 we recall the GIT-quotient had both Δ and D as compactification divisor. Consider its (blown-up) Tits building.

PROPOSITION 4.3.4. *The* GIT-*quotient's (blown-up) Tits building coincides with the* Tbws *mentioned in* 4.3.3. *The scaffolding of this* GIT-*quotient is trivial.*

Hence we see the Tits building of the GIT-quotient specialising to a Tbws in 2 different manners (dual to each other).

5. Modular interpretation

5.1. Families of marked lines

Let $S = \{1, 2, 3, 4, 5, 6\}$ and $\{x_i\}_{i \in S}$ a set of 6 points on $\mathbb{P}^1$, i.e. a map $\phi: S \to \mathbb{P}^1$. The set of all such maps is naturally isomorphic to $(\mathbb{P}^1)^6$. The "diagonals" Δ are the divisors $\{x_i = x_j, i \neq j = 1, \ldots, 6\}$ and their intersections. PGL(2) acts on $X := (\mathbb{P}^1)^6 - \Delta$ and the quotient, which is the GIT quotient discussed in 3.3, is isomorphic to $\mathbb{P}^3 - H$. H the arrangement consisting of the 10 planes of 2.5. In other words we have a fibre space

$$(\mathbb{P}^1)^6 - \Delta \to \mathbb{P}^3 - H$$

with fibre PGL(2). The explicit form of this map is given in [LW, §6], as mentioned above in Section 3. Here we are using the identification of $\mathbb{P}^3 - H$ with $S - D$, S the Segre cubic, derived in 3.3. We let $\mathscr{F}_0$ denote this family of marked $\mathbb{P}^1$'s over $\mathbb{P}^3 - H$.

5.2. Hyperelliptic curves

Let the $\{x_i\}_{i \in S}$ be as in 5.1, and consider the hyperelliptic curve

$$y^2 = \prod_1^6 (t - x_i) \tag{5.2.1}$$

which is a branched cover of $\mathbb{P}^1$, branched at each of the x_i. We let $\mathscr{M}_0 \to \mathscr{F}_0$ be the double cover of $\mathscr{F}_0$ branched at the $\{\phi_i\}$. To be precise, this object only exists in the category of algebraic stacks, i.e., the universal curve only exists locally, and cannot be globally constructed as $-1 \in \Gamma(2)$. However this is sufficient for our purposes. Hence, for each $\{x_i\}$ mod PGL(2) $\in \mathbb{P}^3 - H$, the fibre $(\mathscr{M}_0)_{\{x_i\}}$ is the double cover of $\mathbb{P}^1$ branched at the $\{x_i\}$ mod PGL(2). Thus we get a "fibre space"

$$\mathscr{M}_0 \to \mathbb{P}^3 - H = X(2) - E,$$

of genus 2 curves.

We now describe the degenerations of 5.2.1 corresponding to the divisors D (compactification divisor, 1.4) and E (Humbert surfaces, 1.5). We shall employ the following notation:

$D_i, i = 1, \ldots, 15$:	2 of the x_n coincide
$D_{ij} = D_i \cap D_j$:	2 pairs of the x_n coincide
$D_{ijk} = D_i \cap D_j \cap D_k$:	3 pair of the x_n coincide
$E_\lambda, \lambda = 1, \ldots, 10$:	3 of the x_n coincide
$D_i \cap E_\lambda$:	1 pair and 1 triple of the x_n coincide
$D_{ij} \cap E_\lambda$:	2 triples of the x_n coincide

D_i: If 2 of the x_n coincide, then we have a double cover with 5 branch points; this is a genus one curve (elliptic) with one double point.

D_{ij}: Reasoning as above we get here a rational curve with 2 double points.

D_{ijk}: Now we have only 3 branching points and at each branch point the degeneration is:

so that the double cover splits into 2 curves which are permuted by the Galois group, and each branch point is now a double point of the covering.

E_λ: when 3 of the x_n coincide, the curve is $y^2 = \Pi_1^3(t - x_n)$, an elliptic curve. To see precisely what is going on, let

$$y^2 = \prod_1^6 (t - x_n)$$

be the original equation, and write it as

$$y^2 = (t - x_1)(t - x_2)(t - x_3)(t - \lambda x_4)(t - \lambda x_5)(t - \lambda x_6)$$

and the degeneration is then given by letting $\lambda \to \infty$. The limit curve is

$$y^2 = \prod_1^3 (t - x_n), \tag{5.2.2}$$

with 4th branch point at infinity, and changing variable to $\tilde{t} = \lambda t$ we get

$$y^2 = (\tilde{t} - \lambda x_1)(\tilde{t} - \lambda x_2)(\tilde{t} - \lambda x_3)(\tilde{t} - x_4)(\tilde{t} - x_5)(\tilde{t} - x_6)$$

which for $\lambda \to \infty$ becomes

$$y^2 = \prod_4^6 (\tilde{t} - x_i), \tag{5.2.3.}$$

with 4th branch point at infinity. Therefore, over $p \in E_\lambda$ the corresponding degeneration consists of 2 elliptic curves 5.2.2 and 5.2.3, meeting at their common branch point.

REMARK. The degeneration just described of course depends on the birational model of $X(2)^\wedge$ used. That described here corresponds to the big resolution described in 2.4. There is also a small resolution of the ordinary double point, and in that case the degeneration is just an elliptic curve.

It is now obvious what the remaining degenerations are. A summary is given below in table 5.4.

5.3. Picard curves

Let $\{x_n\}_{n\in S}$ be again as in 5.1 and consider the trigonal curve

$$y^3 = \prod_1^6 (t - x_n). \tag{5.3.1}$$

This curve is called a Picard curve, being studied by Picard more than a century ago (actually it was the genus 3 curve he studied [P]). It is a genus 4 curve, whose Jacobian has complex multiplication by a cube root of unity, coming from the Galois action on the curve. As above we construct the triple cover $\mathscr{C}_0 \to \mathscr{F}_0$ whose fibre $(\mathscr{C}_0)_{\{x_n\}}$ is the curve 5.3.1. We get a fibering

$$\mathscr{C}_0 \to \mathbb{P}^3 - H = Y(1-\rho) - D,$$

of genus 4 curves. This result was first proven by Shimura [S].

We now describe the degenerations of 5.3.1. corresponding to the divisors D (modular subvarieties, 2.5) and E (compactification divisors, 2.4). We use the same notations as in 5.2.

D_i: If one pair of the x_n coincide, we get a 3-fold cover, branched at 5 points, which is a smooth, genus 3 Picard curve (see the introduction and 2.5.)

D_{ij}: Now the equation becomes $y^3 = \prod_1^4 (t - x_n)$, and each branch point induces:

so after 2 branch points we see that the cover splits into one elliptic and one rational component.

D_{ijk}: the picture now becomes:

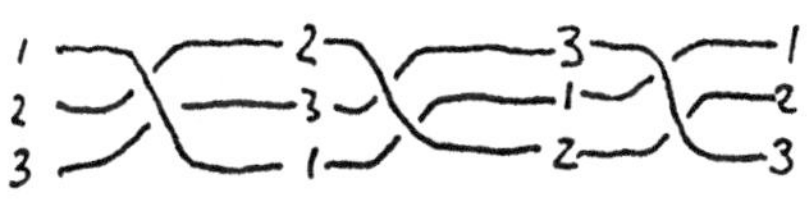

so the cover splits into 3 rational curves.

E_λ: At a point where 3 of the x_n coincide, the action of the Galois group is $\substack{1\\2\\3}$--$\substack{1\\2\\3}$ at that branch point, so it is a double point of the curve, and checking euler numbers it has genus 2.

$E_\lambda \cap D_i$: As above, where the pair of $\{x_n\}$ coincide the cover is still smooth, with the double point of E_λ (from the triple of the $\{x_n\}$ which coincide), and an euler number calculation shows it is elliptic.

$E_\lambda \cap D_{ij}$: Here one gets a rational curve with 2 double points.

We can extend the family $\mathscr{C}_0 \to Y(1-\rho) - D$ to all of $Y(1-\rho)^\wedge$ by adding in the degenerations just described. We denote this by $\mathscr{C}^\wedge \to Y(1-\rho)^\wedge$. We note that this is not a semi-stable family of curves (the arithmetic genus changes).

5.4. Totally degenerate stable curves of genus 3.

There is (at least) yet another moduli interpretation of $\hat{\mathbb{P}}^3$, described in detail in [GHv]. I am indebted to F. Herrlich and L. Gerritzen for this.

DEFINITION 5.4.1. A connected projective curve C, of arithmetic genus g, is called a *totally degenerate stable curve of genus g* if:

(a) every irreducible component of C is a rational curve.
(b) every singular point of C is an ordinary doublepoint, and
(c) every non-singular component C_i of C meets $C - L$ in at least 3 points.

Naturally associated with such a degenerate curve is a more combinatorial object, a tree;

DEFINITION 5.4.2. A connected set of mutually intersecting $\mathbb{P}^1$'s is called an *n-pointed tree of projective lines* if;

(i) the components intersect in ordinary double points
(ii) The intersection graph is a tree, and
(iii) given is a set $\{p_1, \ldots, p_n\}$ of distinct ("marked") points on the components. It is called *stable* if, in addition,
(iv) on every component there are at least 3 points which are either singular or marked.

Totally degenerate curves are also parameterised by the $(x_i) \in \hat{\mathbb{P}}^3$. The generic curve is as follows:

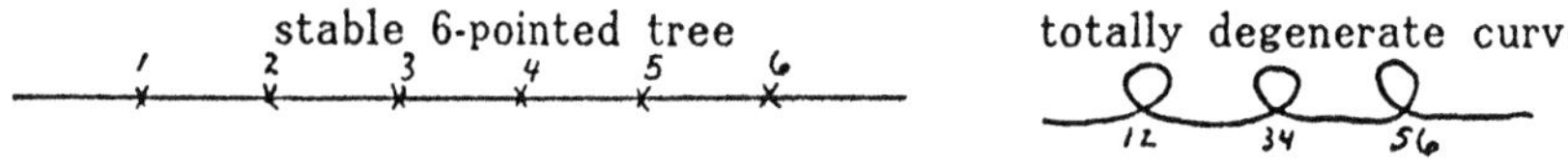

In the following we describe the 6-pointed trees corresponding to the loci D_i and F_λ as above:

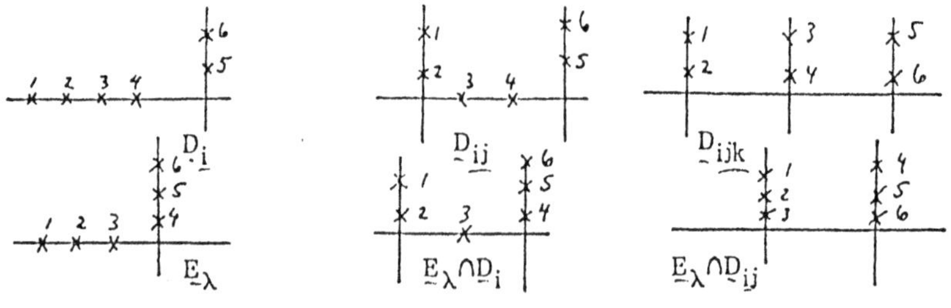

The corresponding degenerate curves depend in addition on the identification of the double points, for example the tree:

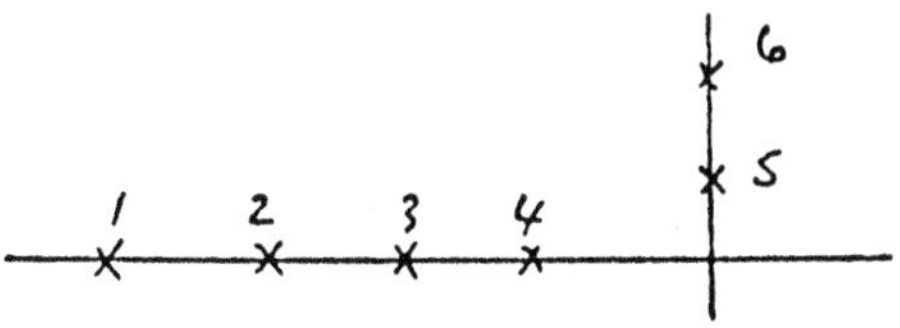

gives both

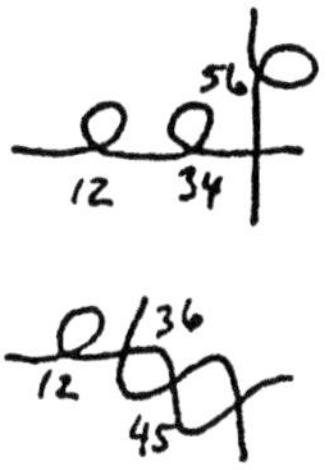

We have included these in the following table.

6. Differential equations

6.1. Picard-Fuchs equations and monodromy

Let $\pi:\mathscr{C}\to S$ be a family of smooth curves over a parameter space S, i.e. $\mathscr{C}$, S are algebraic varieties, and π is a holomorphic map. We consider the sheaf $R^1\pi_*\mathbb{C}$, which is a locally constant sheaf whose stalk at $x\in S$ is just $H^1(F_x,\mathbb{C})$, where F_x is the fibre at x. Since F_x is a Riemann surface this stalk splits according to the Hodge decomposition:

$$H^1(F_x,\mathbb{C}) = H^{1,0}(F_x)\oplus H^{0,1}(F_x).$$

$H^{1,0}(F_x)$ is the vector space of holomorphic 1-forms, and the decomposition is such that the position of $H^{1,0}(F_x)$ in $H^1(F_x,\mathbb{C})=H^1(F,\mathbb{C})$ depends holomorphically on x (here F is a typical fibre). Fix a base $\omega_1(x),\ldots,\omega_g(x)$ of $H^{1,0}(F_x)$, and let $\delta_1,\ldots,\delta_{2g}$ be 1 cycles forming a basis for $H_1(F,\mathbb{Z})$. The period matrix of S is the $g\times 2g$ matrix

$$\Omega(x)=\begin{pmatrix} \int_{\delta_1}\omega_1(x)\cdots\int_{\delta_{2g}}\omega_1(x) \\ \cdots\cdots\cdots\cdots \\ \int_{\delta_1}\omega_g(x)\cdots\int_{\delta_{2g}}\omega_g(x)\end{pmatrix}$$

which can be put in a normalised form $\Omega(x)=(1_g,Z(x))$, where $Z(x)\in\mathbb{S}_g$.

Table 5.4.

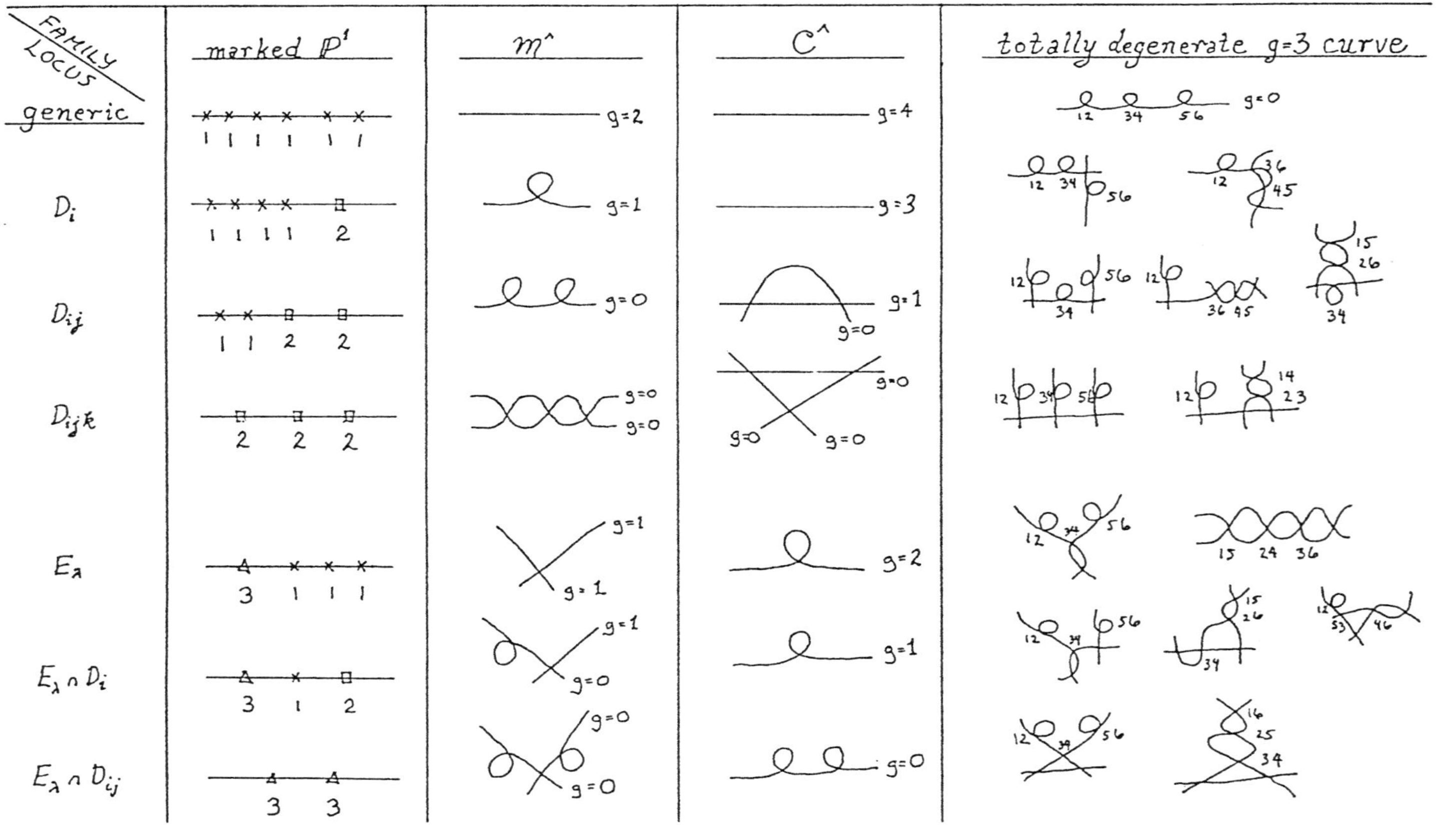
FAMILY
LOCUS
generic
D_i
D_{ij}
D_{ijk}
E_λ
$E_\lambda \cap D_i$
$E_\lambda \cap D_{ij}$
marked $\mathbb{P}^1$
1 1 1 1 1 1
1 1 1 1 2
1 1 2 2
2 2 2
3 1 1 1
3 1 2
3 3
$m^\wedge$
g=2
g=1
g=0
g=0
g=0
g=1
g=1
g=1
g=0
g=0
g=0
$C^\wedge$
g=4
g=3
g=1
g=0
g=0
g=0
g=0
g=2
g=1
g=0
totally degenerate g=3 curve
12 34 56 g=0
12 34 56
12 36 45
12 34 56
12 36 45
15 26 34
12 34 56
12 14 23
12 34 56
15 24 36
12 34 56
15 26 34
12 53 46
12 34 56
16 25 34

Associated with the family $\mathscr{C} \to S$ one then gets a holomorphic *period map*:

$$\begin{aligned} \omega: S &\to \mathbb{S}_g \\ x &\mapsto Z(x). \end{aligned} \tag{6.1.1}$$

Furthermore the Torelli theorem tells us that from $Z \in \mathbb{S}_g$ (mod $\mathrm{Sp}(g, \mathbb{Z})$) we can recover the curve F_x such that $Z = Z(x)$. This is the general picture, and well-known.

Now fix a section of the (1, 0) part of $R^1\pi_*\mathbb{C}$, in other words, fix a holomorphic 1-form $\omega(x)$ on each fibre, depending smoothly on x. Then the periods

$$\int_{\delta_1} \omega, \ldots, \int_{\delta_{2g}} \omega$$

are functions of the parameters, i.e. of x. It was first observed by Fuchs that, given ω, the $2g$ periods $\int_{\delta_i} \omega$ are all solutions of a linear differential equation of degree $2g$, the *Picard-Fuchs* equation. See Katz [Ka] for general remarks and higher-dimensional analogues.

Getting back to our family $\mathscr{C} \to S$ it is intuitively clear that if the fibres are "special" this leads to "special" Picard-Fuchs equations. For example, the special case of cycloelliptic curves

$$y^n = \Pi(t - x_i)^{\beta_i} \tag{6.1.2.}$$

is discussed in detail in part II of Holzapfel's book [Ho1], yielding equations he calls of Euler-Picard type. Both of our families are in fact special cases of these curves (§5.2: hyperelliptic, §5.3.:trigonal). Our special Picard-Fuchs equations will turn out to be hypergeometric differential equations.

We now just sketch how the monodromy of an algebraic differential equation relates to the geometry of the base space. First of all, the Picard-Fuchs equations have regular singular points (see e.g. [Gr] for precise definitions and statement of results). The singular locus $\Sigma \subset S$ is a divisor, which may be taken as normal crossings. For any $x \in S - \Sigma$, local solutions $\phi(x)$ will be single-valued. If the space of solutions of the differential equation has dimension d, let $\phi_1, \ldots, \phi_d$ be a local base at some fixed point $_* \in S - \Sigma$. For $\gamma \in \pi_1(S - \Sigma,_*)$ we can consider analytic continuation of the ϕ_i, and the continued solution, upon returning to $_*$, can be written as a linear combination of the ϕ_i's one started with, yielding a representation

$$\rho: \pi_1(S - \Sigma) \to \mathrm{GL}(d, \mathbb{C}) \tag{6.1.3}$$

called the *monodromy representation*. Of course, interesting things might happen if the image of ρ lies in some particular subgroup. For example, suppose $\mathscr{C} \to S$ is a family of elliptic curves over a curve, i.e. $\mathscr{C}$ is an elliptic surface. The Picard-Fuchs equation here is a second-order linear differential equation (see [St])

$$(D^2 - P(x)D + Q(x))F = 0 \tag{6.1.4}$$

with P and Q being singular at the discriminant of the family $\mathscr{C} \to S$. One then has the theorem [St, §3]:

The differential equation 6.1.4 coming from a family $\mathscr{C} \to S$ of elliptic curves is characterised as follows:

(1) The dimension of the solution space is 2.
(2) If ω_1 and ω_2 are 2 linearly independent solutions, then $\omega_1(x)/\omega_2(x) \in \mathbb{S}_1$ for $x \in S - \Sigma$.
(3) The monodromy lies in $\mathrm{SL}(2, \mathbb{Z})$.

If $S = \mathbb{P}_1$, $\Sigma = \{0, 1, \infty\}$, the resulting equation 6.1.4 is the hypergeometric differential equation (in 1 variable). In that case the curve $\mathscr{C}_x$ is given by the equation

$$y^2 = (t-1)t(t-x), \tag{6.1.5}$$

i.e. a double cover of $\mathbb{P}^1$ branched at $\{0, 1, \infty \text{ and } x\}$. The differential in this case is

$$\omega = \frac{dx}{y} = \frac{dx}{\sqrt{(t-1)t(t-x)}},$$

and the solutions to the HGDE are $\omega_i = \int_{\gamma_i} \omega$, γ_1, γ_2 a basis of $H_1(\mathscr{C}_x, \mathbb{Z})$. Taking their quotient we get a many-valued function

$$\omega(x) := \omega_1(x)/\omega_2(x) = \int_{\gamma_1} \omega \Big/ \int_{\gamma_2} \omega$$

which takes values in the upper half plane (for $x \in S - \Sigma$), yielding a diagram:

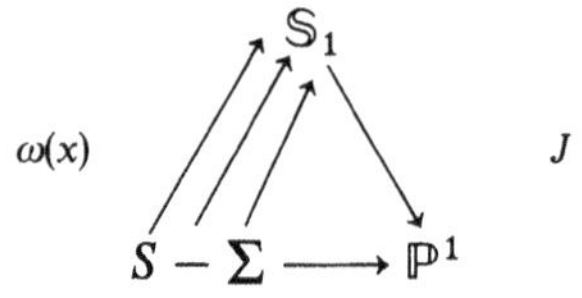

(6.1.6)

The composition of these two maps is now a well defined (single-valued) function from $S - \Sigma$ to $\mathbb{P}^1$, since the many-valuedness of $\omega(x)$ is precisely offset by the

SL(2, $\mathbb{Z}$)-invariance of the modular function J. For $S = \mathbb{P}^1$, $\Sigma = \{0, 1, \infty\}$, this map is an isomorphism onto $\mathbb{P}^1$- $\{0, 1, \infty\}$. For a complete list of examples in this case, i.e. elliptic surfaces over $\mathbb{P}^1$ with 3 singular fibres see [SH].

In the upcoming sections we will find diagrams analogous to 6.1.6 for the families $\mathcal{M}$ and $\mathcal{C}$ considered in Section 5. Before we go into some of the details, let us begin by describing some of the very detailed results of Holzapfel for the analogous case where S is 2-dimensional (in our cases it will be 3-dimensional). One might start by considering the simplest 2-dimensional analogue of 6.1.5.

$$y^2 = (t-1)t(t-x_1)(t-x_2),$$

however this of course doesn't define a family of smooth curves. Instead, consider

$$y^3 = (t-1)t(t-x_1)(t-x_2), \tag{6.1.7}$$

branched also at ∞. This is a genus 3 curve. Let $S = \mathbb{P}^2, \Sigma = l_1 \cup \cdots \cup l_6$, the six lines forming the arrangement 1.4.2. Then 6.1.7. defines a smooth family over $S - \Sigma$, and its period matrix has the form ($\eta_1 = dx/y, \eta_2 = dx/y^2, \eta_3 = x\,dx/y^2$)

$$\prod = \begin{pmatrix} A_1 & A_2 & -\rho^2 A_1 & A_3 & \rho^2 A_2 & \rho A_3 \\ B_1 & B_2 & -\rho B_1 & B_3 & \rho B_2 & \rho^2 B_3 \\ C_1 & C_2 & -\rho C_1 & C_3 & \rho C_2 & \rho^2 C_3 \end{pmatrix}$$

where

$$A_i = \oint_{\alpha_i} \eta_1, \qquad B_i = \oint_{\alpha_i} \eta_2, \qquad C_i = \oint_{\alpha_i} \eta_3, \quad i = 1, 2, 3.$$

from which it follows that $\prod$ is *already determined by the first row*: $x \mapsto (\int_{\alpha_1}\eta_1 : \int_{\alpha_2}\eta_1 : \int_{\alpha_3}\eta_1)$ is a many valued map from $S - \Sigma$ into $\mathbb{P}^2$, whose image lands in $\mathbb{B}^2 \subset \mathbb{P}^2$. Hence, in analogy to 6.1.6 above, we get a diagram

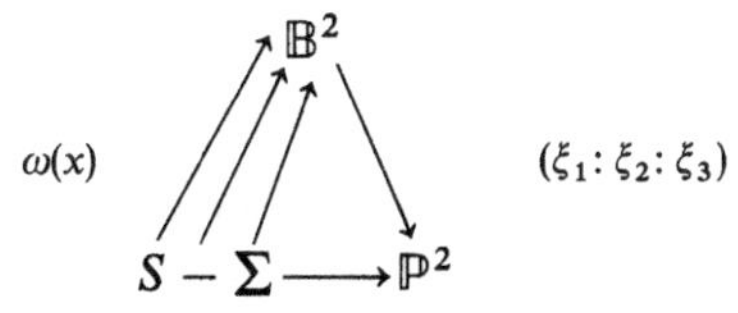

(6.1.8.)

where $\omega(x) = (\int_{\alpha_1}\eta_1(x) : \int_{\alpha_2}\eta_1(x) : \int_{\alpha_3}\eta_1(x))$, and the $(\xi_1 : \xi_2 : \xi_3)$ are automorphic forms of weight 1 ([Hol, pp. 20–21], see also 3.4.1 above).

6.2. The Hypergeometric Differential Equation (HDE)

A general reference for this section is Terada [T]. Let $\mathbb{P}^n$ be given synthetic coordinates $\{(x_0, \dots, x_{n+1}) \mid \Sigma x_i = 0\}$. The natural way to get these coordinates is to consider the $\mathbb{C}^{n+1} \subset \mathbb{C}^{n+2}$:

$$\mathbb{C}^{n+1} = \{(x_i) \in \mathbb{C}^{n+2} \mid \Sigma x_i = 0\},$$

and we take $\mathbb{P}^n = \mathbb{P}(\mathbb{C}^{n+1})$ with the induced homogenous coordinates. Since Σ_{n+2} acts naturally on $\mathbb{C}^{n+2}$ by permuting the coordinates and $\mathbb{C}^{n+1}$ is an invariant subspace, we get the so-called *projective symmetric representation* of Σ_{n+2} (cf. [Ho2]). In different language Σ_{n+2} is a unitary reflection group, and as such defines an arrangement H of planes in $\mathbb{P}^n$ (cf. [OS]). The first 3 such arrangements are:

$$n = 1. \quad H = \{x_0 x_1 (x_0 - x_1) = 0\},$$
$$n = 2. \quad H = \{x_0 x_1 x_2 (x_0 - x_1)(x_1 - x_2)(x_2 - x_0) = 0\} = 1.4.2.$$
$$n = 3. \quad H = \{x_0 x_1 x_2 x_3 (x_0 - x_1) \dots (x_3 - x_0) = 0\} = 2.5.1.$$

The hypergeometric differential equation is:

$$\begin{cases} (x_i - x_j)\partial_i\partial_j F + (\lambda_j - 1)\partial_i F - (\lambda_i - 1)\partial_j F = 0, \quad 1 \leqslant i < j \leqslant n, \\ x_i(x_i - 1)\partial_i^2 F + \Bigg[x_i(x_i - 1) \sum_{1 \leqslant \alpha \leqslant n, \alpha \neq i} (1 - \lambda_\alpha)/(x_i - x_\alpha) + \lambda_0 + \lambda_i - 2 + \\ \quad + (4 - 2\lambda_i - \lambda_0 - \lambda_{n+1})x_i \Bigg] \partial_i F + (\lambda_i - 1) \sum_{1 \leqslant \alpha \leqslant n, \alpha \neq n} x_\alpha(x_\alpha - 1)\partial_\alpha F / \\ \quad (x_i - x_\alpha) + \lambda_\infty(1 - \lambda_i)F = 0 \quad 1 \leqslant i \leqslant n. \end{cases} \tag{6.2.1}$$

where the λ_i are rational numbers with $\Sigma\lambda_i = n + 1$. This is an algebraic differential equation on $\mathbb{P}^n$ with regular singular points, non-singular off the arrangement $H \subset \mathbb{P}^n$ defined by Σ_{n+2}. A *solution* of 6.2.1 is the period of a holomorphic 1-form on the curve

$$y^\nu = t\mu_0(t - 1)\mu_{n+1}(t - x_1)\mu_1 \dots (t - x_n)\mu_n$$

for parameters $\mu_i/\nu = 1 - \lambda_i$. This is part of more general results of [DM]. The exact conditions on the parameters μ_i are known for the differentials $\omega_i = \int_{\gamma_i} y^{-1}\,dx$ to give a uniformisation into the ball $\mathbb{B}^n$. If this is the case, the monodromy group is a discrete subgroup in $PU(n, 1)$, and the parameter space therefore is locally symmetric. These conditions are:

$$(1 - \mu_i - \mu_j)^{-1} \subset \mathbb{Z} \cup \{\infty\}, \Sigma\mu_i = 2. \tag{INT}$$

There are a finite number of parameter values $\{\mu_i\}$ satisfying INT, listed in [T] and [DM, §14].

6.3. Picard Curves

In this section we consider Picard curves as in Section 5,

$$y^3 = t(t-1)(t-x_1)(t-x_2)(t-x_3), \tag{6.3.1}$$

which are curves as considered in 6.2 with parameter values

$$(\mu_i) = (\tfrac{1}{3}, \ldots, \tfrac{1}{3})$$

yielding a particular HDE, the one listed as #1 in [DM]. Let $\rho: \pi_1(S-\Sigma) \to \mathrm{GL}(3, \mathbb{C})$ be the monodromy representation.

THEOREM 6.3.2. *The monodromy group of the* HDE *associated with the family* 6.3.1. *is the Picard lattice* $\Gamma(1-\rho)$.

Proof. There are several ways to prove this. The most straightforward is analogous to Holzapfels proof of the surface case. [Ho pp. 120–125]. We sketch this argument, in a series of lemmas.

LEMMA 6.3.3. *Let* Υ *be the monodromy group of the Picard family* 6.3.1., $\pi_1(S-\Sigma)$ *the fundamental group,* $\gamma_1, \ldots, \gamma_{10} \in \pi_1(S-\Sigma)$ *generators. Then* $\rho(\gamma_i) \in \mathrm{U}(3, 1; \mathcal{O}_K(1-\rho))$.

COROLLARY 6.3.4. $\Upsilon \subset \mathrm{U}(3,1; \mathcal{O}_K(1-\rho))$ *has finite index.*

This follows from [DM]: Υ is an arithmetic lattice in $\mathrm{PU}(3, 1)$. From this one gets a diagram

$$\begin{array}{ccc} & \mathbb{B}^3 & \\ \swarrow & & \searrow \\ (\Upsilon\backslash\mathbb{B}^3)^* & \xrightarrow{\chi} & \Gamma(1-\rho)\backslash\mathbb{B}^3)^* \end{array}$$

and χ is a finite branched cover. 6.3.2 then follows from

LEMMA 6.3.5. χ *is unbranched.*

In fact, since $(\Gamma(1-\rho)\backslash\mathbb{B}^3)^*$ is simply connected, it has no unbranched covers, and 6.3.5 implies χ is an isomorphism. 6.3.5 could be proved in a manner similar to [Ho1], (his 6.3.11), which is a detailed analysis of χ near the cusps.

There is a somewhat easier proof of 6.3.2, which we now sketch. The idea is simple: we will identify fundamental domains of $\Gamma(1-\rho)$ and Υ in $\mathbb{B}^3$. First of all, both groups have 10 cusps. For $\Gamma(1-\rho)$ this was shown in 2.2, and for Σ it can be proved using the cover $Y \to \hat{\mathbb{P}}^3$ described in 2.7. Secondly, both have a scaffolding consisting of 15 modular subvarieties. Proof of this is the same. Now utilising the map

$$\chi\colon (\Sigma\backslash\mathbb{B}^3)^* \to (\Gamma(1-\rho)\backslash\mathbb{B}^3)^*$$

from above, we see both groups have the same fundamental domain in $\mathbb{B}^3$: the fundamental domain of Υ would consist of deg χ-copies of the fundamental domain of $\Gamma(1-\rho)$ if $\Gamma(1-\rho)$ acted freely. But since $\Gamma(1-\rho)$ and Υ have the same elliptic points (see the discussion in 2.8) this is sufficient.

Putting all this together, we get the following diagram:

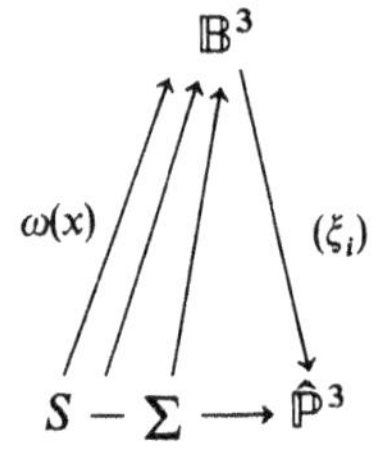

(6.3.6)

where $\omega(x)$ is the map given by the periods giving solutions of 6.2.1, and (ξ_i) are the modular forms discussed in 3.3.

6.4. Hyperelliptic curves

In this section we consider the family of Section 5.1.

$$y^2 = t(t-1)(t-x_1)(t-x_2)(t-x_3). \tag{6.4.1}$$

Here, once again, we have exponents, this time

$$(\mu_i) = (\tfrac{1}{2}, \ldots, \tfrac{1}{2})$$

and a corresponding HDE. However this set of (μ_i) does not fulfill the condition INT, and so the situation here is different than that considered in 6.3 (as it should be). Sasaki and Yoshida [HSY] have succeeded in figuring this example out, yielding a diagram analogous to 6.3.6 in this case. This goes as follows:

THEOREM 6.4.2. *Let $E(\frac{1}{2})$ be the* HDE *in 3 variables with all parameters* $=\frac{1}{2}$. *Let $E \wedge E$ be the wedge product, a 6-dimensional system. Then for these parameter values, the solution space of $E \wedge E$ splits into an irreducible 5-dimensional space + complement. This 5-dimensional space is spanned by the 2×2 minors of* *the periods:*

$$\begin{pmatrix} \int_{\gamma_1} \frac{x\,dx}{y} \cdots \int_{\gamma_4} \frac{x\,dx}{y} \\ \\ \int_{\gamma_1} \frac{dx}{y} \cdots \int_{\gamma_4} \frac{dx}{y} \end{pmatrix}$$

of the family 6.4.1.

Let us denote this solution by $\Omega = \{\sigma_1, \ldots, \sigma_5\}$. Then we get a diagram

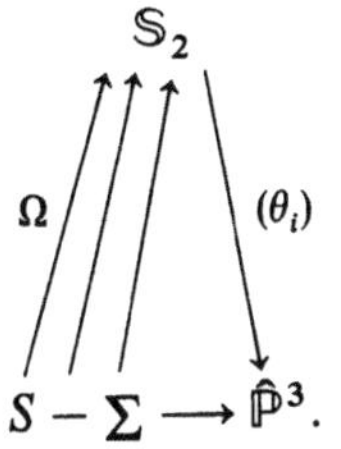

(6.4.3)

analogous to 6.3.6. Here we identify $\mathbb{S}_2$ with the non-compact dual of the hyperquadric $\mathbb{Q}_3$ in $\mathbb{P}^4$, and Ω maps onto projective coordinates.

There is the notion of dual differential equation, [HSY, §4], and one of the things proven in [HSY] is that $E(\frac{1}{2})$ is the unique HDE which is self-dual. This raises

PROBLEM. Is the self-duality of $E(\frac{1}{2})$ related to the double structure of $\hat{\mathbb{P}}^3$ as locally symmetric space?

References

[B] Baker, H.F., Principles of Geometry, Vol. IV, Ch. 5, Cambridge University Press, London, 1940.

[C] Christian, U., Über die Modulgruppe zweiten grades I, *Math. Zeit.* 85, (1964) 1–28.

[DM] Deligne, P. and Mostow, G., Monodromy of hypergeometric functions and non-integral lattice monodromy, *Pupl. Math. IHES*, 63 (1986) 113–140.

[F] Feustel, J.M., Ringe automorpher Formen auf der komplexen Einheitskugel und ihre Erzeugung durch Theta-Konstanten. Preprint P-Math-13/86, Akademie der Wissenschaften, Berlin.

[GHv] Gerritzen, L., Herrlich, F. and van der Put, M., Stable *n*-pointed trees of projective lines. To appear in Indagationes.

[Gr] Griffiths, P., Periods of integrals on algebraic manifolds. Bull. AMS 76, (1970) 228–296.

[GS] Griffiths, P. and Schmid, W., Locally homogenous complex manifolds. *Acta Math.* 123 (1969) 253–302.

[HK] Hashimoto, K. and Koseki, H., On class numbers of positive definite binary and ternary unimodular hermitian forms. *Math. Gottingensis* (1987) 21–22.

[He] Hemperly, J.C., The parabolic contribution to the number of linearly independent automorphic forms on a certain bounded domain. *Amer. J. Math.*, 94 (1972) 1078–1100.

[Hi] Hirzebruch, F., Automorphe Formen und der Satz von Riemann Roch. in: Symp. Inter. de Top. Alg., Unesco 1958.

[Ho1] Holzapfel, R.-P., Geometry and Arithmetic around Euler partial differential equations. D. Reidel Publishing Company: Boston 1986.

[Ho2] Holzapfel, R.-P., On the nebentypus of Picard modular forms. To appear in Math. Nach.

[HSY] Hara, M., Sasaki, T. and Yoshida, M., Tensor products of Linear Differential Equations. Preprint Kyushu University.

[Hu] Hunt, B., Coverings and Ball Quotients. *Bonn. Math. Schrif.* 174, Bonn, 1986.

[I1] Igusa, J., On Siegel modular forms of genus two. *Am. J. Math.* 84 (1962) 175–200 and 86 (1964) 392–412.

[I2] Igusa, J., On the graded ring of Theta Constants. *Am. J. Math.* 86 (1964) 219–246 and 88 (1966) 221–236.

[I3] Igusa, J., On a desingularisation problem in the theory of Siegel modular forms. *Math. Ann.* 168 (1967) 228–260.

[Ka] Katz, N., On the differential equations satisfied by period matrices. Publ. Math. IHES 35, 71–106 (1968)

[KLW] Kirwan, F., Lee, R. and Weintraub, S., Quotients of the complex ball by discrete groups. Pac. J. Math. 130, (1987) 115–141.

[K] Kodaira, K., On Kähler varieties of restricted type. Ann. of Math. 60, (1954) 28–48.

[LW] Lee, R. and Weintraub, S., Cohomology of $Sp_4(\mathbb{Z})$ and related groups and spaces. Topology, 24, (1985) 391–410.

[LW1] Lee, R. and Weintraub, S., Moduli spaces of Riemann surfaces of genus two with level structures. I. Math. Gottingensis. 66 (1986), to appear in Trans. AMS, 1988.

[OS] Orlik, P. and Solomon, L.: Arrangements defined by unitary reflection groups., Math. Ann. 261, (1982) 339–357.

[P] Picard, E., Sur les fonctions hyperfuchsiennes provenant des series hypergeometriques de deux variables. Ann. Ecole Norm. Sup. 3ᵉ ser. 62, (1885) 357–384.

[SH] Schmickler-Hirzebruch, U., Elliptische Flächen über $\mathbb{P}^1$ mit drei Ausnahmefasern und die hypergeometrische Differentialgleichung. Diplomarbeit, Bonn, 1978.

[S] Shimura, G., On purely transcendental fields of automorphic functions of several variables. Osaka J. Math. 1, (1964) 1–14.

[St] Stiller, P., The differential equation associated with elliptic surfaces. J. Math. Soc. Japan 24, (1972) 20–59.

[T] Terada, T., Fonctions hypergeometriques F_1 et fonctions automorphes I. J. Math. Soc. Japan 35, (1983) 451–475.

[V] van der Geer, G., On the geometry of a Siegel modular threefold. Math. Ann. 260, (1982) 317–350.

[Y] Yoshida, M., Fuchsian differential equations. Viehweg, Braunschweig/Wiesbaden 1987

[Z] Zeltinger, H., Volumina und Spitzenanzahlen Picardscher Modulvarietaeten. Bonn. Math. Schrif. 136, (1981).

Compositio Mathematica **76**: 243–245, 1990.

Linearizing some $\mathbb{Z}/2\mathbb{Z}$ actions on affine space

JERZY JURKIEWICZ
Institute of Mathematics, University of Warsaw, P.K.i. N. 9p., 00–901, Warsaw, Poland

Received 3 December 1988; accepted 20 July 1989

Let V be the affine space k^n over an algebraically closed field k, G a linearly reductive group and $A\colon G \times V \to V$ a group action with a fixed point, say the origin. Then for all $g \in G$ let me denote by $A(g)$ the corresponding automorphism of V. We have

$$A(g) = L(g) + D(g)$$

where $L(g), D(g) \in \operatorname{End} V$, $L(g)$ linear and $D(g)$ the sum of terms of higher degrees. Let me recall the well known linearization problem: is the action A linearizable, i.e. conjugated to the linear action $L\colon G \times V \to V$ (see e.g. [B] and [K])? Recently counter-examples have been found, see [S] and [K + S], so it is reasonable to study additional assumptions on the action A. One of them is considered in the present paper.

First I want to define some morphism $\sigma_A\colon V \to V$ which turns out to be a conjugating automorphism for A, provided σ_A is invertible. It will be done using the Reynolds operator i.e. the equivariant projection $\rho\colon \mathcal{O}(G) \to k$. For a finite dimensional k-space W we have the unique linear map $\int_G\colon \operatorname{Mor}(G, W) \to W$ such that for all linear maps $f\colon W \to k$ the induced diagram

$$\begin{array}{ccc} \int_G\colon \operatorname{Mor}(G, W) & \longrightarrow & W \\ \downarrow f_* & & \downarrow f \\ \rho\colon \mathcal{O}(G) & \longrightarrow & k \end{array}$$

is commutative. Now let $\phi\colon G \to \operatorname{End}(V)$ be such a map that the induced map $G \times V \to V$ is an algebraic morphism. Then $W := \text{lin hull}\,(\phi(G))$ is finite dimensional, hence $\int_G \phi$ is a well-defined element of $\operatorname{End}(V)$. Let us apply the above to the map $\phi\colon G \ni g \mapsto L(g^{-1})A(g) \in \operatorname{End}(V)$ and set $\sigma = \sigma_A = \int \phi$ (compare [J]). We have

$$L(h)\sigma = \int_{g \in G} L(h)L(g^{-1})A(g) = \left(\int_{g \in G} L(hg^{-1})A(gh^{-1})\right)A(h) = \sigma A(h)$$

for all $h \in G$.

So σ invertible implies that $A(h) = \sigma^{-1}L(h)\sigma$. In particular the action A is linearizable. Later we will give an example of an action A which can be linearized but for which σ_A is not invertible.

As mentioned in [J], the morphism σ_A can be interpreted as an average deviation of A from being linear.

CONJECTURE (Kraft, Procesi). Assume for some $d \geqslant 2$

$$A(g) = L(g) + H_d(g) + H_{d+1}(g) + \cdots + H_{2d-2}(g), \quad \text{for all } g,$$

where $H_m(g)$ is a homogeneous endomorphism of V of degree m. Then σ_A is invertible. In particular the action A is linearizable.

THEOREM. *The above conjecture is true in the following cases*

1. *G linearly reductive, $d = 2$ and* char $k \neq 2$,
2. *G diagonalizable, $d = 2$ and* char k *arbitrary,*
3. *$G = \mathbb{Z}/2\mathbb{Z}$, d arbitrary and* char $k = 0$.

Cases 1 *and* 2 *are the objects of* [J].

Proof for the case 3. Let I denote the identity map of V. We can write: $G = \{I, L + D\}$, where L and D are endomorphisms of V, L linear and $D = H_d + \cdots + H_{2d-2}$. We have $L^2 = (L + D)^2 = I$. It follows that

$$LD + D(L + D) = 0. \tag{1}$$

Let me denote by $\tilde{H}_d$ the d-linear symmetric map from V^d to V corresponding to H_d. Then we have

$$D(L + D) = DL + d\tilde{H}_d(L, \ldots, L, H_d) + \cdots$$

where the first summand consists of terms of degrees $d, \ldots, 2d - 2$, the second is of degree $2d - 1$ and all further summands have higher degrees. Considering the possible cancellations in (1) we obtain:

$$-LD = DL = D(L + D). \tag{2}$$

By defnition $\sigma = \frac{1}{2}(I + (I + LD)) = I - \frac{1}{2}DL$. We will prove that $I + \frac{1}{2}DL$ is the inverse of σ.

LEMMA. $D(I + mDL) = D$ *for* $m = 0, 1, 2, \ldots$

Proof. Suppose the above holds for some $m - 1, m > 0$. By (2), $D = D(I + DL)$. Therefore

$$D = D(I + (m - 1)DL)(I + DL) = D(I + DL + (m - 1)DL(I + DL)).$$

On the other hand $DL(I + DL) = -LD(I + DL) = -LD = DL$, and we are done.

Since $\mathrm{char}(k) = 0$ the Lemma implies that $D(I + rDL) = D$ for all $r \in k$. Then taking $r = \frac{1}{2}$ we have

$$(I - \tfrac{1}{2}DL)(I + \tfrac{1}{2}DL) = I + \tfrac{1}{2}DL + \tfrac{1}{2}LD(I + \tfrac{1}{2}DL) = I,$$

and the same applies if we interchange the order of factors at the left hand side. Q.E.D.

EXAMPLE OF A NON INVERTIBLE σ. Let the linear endomorphism L of k^2 be given by $L(x, y) = (x, -y)$ and an automorphism τ by $\tau(x, y) = (x - (x + y)^2, y + (x + y)^2)$ so that $\tau^{-1}(x, y) = (x + (x + y)^2, y - (x + y)^2)$. The automorphism $\tau^{-1}L\tau$ has order two, so it defines an action of the group of order two on k^2. The corresponding endomorphism $\sigma = \frac{1}{2}(I + L\tau^{-1}L\tau)$ takes (x, y) to $(x - u + v, y + u + v)$, where $u = \frac{1}{2}(x + y)^2$, $v = \frac{1}{2}(x - y - 2(x + y)^2)^2$. Direct computation shows that the Jacobian determinant of σ is

$$J(\sigma) = 1 - 4(x^2 + y^2) + 8(x^3 + y^3) + 24(x^2y + xy^2).$$

Therefore the endomorphism σ is not invertible, while the considered group action can obviously be linearized.

References

[B] Bass, H., Algebraic group actions on affine spaces. Group actions on rings. Contemporary Mathematics, Vol. 43 (1985), AMS, 1–24.

[J] Jurkiewicz, J., On the linearization of actions of linearly reductive groups; Commentari Mathematici Helvetici 64 (1989) 508–513.

[K] Kraft, H., Algebraic group actions on affine spaces. Geometry today, Progress in Mathematics. Birkhäuser 1984, 251–266.

[S] G.V. Schwarz, Exotic algebraic group actions, preprint.

[K + F] Kraft, H. and Schwarz, G. V., Reductive group actions on affine spaces with one-dimensional quotient. To appear in Contemporary Mathematics, Proceedings of the Conference on Group Actions and Invariant Theory, Montreal 1988.

Compositio Mathematica **76**: 247–264, 1990.

Normal forms and moduli spaces of curve singularities with semigroup $\langle 2p, 2q, 2pq + d\rangle$

IGNACIO LUENGO[1] & GERHARD PFISTER[2]

[1]*Dto. de Algebra, Fac. de Matematicas, Univ. Complutense, 28040-Madrid, Spain;* [2]*Sektion-Mathematik, Humboldt-Universitaet zu Berlin, 1086 Berlin, Unter der Linden 6, DDR*

Received 23 September 1988; accepted in revised form 20 September 1989

This work has been possible thanks to a Scientific Agreement between the Universidad Complutense and the Humboldt Universitaet. This cooperation agreement supported our stay in the Bereich Algebra and Departamento de Algebra respectively.

The aim of this paper is to classify map germs $(\mathbb{C}^2, 0) \to \mathbb{C}$ and germs of curve singularities in $\mathbb{C}^2$ given by an equation of the type $f = (x^p + y^q)^2 + \sum_{iq+jp>2pq} a_{ij}x^i y^j = 0$ with a fixed Milnor number $\mu(f) = \dim_{\mathbb{C}} \mathbb{C}[[x, y]]/(\partial f/\partial x, \partial f/\partial y)$. Here we always suppose $p < q$ and $\gcd(p, q) = 1$.

The moduli space $M_{p,q,\mu}$ of the map germs described above is an affine Zariski-open subset of $\mathbb{C}^{2(p-1)(q-1)-p-q+2+[q/p]}$ devided by a suitable action of μ_{2pq} (the group of $2pq$-roots of unity) depending on $\mu(f)$.

The moduli space $T_{p,q,\mu}$ of all plane curve singularities described above (which is the moduli space of all plane curve singularities with the semigroup $\langle 2p, 2q, \mu - 2(p-1)(q-1)+1\rangle$ if μ is even) is $\mathbb{C}^{(p-2)(q-2)+[q/p]-1}$ devided by a suitable action of μ_d, $d = \mu - (2p-1)(2q-1)$.

In both cases we also get an algebraic universal family. It turns out that the Tjurina-number $\tau(f) = \dim_{\mathbb{C}} \mathbb{C}[[x, y]]/(f, \partial f/\partial x, \partial f/\partial y) = \mu(f) - (p-1)(q-1)$ depends only on $\mu(f)$ and p and q.

Constructing the moduli spaces we use the graduation of $\mathbb{C}[[x, y]]$ defined by p, q: $\deg x^i y^j = iq + jp$.

We use the following idea to construct the moduli spaces: Let $\mu = (2p-1)(2q-1) + d$. We prove that for all f of the above type we can choose the same monomial base of $\mathbb{C}[[x, y]]/(\partial f/\partial x, \partial f/\partial y)$ (Lemma 2). We choose α, β and that $\alpha < p$, $\alpha q + \beta p = 2pq + d$. Hence $\mu(f_0) = \mu$ with $f_0 = (x^p + y^q)^2 + x^\alpha y^\beta$. Then we consider a universal μ-constant unfolding of f_0 as a "global" family (Lemma 3). The parameter space U of that unfolding is an affine open subset of $\mathbb{C}^{2(p-1)(q-1)-p-q+2+[q/p]}$. The group μ_{2pq} acts on U and $M_{p,q,\mu} = U/\mu_{2pq}$.

To construct $T_{p,q,\mu}$ we consider the Kodaira-Spencer map of the universal μ-constant unfolding. The Kernel of the Kodaira-Spencer map is a Lie-algebra acting on U. The integral manifolds of that Lie-algebra are the analytically trivial subfamilies of the unfolding.

We choose a suitable section transversal to those integral manifolds, which turns out to be isomorphic to $\mathbb{C}^{(p-2)(q-2)+[q/p]-1}$. The group μ_d acts on the corresponding family and we prove that $T_{p,q,\mu} = \mathbb{C}^{(p-2)(q-2)+[q/p]-1}/\mu_d$.

1. A normal form for map germs $(\mathbb{C}^2, 0) \to \mathbb{C}$ with initial term $(x^p + y^q)^2$

LEMMA 1. *Let*

$$f = \left(x^p + y^q + \sum_{iq+jp>pq} h_{ij}x^iy^j\right)^2 + \sum_{iq+jp\geqslant 2pq+d} w_{ij}x^iy^j$$

then $\mu(f) \geqslant (2p-1)(2q-1)+d$, *and* $\mu(f) = (2p-1)(2q-1)+d$ *iff*

$$f_d := \sum_{iq+jp=2pq+d} (-1)^{[i/p]}w_{ij} \neq 0.$$

Proof. Either f is irreducible or the components of f have the same tangent direction. This implies that

$$\mu(\tilde{f}) = \mu(f) - 2p(2p-1),$$

where $\tilde{f}$ is the blowing up

$$\tilde{f} = \frac{f(xy, y)}{y^{2p}} = \left(x^p + y^{q-p} + \sum h_{ij}x^iy^{i+j-p}\right)^2 + \sum w_{ij}x^iy^{i+j-2p}$$

$$= \left(x^p + y^{q-p} + \sum_{i(q-p)+jp>(q-p)p} h_{i,j-i+p}x^iy^j\right)^2 +$$

$$+ \sum_{i(q-p)+jp\geqslant 2(q-p)p+d} w_{i,j-i+2p}x^iy^j.$$

Using induction we may assume that

$$\mu(\tilde{f}) \geqslant (2p-1)(2(q-p)-1)+d$$

and

$$\mu(\tilde{f}) = (2p-1)(2(q-p)-1) + d \text{ iff } 0 \neq \sum_{iq+jp=2pq+d} (-1)^{[i/p]} w_{ij}$$

$$= \sum_{i(q-p)+jp=2(p-q)p+d} (-1)^{[i/p]} w_{i,j-i+2p}.$$

This yields the if part of the result. Now if f is as above and $\mu(f) > (2p-1)(2q-1)+d$, the condition $f_d = 0$ says that $(x^p + y^q)$ divides $\sum_{iq+jp=2pq+d} w_{ij}x^iy^j$, and adding $-\frac{1}{2}\sum_{iq+jp=2pq+d} w_{ij}x^iy^j$ to the first part of f one gets $f = (x^p + y^q + \cdots)^2 +$ terms of degree greater than $2pq+d$. Continuing this way, we get the result.

LEMMA 2. *Let* $f = (x^p + y^q)^2 + \sum_{iq+jp>2pq} h_{ij}x^iy^j$ *and* $\mu(f) = (2p-1)(2q-1)+d$. *Let* γ, δ *such that* $\gamma q + \delta p = 3pq - q - p + d$, $\gamma < p$. *Let* $B = \{(i,j) \in N^2/i < 2p-1, j < q-1\} \cup \{(i,j) \in N^2/i < p, j < q\} \cup \{(i,j), (i,j) \in N^2/i < \gamma, j < \delta + q\}$. *Then* $\{x^iy^j\}_{(i,j)\in B}$ *is a base of* $\mathbb{C}[[x,y]]/(\partial f/\partial x, \partial f/\partial y)$.

Proof. We use the algorithm of Mora (cf. [3]) to compute a Groebner base of the ideal $(\partial f/\partial x, \partial f/\partial y)$. We consider $\mathbb{C}[[x,y]]$ as a graded ring with $\deg x = q$, $\deg y = p$. Let $f_1 = 1/2p(\partial f/\partial x)$ and $f_2 = 1/2q(\partial f/\partial y)$.

Consider $s(f_1, f_2) = y^{q-1}f_1 - x^{p-1}f_2$ and let f_3 be the reduction of $s(f_1, f_2) = y^{q-1}f_1 - x^{p-1}f_2$ with respect to the initial terms x^{2p-1} resp. x^py^{q-1} of f_1 resp. f_2, i.e.

$$s(f_1, f_2) = f_3 + h_1f_1 + h_2f_2$$

$$f_3 = \sum_{\gamma_i < p} l_i x^{\gamma_i} y^{\delta_i}$$

$$q\gamma_i + p\delta_i = 3pq - q - p + i$$

and the initial terms of h_1 resp. h_2 have degree $> pq - p$ resp. $> pq - q$. $f_3 \neq 0$ because of $\mu(f) < \infty$. Let k be the minimal such that $l_k \neq 0$, i.e. $l_k x^{\gamma_k} y^{\delta_k}$ is the initial term of f_3. Consider now

$$s(f_2, f_3) = l_k y^{\delta_k - q + 1} f_2 - x^{p-\gamma_k} f_3$$

$$= l_k y^{\delta_k + q} + \text{terms of degree} > p\delta_k + pq$$

$$=: f_4.$$

It is not difficult to see that the reductions of $s(f_1, f_3)$ and $s(f_i, f_4)$ $i = 1, 2, 3$ with respect to the initial terms of f_1, f_2, f_3, f_4 are zero, i.e. f_1, f_2, f_3, f_4 is a

Groebner base of $(\partial f/\partial x, \partial f/\partial y)$. This implies that

$$\{x^i y^j\}_{(i,j)\in B'}, B' = \{(i,j), i < 2p-1, j < q-1\} \cup \{(i,j), i < p, j < \delta_k\} \cup \{(i,j), i < \gamma_k, j < \delta_k + q\},$$

is a base of $\mathbb{C}[[x,y]]/(\partial f/\partial x, \partial f/\partial y)$.

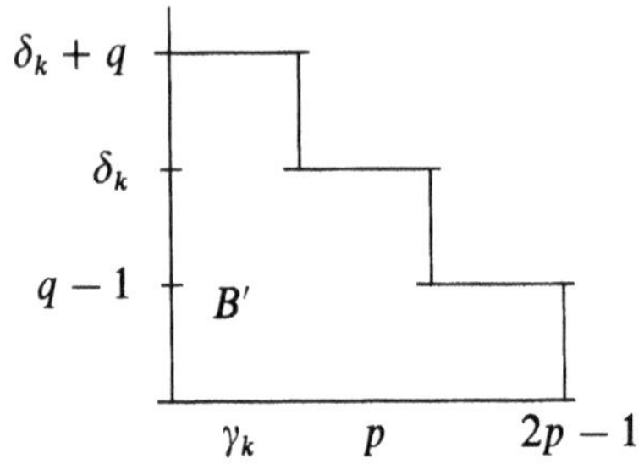

This implies $\mu(f) = (p-1)(q-1) + q\gamma_k + p\delta_k$ and therefore $\gamma = \gamma_k$ and $\delta = \delta_k$ and $B = B'$. □

LEMMA 3. *Let* $f = (x^p + y^q)^2 + \sum_{iq+jp>2pq} a_{ij} x^i y^j$ *and* $\mu(f) = (2p-1)(2q-1) + d$.

Let γ, δ *be defined by*

$$\gamma < p \quad \textit{and} \quad \gamma q + \delta p = 3pq - q - p + d$$

Let $B_0 = \{(i,j), iq + jp > pq, i \leqslant p-2, j \leqslant q-2\}$;

$$B_1 = \{(i,j), iq + jp \geqslant 2pq + d, i < p, j < \delta\} \cup \{(i,j), iq + jp \geqslant 2pq + d, i < \gamma, j < \delta + q\}.$$

There is an automorphism $\varphi: \mathbb{C}[[x,y]] \to \mathbb{C}[[x,y]]$ *such that*

$$f(\varphi) = \left(x^p + y^q + \sum_{(i,j)\in B_0} h_{ij} x^i y^j\right)^2 + \sum_{(i,j)\in B_1} w_{ij} x^i y^j$$

for suitable $h_{ij}, w_{ij} \in \mathbb{C}$.

Proof. Using Lemma 1 we may assume that

$$f = \left(x^p + y^q + \sum_{iq+jp>pq} b_{ij} x^i y^j\right)^2 + \sum_{iq+jp\geqslant 2pq+d} c_{ij} x^i y^j.$$

Assume that there is an automorphism $\varphi^{(k)}$ such that

$$f(\varphi^{(k)}) = \Big(x^p + y^q + \sum_{(i,j)\in B_0} h_{ij}^{(k)} x^i y^j + \sum_{iq+jp \geqslant pq+k} b_{ij}^{(k)} x^i y^j\Big)^2 + \\ + \sum_{(i,j)\in B_1} w_{ij}^{(k)} x^i y^j + \sum_{iq+jp\geqslant 2pq+k} c_{ij}^{(k)} x^i y^j,$$

$\varphi^{(1)} =$ identity.

Now

$$\sum_{iq+jp=2pq+k} c_{ij}^{(k)} x^i y^j = (x^p + y^q)H + \sum_{iq+jp=2pq+k} (-1)^{[i/p]} c_{ij}^{(k)} x^{i_0} y^{j_0}$$

for a suitable homogeneous H of degree $pq + k$ and $i_0 < p, i_0 q + j_0 p = 2pq + k$.

If $\Sigma_{iq+jp>2pq+k}(-1)^{[i/p]}c_{ij}^{(k)} \neq 0$ and $(i_0, j_0) \notin B_1$ then $k \geqslant pq - q - p + d$ (Lemma 1) and $j_0 \geqslant \delta, i_0 \geqslant \gamma$ or $j_0 \geqslant \delta + q$.

Let α, β be defined by $q\alpha + p\beta = 2pq + d, \alpha < p$, then $w_{\alpha\beta} \neq 0$. Notice that $\alpha - 1 \equiv \gamma \bmod p$ and $\beta - 1 \equiv \delta \bmod q$.

Let

$$g := x^p + y^q + \sum_{(i,j)\in B_0} h_{ij}^{(k)} x^i y^j + \sum_{iq+jp\geqslant pq+k} b_{ij}^{(k)} x^i y^j$$

and

$$\omega := e \cdot x^{\xi} y^{\eta} \left(\frac{\partial g}{\partial y}\frac{\partial}{\partial x} - \frac{\partial g}{\partial x}\frac{\partial}{\partial y}\right), \ \xi q + \eta p = k - pq + p + q - d$$

$$e := \frac{1}{(\alpha q + \beta p) w_{\alpha\beta}} \cdot \sum_{ip+jq=2pq+k} (-1)^{[i/p]-[\alpha-1+\xi/p]+1} c_{ij}^{(k)}$$

$$\text{with } (\xi, \eta) = \begin{cases} (i_0 - \gamma, j_0 - \delta) & \text{if } j_0 \geqslant \delta, \quad i_0 \geqslant \gamma \\ (i_0 - \gamma + p, j_0 - \delta - q) & \text{if } j_0 \geqslant \delta + q. \end{cases}$$

Let $\psi: \mathbb{C}[[x, y]] \to \mathbb{C}[[x, y]]$ the automorphisms corresponding to the vector field ω, then $g(\psi) = g$.

Hence,

$$f(\psi \circ \varphi^{(k)}) = g^2 + \sum_{(i,j)\in B_0} w_{ij}^{(k)} x^i y^j + \sum_{iq+jp\geqslant 2pq+k} \bar{c}_{ij}^{(k)} x^i y^j$$

and

$$\sum_{iq+jp=2pq+k} (-1)^{[i/p]}\bar{c}_{ij}^{(k)}$$
$$= \sum_{iq+jp=2pq+k} (-1)^{[i/p]}c_{ij}^{(k)} + (-1)^{[\alpha-1+\xi/p]}(\alpha q + \beta p)w_{\alpha\beta}\cdot e.$$

If $(i_0, j_0) \notin B_1$ we may assume now that $\sum_{iq+jp=2pq+k}(-1)^{[i/p]}c_{ij}^{(k)} = 0$.

Let $g_1 := g + \frac{1}{2}H$ and

$$\sum_{iq+jp=pq+k} b_{ij}^{(k)}x^i y^j + \tfrac{1}{2}H + m_k\frac{\partial g_1}{\partial x} + n_k\frac{\partial g_1}{\partial y} = \sum_{(i,j)\in B_0} d_{ij}^{(k)}x^i y^j.$$

The degree of the initial part of m_k resp. n_k is $q + k$ resp. $p + k$.

We define $\varphi^{(k+1)}$ by

$$\varphi^{(k+1)}(x) = \varphi^{(k)}(x) + m_k$$
$$\varphi^{(k+1)}(y) = \varphi^{(k)}(y) + n_k$$

and

$$h_{ij}^{(k+1)} = h_{ij}^{(k)} + d_{ij}^{(k)}$$
$$w_{ij}^{(k+1)} = w_{ij}^{(k)} \quad \text{if } (i,j) \neq (i_0, j_0)$$
$$w_{i_0,j_0}^{(k+1)} = w_{i_0,j_0}^{(k)} + \sum_{iq+jp=2pq+k} (-1)^{[i/p]}c_{ij}^{(k)} \quad \text{if } (i_0, j_0)\in B_1.$$

Then

$$f(\varphi^{(k+1)}) = \left(x^p + y^q + \sum_{(i,j)\in B_0} h_{ij}^{(k+1)}x^i y^j + \sum_{iq+jp\geqslant pq+k+1} b_{ij}^{(k+1)}x^i y^j\right)^2 +$$
$$+ \sum_{(i,j)\in B_1} w_{ij}^{(k+1)}x^i y^j + \sum_{iq+jp\geqslant 2pq+k+1} c_{ij}^{(k+1)}x^i y^j$$

for suitable $b_{ij}^{(k+1)}, c_{ij}^{(k+1)}$. □

LEMMA 4. *Let*

$$f_t = (x^p + y^q)^2 + \sum_{iq+jp>2pq} a_{ij}(t)x^i y^j, a_{ij}(t)\in \mathbb{C}[t]$$

and $\mu(f_t) = (2p-1)(2q-1) + d$ *for* $t\in\mathbb{C}$.

Let γ, δ, B_0, B_1 *be as in Lemma* 3. *There is a* $\mathbb{C}[t]$*-automorphism* $\varphi_t\colon \mathbb{C}[t][[x, y]] \to \mathbb{C}[t][[x, y]]$ *such that*

$$f_t(\varphi_t) = \left(x^p + y^q + \sum_{(i,j)\in B_0} h_{ij}(t)x^i y^j\right)^2 + \sum_{(i,j)\in B_1} w_{ij}(t)x^i y^j$$

for suitable $h_{ij}(t), w_{ij}(t) \in \mathbb{C}[t]$.

The proof is similar to that of Lemma 3. □

Let us consider the family

$$F(x, y, H, W) = \left(x^p + y^q + \sum_{(i,j)\in B_0} H_{ij}x^i y^j\right)^2 + \sum_{(i,j)\in B_1} W_{ij}x^i y^j$$

depending on the parameters $H = (H_{ij})_{(i,j)\in B_0}$, $W = (W_{ij})_{(i,j)\in B_1}$ and define $N = \# B_0 + \# B_1$, then $\mu(F) = (2p-1)(2q-1) + d$ on the open set U defined by $W_{\alpha\beta} \neq 0$, $\alpha q + \beta p = 2pq + d$, in $\mathbb{C}^N = \operatorname{Spec}\mathbb{C}[H, W]$. Notice that $N = 2(p-1)(q-1) - p - q + 2 + [q/p]$ is not depending on d!

The group of $2pq$-roots of unity acts on U:

$$\lambda \in \mu_{2pq}, \lambda \circ ((h_{ij}), (w_{ij})) := ((\lambda^{iq+jp-pq}h_{ij}), (\lambda^{iq+jp-2pq}w_{ij})).$$ □

THEOREM 1. U/μ_{2pq} *is the moduli space of all functions*

$$f = (x^p + y^q)^2 + \sum_{iq+jp>2pq} a_{ij}x^i y^j$$

with $\mu(f) = (2p-1)(2q-1) + d$ *and* F *is the universal family.*

Proof. Using Lemma 3 we have to prove the following

LEMMA 5. *Let* φ *be an automorphism* $\varphi\colon \mathbb{C}[[x, y]] \to \mathbb{C}[[x, y]]$ *such that*

$$F(\varphi(x), \varphi(y), \bar{h}, \bar{w}) = F(x, y, h, w) \tag{$*$}$$

for $(\bar{h}, \bar{w}), (h, w) \in U \subseteq \mathbb{C}^N$ *then* $\lambda \cdot (\bar{h}, \bar{w}) = (h, w)$ *for a suitable* $\lambda \in \mu_{2pq}$.

Proof. Let $\bar{x} := \varphi(x)$, $\bar{y} := \varphi(y)$ then grouping the squared part of $(*)$ one gets:

$$\left(x^p + y^q + \bar{x}^p + \bar{y}^q + \sum \bar{h}_{ij}\bar{x}^i\bar{y}^j + \sum h_{ij}x^i y^j\right) \times$$

$$\times \left(x^p + y^q - \bar{x}^p - \bar{y}^q - \sum \bar{h}_{ij}\bar{x}^i\bar{y}^j + \sum h_{ij}x^i y^j\right)$$

$$= \sum \bar{w}_{ij}\bar{x}^i\bar{y}^j - \sum w_{ij}x^i y^j.$$

This equation implies obviously that the degree of the initial term of $\varphi(x)$ is $\geqslant q$ and

$$\bar{x} = \lambda^q\left(x + \sum_{iq+jp>q} a_{ij}^{(1)}x^iy^j\right) \qquad \bar{y} = \lambda^p\left(y + \sum_{iq+jp>p} a_{ij}^{(2)}x^iy^j\right), \lambda \in \mu_{2pq}.$$

We may assume that $\lambda = 1$ and prove $(\bar{h}, \bar{w}) = (h, w)$.

Let the degree of the leading parts of both sides of the above equation be $2pq+m$ and let r be the degree of φ, i.e. $a_{ij}^{(1)} = 0$ if $iq + jp < q + r$, $a_{ij}^{(2)} = 0$ if $iq + jp < p + r$ and $a_{ij}^{(1)} \neq 0$ or $a_{ij}^{(2)} \neq 0$ for suitable i, j with $iq + jp = q + r$ resp. $iq + jp = p + r$.

1. Step. We prove that

(a) $r \geqslant pq - p - q$

$$\sum_{iq+jp=q+r} a_{ij}^{(1)}x^iy^j = \frac{1}{p}y^{q-1}\cdot k, \qquad \sum_{iq+jp=p+r} a_{ij}^{(2)}x^iy^j = -\frac{1}{q}x^{p-1}\cdot k$$

(b) $h = \bar{h}$ and $w_{ij} = \bar{w}_{ij}$ if $iq + jp < 3pq - p - q + d$.

First of all $m \geqslant d + r$ because the leading part of the left side of the equation is divisible by $x^p + y^q$ and $m < d + r$ would imply that the leading part of the right side is a monomial. This implies $w_{ij} = \bar{w}_{ij}$ if $iq + jp < 2pq + d + r$. Now $h_{ij} = \bar{h}_{ij}$ if $iq + jp < pq + r$. Otherwise the leading part of the left side of the equation would be $2(x^p + y^q)(h_{ij} - \bar{h}_{ij})x^iy^j$ for some i, j with $iq + jp < pq + r$ and therefore of degree $2pq + r < 2pq + m$.

Now suppose $r < pq - p - q$. Then there is at most one monomial of degree $p + r$ resp. $q + r$.

If $iq + jp = pq + r$ for some $(i, j) \in B_0$ and

$$qi_0 + pj_0 = q + r$$

$$qi_1 + pj_1 = p + r$$

then

$$(h_{ij} - \bar{h}_{ij})x^iy^j - pa_{i_0j_0}^{(1)}x^{i_0+p-1}y^{j_0} - qa_{i_1j_1}^{(2)}x^{i_1}y^{j_1+q-1} = 0$$

otherwise the leading part of the left side of the equation would have degree $2pq + r < 2pq + m$.

But $(i, j) \in B_0$, i.e. $i < p - 1$ and $j < q - 1$. This implies $h_{ij} = \bar{h}_{ij}$, $a_{i_0j_0}^{(1)} = a_{i_1j_1}^{(2)} = 0$ (because of $r < pq - p - q$ we have $i_1 < p - 1$). This is a contradiction since $a_{i_0j_0}^{(1)} \neq 0$ or $a_{i_1j_1}^{(2)} \neq 0$ by the definition of r.

Similarly one gets a contradiction if there is no $(i, j) \in B_0$ with $qi + pj = p + r$, resp. no i_0, j_0 with $qi_0 + pj_0 = q + r$ resp. no i_1, j_1 with $qi_1 + pj_1 = p + r$.

This proves that $r \geqslant pq - p - q$. With the same method we obtain

$$\sum_{iq+jp=p+r} a_{ij}^{(2)} x^i y^j = -\frac{1}{q} x^{p-1} k \quad \text{and} \quad \sum_{iq+jp=q+r} a_{ij}^{(1)} x^i y^j = \frac{1}{p} y^{q-1} k.$$

(b) is clear now by the choice of B_0 and the fact that $r \geqslant pq - p - q$.

2. *Step.* We prove that $r \geqslant 2pq - p - q$.

Assume that $r < 2pq - p - q$. Then $\deg k < pq$, i.e., k is a monomial. The leading part of the left side of the above equation is divisible by $x^p + y^q$. The leading part L of the right side is

$$(\bar{w}_{ij} - w_{ij}) x^i y^j + \bar{w}_{\alpha\beta} \cdot k \left(\frac{\alpha}{p} x^{\alpha-1} y^{\beta+q-1} - \frac{\beta}{q} x^{\alpha+p-1} y^{\beta-1} \right)$$

if $iq + jp = 2pq + m$ for some $(i, j) \in B_1$ or

$$\bar{w}_{\alpha\beta} \cdot k \left(\frac{\alpha}{p} x^{\alpha-1} y^{\beta+q-1} - \frac{\beta}{q} x^{\alpha+p-1} y^{\beta-1} \right)$$

if $iq + jp \neq 2pq + m$ for $(i, j) \in B_1$.

Let $k = \kappa \cdot x^\xi y^\eta$. If $\alpha + \xi - 1 < p$, then $i = \alpha + \xi - 1$ and $j = \beta + \eta + q - 1$. If $\alpha + \xi - 1 \geqslant p$, then $i = \alpha + \xi - 1 - p$ and $j = \beta + \eta + 2q - 1$. But $(\alpha + \xi - 1, \beta + \eta + q - 1) \notin B_1$ and $(\alpha + \xi - 1 - p, \beta + \eta + 2q - 1) \notin B_1$. This implies $L = \bar{w}_{\alpha\beta} \cdot k x^{\alpha-1} y^{\beta-1}((\alpha/p) y^q - (\beta/p) x^p)$ which is not divisible by $x^p + y^q$. This is a contradiction and therefore $r \geqslant 2pq - p - q$.

Now $iq + jp \leqslant 4pq - 2p - 2q + d$ for $(i, j) \in B_1$ then $w_{ij} = \bar{w}_{ij}$ for all $(i, j) \in B_1$. □

2. The construction of the moduli space

We will construct the moduli space of all plane curve singularities given by an equation $(x^p + y^q)^2 + \sum_{iq+jp>2pq} a_{ij} x^i y^j = 0$ with fixed Milnor number μ.

For μ being even we get especially the moduli space for all irreducible plane curve singularities with the semigroup $\Gamma = \langle 2p, 2q, \mu - 2(p-1)(q-1) + 1 \rangle$.

We use the family

$$V(F) \subseteq U \times \mathbb{C}^2 \to U$$

constructed in Theorem 1.

U admits a $\mathbb{C}^*$-action defined by

$$\lambda \circ ((h_{ij}), (w_{ij})) := ((\lambda^{iq+jp-pq} h_{ij}), (\lambda^{iq+jp-2pq} w_{ij})).$$

We get

$$F(\lambda^q x, \lambda^p y, h, w) = \lambda^{2pq} F(x, y, \lambda \circ (h, w)).$$

If $\mu = (2p-1)(2q-1) + d$ and $\alpha q + \beta p = 2pq + d, \alpha < p$, then $U \subseteq \mathbb{C}^N$ was defined by $W_{\alpha,\beta} \neq 0$.

For the construction of the moduli space it is enough to consider the restriction of our family to the transversal section to the orbits of the C^*-action defined by $W_{\alpha,\beta} = 1$.

Let W' be defined by $W = (W_{\alpha,\beta}, W')$ and $G(x, y, H, W') = F(x, y, H, 1, W')$. The parameter space of G is $\mathbb{C}^{N-1} = \operatorname{Spec} \mathbb{C}[H, W']$.

The group μ_d of dth roots of unity acts on the family

$$V(G) \subseteq \mathbb{C}^2 \times \mathbb{C}^{N-1} \to \mathbb{C}^{N-1}$$

induced by the above C^*-action

$$G(\lambda^q x, \lambda^p y, h, w') = \lambda^{2pq} G(x, y, \lambda \circ (h, w)) \quad \lambda \in \mu_d.$$

LEMMA 6. *Let $\varphi: \mathbb{C}[[x, y]] \to \mathbb{C}[[x, y]]$ be an automorphism and $u \in \mathbb{C}[[x, y]]$ a unit such that*

$$u \cdot G(\varphi(x), \varphi(y), h, w') = G(x, y, \bar{h}, \bar{w}')$$

then there is a $\lambda \in \mu_d$ such that (h, w') and $\lambda \circ (\bar{h}, \bar{w}')$ are contained in an analytically trivial subfamily of $V(G) \to \mathbb{C}^{N-1}$.

Proof. Let

$$\varphi(x) = \sum a_{ij}^{(1)} x^i y^j \quad \text{and} \quad \varphi(y) = \sum a_{ij}^{(2)} x^i y^j, \quad u = \sum u_{ij} x^i y^j.$$
$$u \cdot G(\varphi(x), \varphi(y), h, w') = G(x, y, \bar{h}, \bar{w}')$$

implies

(1) $a_{ij}^{(1)} = 0$ if $iq + jp < q$
(2) $a_{1,0}^{(1)2p} = a_{0,1}^{(2)2q} = a_{1,0}^{(1)p} a_{0,1}^{(2)q} = u_{0,0}^{-1}$.

Let $a_{1,0}^{(1)} = \lambda^q$ and $a_{0,1}^{(2)} = \lambda^p$.
We will prove later that $\lambda^d = 1$.

Now we may assume that $\lambda = 1$ and prove that (h, w') and $(\bar{h}, \bar{w}')$ are contained in an analytically trivial subfamily of $V(G) \to \mathbb{C}^{N-1}$.

We choose

(1) $u(t) \in \mathbb{C}[t][[x, y]]$ with the following properties $u(0) = 1$, $u(1) = u$ and u is a unity for all $t \in \mathbb{C}$.

(2) φ_t: $\mathbb{C}[t][[x, y]] \to \mathbb{C}[t][[x, y]]$ with the following properties $\varphi_0 =$ identity, $\varphi_1 = \varphi$ and φ_t is an automorphism of positive degree for all $t \in \mathbb{C}$.

Let $H(t) := u(t)G(\varphi_t(x), \varphi_t(y), h, w')$ and apply Lemma 4. There is an $\mathbb{C}[t]$-automorphism Φ_t: $\mathbb{C}[t][[x, y]] \to \mathbb{C}[t][[x, y]]$ such that

$$H(\Phi_t) = F(x, y, h(t), w(t))$$

for suitable $h_{ij}(t), w_{ij}(t) \in \mathbb{C}[t]$ with the property

$$h(0) = h$$
$$w(0) = (1, w').$$

$H(\Phi_t)$ has a constant Milnor number, i.e. $w_{\alpha,\beta}(t)$ has to be constant.

This implies

$$H(\Phi_t) = G(x, y, h(t), w'(t)).$$

But,

$$G(x, y, h(1), w'(1)) = H(\Phi_1) = G(\Phi_1(x), \Phi_1(y), \bar{h}, \bar{w}').$$

Using Lemma 5 and the fact that Φ_1 has positive degree we get

$$\bar{h} = h(1)$$
$$\bar{w}' = w'(1),$$

i.e. (h, w') and $(\bar{h}, \bar{w}')$ are in the trivial family

$$G(x, y, h(t), w'(t)) = u(\Phi_t)G(\Phi_t \varphi_t, h, w').$$

To finish the proof of Lemma 6 we have to prove

LEMMA 7. *Let*

$$f_k = \left(x^p + y^q + \sum_{iq+jp>pq} a_{ij}^{(k)} x^i y^j\right)^2 + x^\alpha y^\beta + \sum_{iq+jp>2pq+d} b_{ij}^{(k)} x^i y^j, \quad k = 1, 2$$

$\alpha < p$, $\alpha q + \beta p = 2pq + d$.

Let $\varphi\colon \mathbb{C}[[x,y]] \to \mathbb{C}[[x,y]]$ *be an automorphism with the property*

$$\varphi(x) = \lambda^q x + \textit{terms of degree} > q$$
$$\varphi(y) = \lambda^p y + \text{terms of degree} > p$$

and u a unit, such that

$$f_1(\varphi) = f_2 \cdot u$$

then $\lambda^d = 1$.

Proof. $u = \lambda^{2pq}$ + terms of higher degree.

$$u \cdot f_2 = \lambda^{2pq}\Bigg(x^p + y^q + \sum_{iq+jp>pq} \bar{a}_{ij}^{(2)} x^i y^j\Bigg)^2 + \lambda^{2pq} x^\alpha y^\beta + \sum_{iq+jp>2pq+d} \bar{b}_{ij}^{(2)} x^i y^j$$

$$f_1(\varphi) = \lambda^{2pq}\Bigg(x^p + y^q + \sum_{iq+jp>pq} \bar{a}_{ij}^{(1)} x^i y^j\Bigg)^2 + $$
$$+ \lambda^{2pq+d} x^\alpha y^\beta + \sum_{iq+jp\geqslant 2pq+d} \bar{b}_{ij}^{(1)} x^i y^j$$

for suitable $\bar{a}_{ij}^{(k)}, \bar{b}_{ij}^{(k)}$.

This implies

$$\Bigg(2x^p + 2y^q + \sum_{iq+jp>pq} (\bar{a}_{ij}^{(1)} + \bar{a}_{ij}^{(2)}) x^i y^j\Bigg) \cdot \sum_{iq+jp>pq} (\bar{a}_{ij}^{(1)} - \bar{a}_{ij}^{(2)}) x^i y^j$$
$$= (1-\lambda^d) x^\alpha y^\beta + \sum_{iq+jp>2pq+d} \lambda^{-2pq}(\bar{b}_{ij}^{(2)} - \bar{b}_{ij}^{(1)}) x^i y^j.$$

Because the leading term of the left side of the equation is divisible by $x^p + y^q$, we get $\lambda^d = 1$. □

We consider now the Kodaira-Spencer map of the family

$V(G) \to \mathbb{C}^{N-1}$:

$$\rho\colon \mathrm{Der}_{\mathbb{C}}\mathbb{C}[H, W'] \to \mathbb{C}[H, W'][[x,y]]\Big/\left(G, \frac{\partial G}{\partial x}, \frac{\partial G}{\partial y}\right)$$

defined by

$$\rho(\delta) = \text{class}(\delta G).$$

The kernel of the Kodaira-Spencer map is a Lie-algebra L and along the integral manifolds of L the family is analytically trivial. We will choose a transversal section to the integral manifolds of L and divide by the action of μ_d to get the moduli space. To describe this transversal section we choose a suitable subset of B_1:

$$B_2 = \{(i,j) \in B_1, i \leqslant \gamma, j \leqslant \delta\} \cup \{(i,j) \in B_1, j \leqslant \delta - q\}.$$

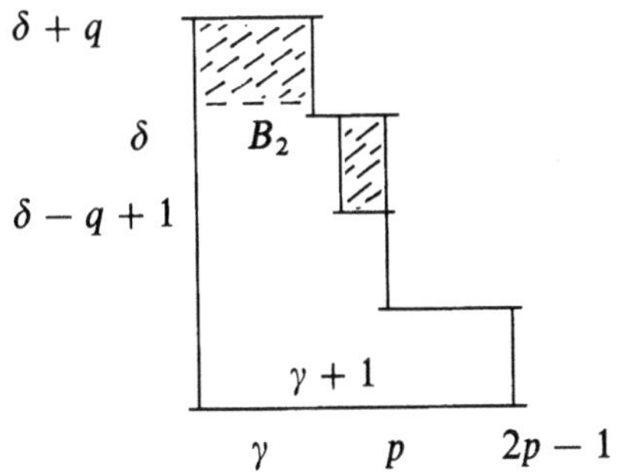

Let $M := \# B_0 + \# B_2 = N - (p-1)(q-1) = (p-2)(q-2) + [q/p] - 1$. Let $W'' := (W_{ij})_{(ij) \in B_2}$ and $\mathbb{C}^M = \operatorname{Spec} \mathbb{C}[H, W''\}$

$$G_u(x, y, H, W'') := \left(x^p + y^q + \sum_{(i,j) \in B_0} H_{ij} x^i y^j\right)^2 + x^\alpha y^\beta + \sum_{(i,j) \in B_2} W_{ij} x^i y^j$$

As before μ_d acts on the family $V(G_u) \subseteq \mathbb{C}^2 \times \mathbb{C}^M \to \mathbb{C}^M$.

THEOREM 2. *$\mathbb{C}^M/\mu_d$ is the moduli space of all plane curve singularities defined by an equation*

$$(x^p + y^q)^2 + \sum_{iq + jp > 2pq} a_{\overline{ij}} x^i y^j = 0$$

with Milnor numbers $\mu = (2p-1)(2q-1) + d$ and G_u is the corresponding universal family.

Especially the Tjurina number $\tau = \mu - (p-1)(q-1)$ only depends on μ for these singularities.

COROLLARY. *Let $\Gamma = \langle 2p, 2q, 2pq + d \rangle$, d odd, a semigroup.*

Then $\mathbb{C}^{(p-2)(q-2)+[q/p]-1}/\mu_d$ is the moduli space of all irreducible plane curve singularities with the semigroup Γ.

G_u is the corresponding universal family.

Proof. To prove the theorem we compute generators of the kernel of the

Kodaira-Spencer map.

Let

$$G^{(0)} = x^p + y^q + \sum_{(i,j)\in B_0} H_{ij}x^iy^j$$

$$G^{(1)} = x^\alpha y^\beta + \sum_{\substack{(i,j)\in B_1\\ iq+jp>2pq+d}} W_{ij}x^iy^j, \quad \text{i.e.}$$

$$G = G^{(0)2} + G^{(1)}.$$

Let $\delta \in \mathrm{Der}_{\mathbb{C}}\, \mathbb{C}[H, W']$ be a vector field which belongs to the kernel of the Kodaira-Spencer map, i.e.

$$\delta G \in \left(G, \frac{\partial G}{\partial x}, \frac{\partial G}{\partial y}\right).$$

Now

$$\delta G = 2G^{(0)} \sum_{(i,j)\in B_0} \delta H_{ij}x^iy^j + \sum_{\substack{(i,j)\in B_1\\ iq+jp>2pq+d}} \delta W_{ij}x^iy^j = S\cdot G \bmod \frac{\partial G}{\partial x}, \frac{\partial G}{\partial y}$$

for a suitable $S \in \mathbb{C}[H, W'][[x, y]]$.

We will associate to any monomial x^ay^b, $(a, b) \neq (0, 0)$, a vector field $\delta_{a,b} \in \mathrm{Der}_{\mathbb{C}}[H, W']$ such that

$$\delta_{a,b}G = x^ay^bG \bmod \left(\frac{\partial G}{\partial x}, \frac{\partial G}{\partial y}\right).$$

Obviously $\{\delta_{a,b}\}$ generate the kernel of the Kodaira-Spencer map as $\mathbb{C}[H, W']$-module.

Now consider

$$x^ay^bG = x^ay^bG^{(0)2} + x^ay^bG^{(1)}.$$

Let $$x^ay^bG^{(0)} = \sum_{(i,j)\in B_0} E^{ab}_{ij}x^iy^j + L_1\frac{\partial G^{(0)}}{\partial x} + L_2\frac{\partial G^{(0)}}{\partial y}$$

for suitable $E^{ab}_{ij} \in \mathbb{C}[H, W']$, $L_1, L_2 \in \mathbb{C}[H, W'][[x, y]]$,

$$L_1 = \frac{1}{p}x^{a+1}y^b + \text{terms of higher degree}$$

$$L_2 = \frac{1}{q}x^ay^{b+1} + \text{terms of higher degree},$$

then

$$x^a y^b G = G^{(0)} \sum_{(i,j)\in B_0} E_{ij}^{ab} x^i y^j + x^a y^b G^{(1)} - \tfrac{1}{2} L_1 \frac{\partial G^{(1)}}{\partial x} -$$

$$-\tfrac{1}{2} L_2 \frac{\partial G^{(1)}}{\partial y} \bmod \left(\frac{\partial G}{\partial x}, \frac{\partial G}{\partial y} \right).$$

The leading term of

$$x^a y^b G^{(1)} - \tfrac{1}{2} L_1 \frac{\partial G^{(1)}}{\partial x} - \tfrac{1}{2} L_2 \frac{\partial G^{(1)}}{\partial y}$$

is $-(d/2pq) x^{\alpha+a} y^{\beta+b}$.

Using Lemma 2 we get

$$x^a y^b G^{(1)} - \tfrac{1}{2} L_1 \frac{\partial G^{(1)}}{\partial x} - \tfrac{1}{2} L_2 \frac{\partial G^{(1)}}{\partial y}$$

$$= \sum_{\substack{(i,j)\in B_1 \\ iq+jp \geqslant 2pq+d+aq+bp}} D_{ij}^{ab} x^i y^j \bmod \left(\frac{\partial G}{\partial x}, \frac{\partial G}{\partial y} \right)$$

for suitable $D_{ij}^{ab} \in \mathbb{C}[H, W']$.

This implies

$$x^a y^b G = G^{(0)} \sum_{(i,j)\in B_0} E_{ij}^{ab} x^i y^j + \sum_{\substack{(i,j)\in B_1 \\ iq+jp \geqslant 2pq+d+aq+bp}} D_{ij}^{ab} x^i y^j \bmod \left(\frac{\partial G}{\partial x}, \frac{\partial G}{\partial y} \right)$$

We define for $(a, b) \neq (0, 0)$

$$\delta_{a,b}(H_{ij}) := \tfrac{1}{2} E_{ij}^{ab}$$

$$\delta_{a,b}(W_{ij}) := D_{ij}^{ab}, \text{ i.e.,}$$

$$\delta_{a,b} = \frac{1}{2} \sum E_{ij}^{ab} \frac{\partial}{\partial H_{ij}} + \sum D_{ij}^{ab} \frac{\partial}{\partial W_{ij}}$$

The vector fields $\delta_{a,b}$ have the following properties:

(1) $\delta_{a,b}$ is zero if $aq + bp > 2pq - 2p - 2q$

(2) $\delta_{a,b}(W_{ij}) = 0$ if $iq + jp < 2pq + d + aq + bp$

(3) $\delta_{a,b}(W_{ij}) = -d/2pq$ if $(i,j) = (\alpha + a, \beta + b)$ or $(i,j) = (\alpha + a - p, \beta + b + q)$ (in this case $iq + jp = 2pq + d + aq + bp$).

(4) $\delta_{a,b}(H_{ij}) = 0$ for all $(i, j) \in B_0$ if $aq + bp \geqslant pq - 2p - 2q$

(5) For $iq + jp \geqslant 3pq + d - q$ and $(i, j) \in B_1$ there is (a', b') such that

$$(i, j) = (\alpha + a', \beta + b') \text{ or } (i, j) = (\alpha + a' - p, \beta + b' + q),$$

i.e. $\delta_{a',b'}(W_{ij}) = -d/2pq$.

(6) For any (a, b), $aq + bp < pq - q$, always $(\alpha + a, \beta + b)$ or $(\alpha + a - p, \beta + b + q) \in B_1$, i.e. $\delta_{a,b}(W_{ij}) = -d/2pq$ for the corresponding $(i, j) \in B_1$.

(1) and (4) hold because of the fact that

$$iq + jp \leqslant 2pq - 2p - 2q \quad \text{if } (i, j) \in B_0$$
$$iq + jp \leqslant 4pq + d - 2p - 2q \quad \text{if } (i, j) \in B_1.$$

(2) and (3) hold because of the fact that the leading term of $G^{(1)}$ has degree $2pq + d$ and because of Lemma 2. To prove (5) we consider two cases

1. *Case* $i \geqslant \alpha - 1$
In this case $(i, j) \in B_1$ implies $i \leqslant p - 1$. But $iq + jp \geqslant 3pq + d - q$ implies $j \geqslant \beta$. Then $a' = i - \alpha, b' = j - \beta$ have the required properties. Notice that $i < \alpha - 1$ and $iq + jp \geqslant 3pq + d - q$ implies $(i, j) \notin B_1$

2 *Case* $i < \alpha - 1$
Now $iq + jp \geqslant 3pq + d - q$ implies $j \geqslant \beta + q$ then $a' = p + i - \alpha$, $b' = j - \beta - q$ have the required properties. (6) is similar to (5):
We may assume that $(\alpha + a, \beta + b) \notin B_1$. This implies $2p - 3 \geqslant \alpha + a \geqslant p$ and $\alpha \geqslant 2$ because $b \leqslant q - 2$, $a \leqslant p - 2$. Suppose $(\alpha + a - p, \beta + b + q) \notin B_1$ then $\beta + b + q \geqslant \beta + 2q - 1$, i.e. $b \geqslant q - 1$, or $\alpha + a - p \geqslant \alpha - 1$, i.e. $a \geqslant p - 1$, but this is not possible. □

For the coefficients to the vectorfields $\delta_{a,b}$ we get, because of (1)–(6), the following matrix:

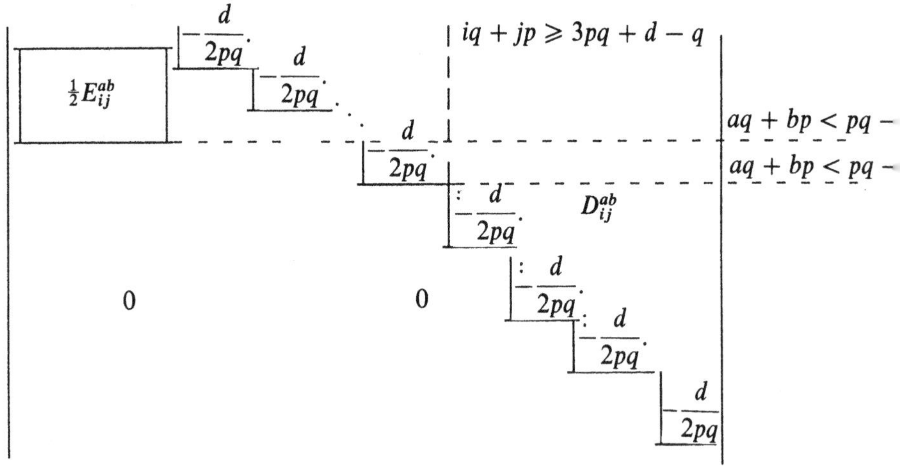

This implies that the kernel of the Kodaira–Spencer map is generated (as $\mathbb{C}[H, W']$-module) by the vector fields

$$\delta'_{ij} := \frac{\partial}{\partial w_{ij}} \quad (i, j) \in B_1, \quad iq + jp \geqslant 3pq + d - q$$

and

$$\delta'_{l,m} = -\frac{pq}{d} \sum_{(i,j)\in B_0} E^{ab}_{ij} \frac{\partial}{\partial H_{ij}} + \frac{\partial}{\partial W_{l,m}} +$$

$$+ \sum_{\substack{(i,j)\in B_1 \\ 2pq+d+aq+bp<iq+jp<3pq+d-q}} \left(-\frac{2pq}{d}\right) D^{ab}_{ij} \frac{\partial}{\partial W_{ij}}$$

$$\times\ (l, m) \in B_1 \backslash (B_2 \cup \{(\alpha, \beta)\}, lq + mp < 3pq + d - q,$$

with $(l, m) = \begin{cases} (\alpha + a, \beta + b) & \text{if } l \geqslant \alpha \\ (\alpha + a - p, \beta + b + q) & \text{else.} \end{cases}$

The vectorfields $\delta'_{l,m}$ act nilpotentely on $\mathbb{C}[H, W']$. Namely, if we consider $\mathbb{C}[H, W']$ as a graded algebra defined by $\deg H_{ij} = pq - iq - jp < 0$, $\deg W_{ij} = 2pq - iq - jp < 0$ then the E^{ab}_{ij} resp. D^{ab}_{ij} are polynomials in $\mathbb{C}[H, W']$ of degree $\geqslant aq + bp + pq - iq - jp$ resp. $\geqslant aq + bp + 2pq - iq - jp$. Notice that their degree is always $\leqslant 0$. Let $A \in \mathbb{C}[H, W']$ be any polynomial of degree $0 \geqslant \deg A = s$ ($\deg A$ = minimum of the degrees of the monomials in A). Then the degree of $\delta'_{lm}(A) > s$. Therefore there is some n with $\delta'^{n}_{lm}(A) = 0$.

LEMMA 8. *Let A be a ring of finite type over a field k. $L \subseteq \mathrm{Der}_k(A)$ a Lie-Algebra.*

Let $\delta_1, \ldots, \delta_r$ vector fields with the following properties:

(1) $\delta_1, \ldots, \delta_r \in L$ *and* $L \subseteq \sum \delta_i A$

(2) $[\delta_i, \delta_j] \in \sum_{k > \max\{i,j\}} \delta_k A$

(3) *There are $x_1, \ldots, x_r \in A$ such that*

$$\delta_i(x_i) = 1 \text{ and } \delta_j(x_i) = 0 \quad j > i$$

(4) *$\delta_1, \ldots, \delta_r$ act nilpotentely on A.*

Then $A^L[x_1, \ldots, x_r] = A$.

The Lemma is not difficult to prove. A similar lemma was used in the construction of the moduli space for curve singularities with the semi-group $\langle p, q\rangle$ (cf. [1], [2]).

Obviously A^L is the ring of all elements of A being invariant under $\delta_1, \ldots, \delta_r$.

Now $A^{\delta_r}[x_r] = A$ and the conditions (2)–(4) of the lemma are satisfied for $\delta_1, \ldots, \delta_{r-1}$ acting on A^{δ_r}.

Now we may apply the Lemma 8 to the kernel of the Kodaira-Spencer map and its generators $\{\delta'_{lm}\}$.

Because of the lemma the geometric quotient of $\mathbb{C}^{N-1} = \operatorname{Spec} \mathbb{C}[H, W']$ by the action of the kernel of the Kodaira-Spencer map exist and is isomorphic to the transversal section to the maximal integral manifolds (which intersect therefore each of these integral manifolds exactly in one point) defined by

$$W_{l,m} = 0, \quad (l,m) \in B_1 \backslash (B_2 \cup \{(\alpha, \beta)\}).$$

Now we use Lemma 6 and get Theorem 2. Notice that

$$\{G^{(0)} x^i y^j\}_{(i,j)\in B_0} \cup \{x^i y^j\}_{(i,j)\in B_2} \cup \{x^i y^j, iq + jp \leqslant 2pq, (i,j) \in B\}$$

is a base of the free $\mathbb{C}[H, W']$-module $\mathbb{C}[H, W'][[x, y]]/(G, \partial G/\partial x, \partial G/\partial y)$. This implies $\mu - \tau = \#(B_1 \backslash B_2) = (p-1)(q-1)$. □

References

[1] Laudal, O.A., Martin, B., and Pfister, G., Moduli of irreducible plane curve singularities with the semi-group $\langle a, b \rangle$, *Proc. Conf. Algebraic Geometry Berlin.* Teubner-Texte 92 (1986).

[2] Laudal, O.A., and Pfister, G., Local moduli and singularities, Springer, *Lecture Notes* 1310 (1988).

[3] Mora, F., A constructive characterization of standard bases, *Boll. U.M.I. sez. D.* 2 (1983) 41–50.

[4] Zariski, O., Le probleme des modules pour les branches planes, Ed. Hermann, Paris 1986.

Compositio Mathematica **76**: 265–275, 1990.

On Castelnuovo's regularity and Hilbert functions

UWE NAGEL

Martin-Luther-University, Dep. of Mathematics, DDR-4010 Halle, German Democratic Republic

Received 3 December 1988; accepted 20 July 1989

Abstract. New bounds for Castelnuovo's regularity are established. As a consequence, we show that a property of Hilbert functions stated by J. Harris and D. Eisenbud in [7], p. 82 is only true for curves and false for higher-dimensional subschemes. The letter of W. Vogel [25] gives rise to study this property again.

1. Introduction

Castelnuovo's regularity was first defined by D. Mumford [10], who attributes the idea to G. Castelnuovo, for coherent sheaves on projective spaces. In a more algebraic setting it was defined by D. Eisenbud and S. Goto [3] and A. Ooishi [13] (see Section 2). It comes out that Castelnuovo's regularity gives an upper bound for the maximal degrees of the syzygies in a minimal free resolution [3]. D. Bayer and M. Stillman [2] showed that an estimate of the regularity of an ideal gives a bound on the complexity of algorithms for computing syzygies. In [3], p. 93 D. Eisenbud and S. Goto stated the following well-known conjecture on such an estimation.

Let $X \subseteq \mathbb{P}^n_K$ (K an algebraically closed field) be a nondegenerate, that is, X is not contained in a hyperplane of $\mathbb{P}^n$, irreducible, reduced subvariety then holds

$$\operatorname{reg}(X) \leqslant \operatorname{degree}(X) - \operatorname{codim}(X) + 1.$$

So far, this conjecture has been proved for curves [6] and, if $\operatorname{char}(K) = 0$, for smooth surfaces [9] (see also [5]), for a large class of smooth threefolds in $\mathbb{P}^5_K$ [18], Theorem 3.3, and if X is arithmetically Buchsbaum or $\operatorname{degree}(X) \leqslant \operatorname{codim}(X) + 2$ [21]. In the other cases only weaker results are known by using certain correction terms (see, e.g., [1], [22], [23], [5], [11], [18]). We also note that [11] and [12] describe applications of Castelnuovo's regularity. The aim of this paper is to describe a new approach for providing Castelnuovo bounds as presented in [23]. This provides new bounds for Castelnuovo's regularity which improve bounds of [23] in some cases. Moreover, we prove special cases of the conjectures of D. Eisenbud and S. Goto and of D. Bayer and M. Stillman [2] (see Theorem 2). In Section 4 we will apply these results in order to show that an assertion of J. Harris and D. Eisenbud [7], p. 82 on the equality of the abstract

Hilbert function and the Hilbert polynomial is not true in general (see Theorem 3). Providing our Theorem 3 we construct our counterexamples.

Acknowledgement

I thank W. Vogel for several helpful discussions and suggestions, especially for showing me the letter [25].

2. Notations and preliminary results

We work over an algebraically closed field K. Let $S = K[x_0, \dots, x_n]$ be a polynomial ring and $\mathfrak{a} \subseteq S$ be a homogeneous ideal. We set $A = S/\mathfrak{a}$., that is, A is a graded K-algebra. We denote by $P_A := \oplus_{n>0} A_n$ the irrelevant ideal of A. When there is no possibility of confusion we will denote P_A simply by P. Let $M = \oplus_{n \in \mathbb{Z}} M_n$ be a graded A-module. The i-th local cohomology module of M with support in P, denoted by $H_P^i(M)$, is also a graded A-module. Let $[M]_i$ denote the i-th graded part of M for $i \in \mathbb{Z}$, i.e. $[M]_i = M_i$. Let j be an integer then let $M(j)$ denote the graded A-module whose underlying module is the same as that of M and whose grading is given by $[M(j)]_i = [M]_{i+j}$ for all $i \in \mathbb{Z}$. We set for an arbitrary A-module M: $e(M) := \sup\{t \in \mathbb{Z} : [M]_t \neq 0\}$

$$\operatorname{sgn}(M_n) := \begin{cases} 1, & \text{if } M_n \neq 0 \\ 0, & \text{otherwise} \end{cases} \quad \text{and}$$

$$r_k(M) := \sup\{i + e(H_P^i(M)) : i \geqslant k\}.$$

For a finitely generated graded A-module M we define Castelnuovo's regularity, denoted by $\operatorname{reg}(M)$, by $\operatorname{reg}(M) := r_0(M) = r_{\operatorname{depth}(M)}(M)$.

Let $\{x_1, \dots, x_m\}$ be a part of a system of parameters for M. It is said to be a filter-regular sequence if

$$e((x_1, \dots, x_{i-1})M :_M x_i/(x_1, \dots, x_{i-1})M) < \infty \quad \text{for } i = 1, \dots, m$$

(see, e.g., [20], appendix for further informations). We set $M^i = M/(x_1, \dots, x_i)M$ $(i = 0, \dots, m)$ for a filter-regular sequence $\{x_1, \dots, x_m\}$ for M. We have the following result.

LEMMA 1. *Let M be a Noetherian graded A-module of dimension $d > 0$. Then there is a filter-regular sequence $\{l_1, \dots, l_d\}$ of forms $\in A_1$.*

Proof. It is sufficient to show that there is a filter-regular element $l \in A_1$ for M.

Let $\{\mathfrak{p}_1, \ldots, \mathfrak{p}_s\}$ be the set of prime ideals $\mathfrak{p}$ of $\mathrm{Ass}_A(M)$ with Krull-dim$(A/\mathfrak{p}) \geqslant d$. Using, for example [16], Theorem 2.3 we can find an element $l \in A_1 \backslash (P \cdot A_1 \cup \mathfrak{p}_1 \cup \cdots \cup \mathfrak{p}_s)$ since $P \not\subset \mathfrak{p}_1 \cup \cdots \cup \mathfrak{p}_s$. It follows from [20], Theorem 7 of the appendix that l is a filter-regular element for M. Q.E.D.

Let M be a Noetherian graded A-module of dimension $d \geqslant 1$. We denote by $h_0(M)$ the $(d-1)!$-fold of the leading coefficient of the Hilbert polynomial $p_M(t)$. We recall that $p_M(t) = \mathrm{rank}_K[M]_t$ for all $t \gg 0$. If $M = \mathrm{A} = S/\mathfrak{a}$ we define

$$\mathrm{degree}(\mathfrak{a}) := \begin{cases} h_0(S/\mathfrak{a}), & \text{if Krull-dim}(S/\mathfrak{a}) > 0 \\ \mathrm{length}(S/\mathfrak{a}), & \text{if Krull-dim}(S/\mathfrak{a}) = 0. \end{cases}$$

Further, we set $\mathfrak{a}: \langle P \rangle = \{x \in S\colon$ there is an integer $m \geqslant 0$ with $P^m \cdot x \subseteq \mathfrak{a}\}$.

Let X be a subscheme of P_K^n. Then we denote by $I(X)$ the defining ideal of X in $S = K[x_0, \ldots, x_n]$. If $\mathfrak{a} \subseteq S$ is a homogeneous ideal let $V(\mathfrak{a})$ be the corresponding subscheme of $\mathbb{P}_K^n$. The ideal $\mathfrak{a}$ is said to be regular if $V(\mathfrak{a})$ is smooth. Note that degree $(X) =$ degree$(I(X))$. We set $r_k(X) := r_{k+1}(I(X)) = r_k(S/I(X)) + 1$ $(k \geqslant 0)$.

For a set B we write card(B) for its cardinality. Finally, we set for integers a, $b \geqslant 0$: $\{a/b\} := \inf\{t \in \mathbb{Z}\colon a \leqslant tb\}$. If $a > b$ we define a sum $\sum_{i=a}^{b} \ldots$ to be zero and a condition, say B_i, for $i = a, a+1, \ldots, b$ to be empty.

3. Castelnuovo bounds

Studying our integers r_k we will prove a generalization of a theorem of D. Mumford [10], p. 99 and A. Ooishi [13], Theorem 2.

THEOREM 1. *Let M be a finitely generated graded A-module of dimension d and let m and $k \geqslant 1$ be integers. Suppose that $[H_P^i(M)]_{m-i} = 0$ for all $i \geqslant k$. Then $r_k(M) \leqslant m-1$. Moreover,* reg$(M) \leqslant m-1$ *provides $A_i M_j = M_{i+j}$ for all $i \geqslant 0$ and $j \geqslant m-1$.*

Proof. We induct on d. In case of $d = 0$ the assertions are trivial since $H_P^0(M) = M$ and $H_P^i(M) = 0$ for all $i > 0$. Let $d > 0$. According to Lemma 1 we can choose a filter-regular element $l \in A_1$ for M. Then we get $H_P^i(M/0{:}\, l) \cong H_P^i(M)$ for $i > 0$ from the long exact cohomology sequence of $0 \to 0{:}_M l \to M \to M/0{:}_M l \to 0$. The exact sequence $0 \to M/0{:}\, l(-1) \xrightarrow{l} M \to M/lM \to 0$ gives rise to the cohomology sequence

$$0 \to 0{:}_M l(-1) \to H_P^0(M)(-1) \xrightarrow{l} H_P^0(M) \to H_P^0(M/lM) \to H_P^1(M)(-1) \to \cdots \qquad (*)$$

because we have $0{:}_M l \subseteq H_P^0(M)$ by the choice of l. Considering the following sequences of (∗)

$$[H^i_P(M)]_{m-i} \to [H^i_P(M/lM)]_{m-i} \to [H^{i+1}(M)]_{m-i-1} \quad (i \geqslant 0)$$

we get $[H^i_P(M/lM)]_{m-1} = 0$ for all $i \geqslant k$ by assumption. Therefore the induction hypothesis provides $[H^i_P(M/lM)]_j = 0$ for all i and j with $i \geqslant k$ and $i + j \geqslant m$. The following sequences of (∗)

$$[H^i_P(M)]_{m-i} \to [H^i_P(M)]_{m-i+1} \to [H^i_P(M/lM)]_{m+1-i} \quad (i \geqslant 0)$$

gives us $[H^i_P(M)]_{m+1-i} = 0$ for all $i \geqslant k$. By induction we therefore have the first assertion. Proving the second assertion we first note that $\operatorname{reg}(M) \leqslant m - 1$ and (∗) yield $\operatorname{reg}(M/lM) \leqslant m - 1$. If we set $M' = M/lM$ and $A' = A/lA$ we have $A'_i M'_j = M'_{i+j}$ for $i \geqslant 0$ and $j \geqslant m - 1$ by induction hypothesis. Hence $A_i M_j + lM_{i+j-1} = M_{i+j}$. It follows from this by induction on i that $M_{i+j-1} = A_{i-1} M_j$. We obtain $M_{i+j} = A_i M_j + lM_{i+j-1} = A_i M_j$. Q.E.D.

Moreover we get from the proof of Theorem 1. (see also [24], Lemma 2.3):

LEMMA 2. *Let M be a finitely generated graded A-module and let $l \in A_1$ be a filter-regular element for M then*

$$e(H^{i+1}_P(M)) < e(H^i_P(M/lM)) \leqslant \max\{e(H^i_P(M)), 1 + e(H^{i+1}_P(M))\} \quad (i \geqslant 0).$$

Proof. The assertions follows from the exact sequence (∗) of the proof of Theorem 1. Q.E.D.

LEMMA 3. *Let $k \geqslant 1$ and c be integers. Then we have for all finitely generated graded A-modules M of dimension d*

$$r_k(M) \leqslant c - 1 + \sum_{i=k}^{d} \operatorname{card}\{t \in \mathbb{Z} : t \geqslant c - i \text{ and } [H^i_P(M)]_t \neq 0\}.$$

Proof. The assertion is trivial for $k > d$. Let $1 \leqslant k \leqslant d$. We have $\sum_{i=k}^{d} \operatorname{sgn}[H^i_P(M)]_{m-i} \geqslant 1$ for all $m \leqslant r_k(M)$ by Theorem 1. Thus it follows from all $c \in \mathbb{Z}$:

$$\begin{aligned} r_k(M) &\leqslant c - 1 + \sum_{m=c}^{r_k(M)} \sum_{i=k}^{d} \operatorname{sgn}[H^i_P(M)]_{m-i} \\ &= c - 1 + \sum_{i=k}^{d} \sum_{t=c-i}^{r_k(M)-i} \operatorname{sgn}[H^i_P(M)]_t \\ &= c - 1 + \sum_{i=k}^{d} \operatorname{card}\{t \in \mathbb{Z} : t \geqslant c - i \text{ and } [H^i_P(M)]_t \neq 0\} \end{aligned}$$

where the last equality follows from the definition of $r_k(M)$. Q.E.D.

In case of $M = A$ we obtain something more.

MAIN LEMMA: *Let $k \geqslant 0$ and c be integers. Let A be a graded K-algebra of Krulldimension d. Then:*

$$r_k(A) \leqslant c - 1 + \sum_{i=k}^{d} \operatorname{card}\{t \in \mathbb{Z}: t \geqslant c - i \text{ and } [H_P^i(A)]_t \neq 0\}.$$

Proof. We consider only the case $k = 0$ according to Lemma 3. Since $r_0(A) = \max\{e(H_P^0(A)), r_1(A)\}$ the assertion follows from Lemma 3 with $k = 1$ in assuming $e(H_P^0(A)) \leqslant r_1(A)$. We therefore suppose that $e(H_P^0(A)) > r_1(A)$. We set $A = S/\mathfrak{a} \cap \mathfrak{q}$ where $S = K[x_0, \ldots, x_n]$ and $\mathfrak{a}, \mathfrak{q} \subseteq S$ are homogeneous ideals such that $\mathfrak{a}: \langle P \rangle = \mathfrak{a}$, $\mathfrak{a} \nsubseteq \mathfrak{q}$ and $\mathfrak{q}$ is a primary ideal belonging to P. If we set $A' = S/\mathfrak{a}$ we get $\operatorname{depth}(A') \geqslant 1$. Hence $\operatorname{reg}(A') = r_1(A') = r_1(A)$, consequently $\operatorname{reg}(\mathfrak{a}) = 1 + r_1(A)$. It follows from the second assertion of Theorem 1. that $\mathfrak{a}$ is generated by forms of degree $\leqslant 1 + r_1(A)$. Since $H_P^0(A) = \mathfrak{a}/\mathfrak{a} \cap \mathfrak{q}$ we can deduce $[H_P^0(A)]_t \neq 0$ for all t with $r_1(A) < t \leqslant e(H_P^0(A))$. This gives us

$$r_0(A) = e(H_P^0(A)) = r_1(A) + \operatorname{card}\{t \in \mathbb{Z}: t > r_1(A) \text{ and } [H_P^0(A)]_t \neq 0\}.$$

Therefore the assertion follows again from Lemma 3 with $k = 1$. Q.E.D.

COROLLARY 1. *Let A be a graded K-algebra of Krulldimension d. Let j and k be integers such that $j + k \leqslant d$ and $\{l_1, \ldots, l_j\}$ be a filter-regular sequence for A. We set $A^j := A/(l_1, \ldots, l_j)A$. Then:*

$$r_k(A) \leqslant c - 1 + \sum_{i=k}^{k+j-1} \operatorname{card}\{t \in \mathbb{Z}: t \geqslant c - i \text{ and } [H_P^i(A)]_t \neq 0\}$$

for all $c > r_k(A^j)$.

Proof. Lemma 2 gives us for $i = k, \ldots, d - j$

$$i + j + e(H_P^{i+j}(A)) \leqslant i + e(H_P^i(A^j)) \leqslant r_k(A^j).$$

Therefore the assertion follows from the Main Lemma. Q.E.D.

REMARKS (i) If we suppose $k \geqslant 1$ in Corollary 1 the above result remains true even for finitely generated graded A-modules because we can apply Lemma 3.

(ii) If we set $k = 1$ and $j = d - 1$ in Corollary 1 we obtain the main lemma of [23]. Hence we could deduce the Castelnuovo bounds of [23]. Here we want to state some new bounds.

THEOREM 2. *Let $\mathfrak{a} = \mathfrak{p}_1 \cap \cdots \cap \mathfrak{p}_m \subset S = K[x_0, \ldots, x_n]$ be an intersection of m equidimensional (homogeneous) prime ideals. Let d be the Krulldimension of $A = S/\mathfrak{a}$. Then we have:*

(i) $\operatorname{reg}(\mathfrak{a}) \leqslant c + \sum_{i=1}^{d-2} \operatorname{card}\{t \in \mathbb{Z}: t \geqslant c - i$ *and* $[H_P^i(A)]_t \neq 0\}$ *for all* $c \geqslant \operatorname{degree}(\mathfrak{a})$.

(ii) *In case of $m = 1$ we get:*

(ii.1) $\operatorname{reg}(\mathfrak{a}) \leqslant c + \sum_{i=1}^{d-1} \operatorname{card}\{t \in \mathbb{Z}: t \geqslant c - i$ *and* $[H_P^i(A)]_t \neq 0\}$ *for all* $c > \{\operatorname{degree}(\mathfrak{a}) - 1/\operatorname{rank}_K[A]_1 - d\}$ *if* $\operatorname{char}(K) = 0$

(ii.2) $\operatorname{reg}(\mathfrak{a}) \leqslant c + \sum_{i=1}^{d-2} \operatorname{card}\{t \in \mathbb{Z}: t \geqslant c - i$ *and* $[H_P^i(A)]_t \neq 0\}$ *for all* $c > \operatorname{degree}(\mathfrak{a}) + d - \operatorname{rank}_K[A]_1$

(ii.3) $\operatorname{reg}(\mathfrak{a}) \leqslant c + \sum_{i=1}^{d-3} \operatorname{card}\{t \in \mathbb{Z}: t \geqslant c - i$ *and* $[H_P^i(A)]_t \neq 0\}$ *for all* $c > \operatorname{degree}(\mathfrak{a}) + d - \operatorname{rank}_K[A]_1$ *if $\mathfrak{a}$ is regular and* $\operatorname{char}(K) = 0$.

Proof. First we show (i) and (ii.2). If $d = 0$ or $d = 1$ then A is Cohen-Macaulay and we get $\operatorname{reg}(\mathfrak{a}) \leqslant \operatorname{degree}(\mathfrak{a}) + d - \operatorname{rank}_K[A]_1 + 1$ by [13], Proposition 13. This proves (i) and even (ii.2) since $\operatorname{rank}_K[A]_1 \geqslant d$, where equality holds if and only if A is isomorphic to a polynomial ring over K. But in this case we have $\operatorname{degree}(\mathfrak{a}) = \operatorname{reg}(\mathfrak{a}) = 1$. Let $d \geqslant 2$. According to H. Flenner [4] there are generic linear forms $l_1, \ldots, l_{d-2} \in S$ such that $\mathfrak{a} + (l_1, \ldots, l_i)S$ is an intersection of m prime ideals of dimension $d - i$ up to a primary component belonging to P for $i = 0, \ldots, d-2$. Therefore $\{l_1, \ldots, l_{d-2}\}$ is a filter-regular sequence for A. If we set $A^{d-2} = A/(l_1, \ldots, l_{d-2})A$ and $\mathfrak{a}' = \mathfrak{a} + (l_1, \ldots, l_{d-2})S : \langle P \rangle$ we obtain $r_1(A^{d-2}) = r_1(S/\mathfrak{a}') = \operatorname{reg}(S/\mathfrak{a}')$. Thus we get $r_1(A^{d-2} \leqslant \operatorname{degree}(\mathfrak{a}') + 2 - \operatorname{rank}_K[S/\mathfrak{a}']_1$ by Theorem 1.1. of [6] for $m = 1$ and $r_1(A^{d-2}) \leqslant \operatorname{degree}(\mathfrak{a}') - 1$ according to the remark after the proof of Theorem 1.1. in [6] for $m \geqslant 1$. Therefore Corollary 1 with $k = 1$ and $j = d - 2$ proves (i) since $\operatorname{degree}(\mathfrak{a}) = \operatorname{degree}(\mathfrak{a}')$ by Bezout's theorem.

Proving (ii.2) we will show that $2 - \operatorname{rank}_K[S/\mathfrak{a}']_1 \leqslant d - \operatorname{rank}_K[A]_1$. Then we can apply Corollary 1. It follows from [22], Lemma 3 that

$$\begin{aligned}
\operatorname{rank}_K[\mathfrak{a}']_1 &= \operatorname{rank}_K[\mathfrak{a} + (l_1, \ldots, l_{d-3})S]_1 + 1 \\
&\leqslant \operatorname{rank}_K[\mathfrak{a} + (l_1, \ldots, l_{d-3})S : \langle P \rangle]_1 + 1 \\
&= \operatorname{rank}_K[\mathfrak{a} + (l_1, \ldots, l_{d-4})S]_1 + 2 \\
&\leqslant \cdots \leqslant \operatorname{rank}_K[\mathfrak{a}]_1 + d - 2.
\end{aligned}$$

The proof of (ii.3) is analogous to the proof of (ii.2). For this we note that we can choose the linear forms l_i according to [4] such that $\mathfrak{a}' := \mathfrak{a} + (l_1, \ldots, l_{d-3})S : \langle P \rangle$ is regular if $\mathfrak{a}$ is regular. Therefore the assertion follows from Corollary 1 with

$k = 1$ and $j = d - 3$ by using

$$
\begin{aligned}
r_1(A/(l_1,\ldots,l_{d-3})A) &= \operatorname{reg}(S/\mathfrak{a}') \\
&\leqslant \operatorname{degree}(\mathfrak{a}') + 3 - \operatorname{rank}_K[S/\mathfrak{a}']_1 \quad \text{by [9] (or [5])} \\
&\leqslant \operatorname{degree}(\mathfrak{a}') + d - \operatorname{rank}_K[A]_1 \quad \text{by [22]},
\end{aligned}
$$

Lemma 3 and Bezout's theorem.

Now we show (ii.1). The assertion is trivial for $d = 0$ and $d = 1$. Let $d \geqslant 2$. Take the linear forms $l_1, \ldots, l_{d-2}$ constructed in our proof of (ii.2). Consider a general linear form l_{d-1} and set $\mathfrak{a}' = \mathfrak{a} + (l_1, \ldots, l_{d-1})S\colon \langle P\rangle$. Then we get

$$
r_1(A/(l_1,\ldots,l_{d-1})A) = \operatorname{reg}(S/\mathfrak{a}') \leqslant \left\{\frac{\operatorname{degree}(\mathfrak{a}') - 1}{\operatorname{rank}_K[S/\mathfrak{a}']_1 - 1}\right\}
$$

(see, e.g., [22], Lemma 1). We have again $\operatorname{degree}(\mathfrak{a}) = \operatorname{degree}(\mathfrak{a}')$ by Bezout's theorem and $\operatorname{rank}_K[\mathfrak{a}']_1 \leqslant \operatorname{rank}_K[\mathfrak{a}]_1 + d - 1$. On the other hand Lemma 3 of [22] gives us

$$
\begin{aligned}
\operatorname{rank}_K[\mathfrak{a}']_1 &= \operatorname{rank}_K[\mathfrak{a} + (l_1,\ldots,l_{d-2})S]_1 + \\
&\quad + 1 \geqslant \operatorname{rank}_K[\mathfrak{a}]_1 + d - 1.
\end{aligned}
$$

Putting all together we obtain

$$
r_1(A^{d-1}) \leqslant \left\{\frac{\operatorname{degree}(\mathfrak{a}) - 1}{\operatorname{rank}_K[A]_1 - d}\right\}. \tag{+}
$$

Consequently Corollary 1 with $k = 1$ and $j = d - 1$ proves the assertion (ii.1). This completes the proof of Theorem 2. Q.E.D.

COROLLARY 2. *Let X be a nondegenerate, irreducible and reduced subscheme of $\mathbb{P}^n$ of dimension d. Then we have:*

$$
d + 1 + e(H_P^{d+1}(S/I(X))) \leqslant \left\{\frac{\operatorname{degree}(X) - 1}{\operatorname{codim}(X)}\right\}.
$$

Proof. The assertion follows from (+) of the proof of Theorem 2(ii.1) and Lemma 2. Q.E.D.

REMARKS. Theorem 2(ii.1) is Theorem 2(ii) of [23]. Theorem 2(i), (ii.2) and (ii.3) improve Theorem 2(i) of [23] in some special cases.

The assumption $\mathrm{char}(K) = 0$ is necessary in Theorem 2(ii.1) because the general position lemma does not remain true if $\mathrm{char}(K) > 0$ (see [17], Example 1.2).

The conjecture of D. Bayer and M. Stillman [2] gives $\mathrm{reg}(\mathfrak{a}) \leqslant \mathrm{degree}(\mathfrak{a})$. Therefore Theorem 2(i) and (ii.2) (see also [22], corollary) prove the conjectures of Bayer and Stillman and of Eisenbud and Goto in case of $\mathrm{depth}(S/\mathfrak{a}) \geqslant d - 1$. This means, for example, that the latter conjecture is true for surfaces in $\mathbb{P}^4$ if the homogeneous coordinate ring has depth $\geqslant 2$. Note that (see the introduction) singular surfaces in $\mathbb{P}^4$ are the simplest varieties such that the conjecture of D. Eisenbud and S. Goto is open.

4. Counterexamples to an assertion of J. Harris and D. Eisenbud

In this section we will apply Theorem 2 in order to study the equality between Hilbert functions and Hilbert polynomials. Let X be a subscheme of P^n and $A = S/I(X)$ be its homogeneous coordinate ring. We recall that the Hilbert function of X is defined by $h_X(t) := \mathrm{rank}_K[A]_t$ for $t \geqslant 0$. The so-called Hilbert polynomial, denoted by $p_X(t)$, is given by $h_X(t)$ for $t \gg 0$. It is well-known that $p_X(t) = \sum_{i\geqslant 0}(-1)^i h^i(X, \mathcal{O}_X(t))$ where $h^i(X, O_X(t))$ is the dimension of $H^i(X, \mathcal{O}_X(t))$. Following [7] the function $h'_X(t) := h^0(X, \mathcal{O}_X(t))$ is said to be the abstract Hilbert function of X. The index of regularity of X, denoted by $r(X)$, is defined as $r(X) := \min\{t \in \mathbb{N} : h_X(i) = p_X(i) \text{ for all } i \geqslant t\}$. Moreover, we set

$$r'(X) := \min\{t \in \mathbb{N} : h'_X(i) = p_X(i) \text{ for all } i \geqslant t\}, \ \mathrm{reg}(X) = \mathrm{reg}(I(X)).$$

LEMMA 4. (i) ([14], *Corollary* 2.2) $r(X) \leqslant \mathrm{reg}(X) - \mathrm{depth}(S/I(X))$,
(ii) $r'(X) \leqslant r_2(X) - \max\{2, \mathrm{depth}(S/I(X))\}$.

Proof. We have $h_X(t) - p_X(t) = \sum_{i\geqslant 0}(-1)^i \mathrm{rank}_K[H^i_P(S/I(X))]_t$ according to [19]. This proves (i). We obtain (ii) from the characterization of $p_X(t)$ as an Euler-Poincaré characteristic and the isomorphisms $H^i(X, \mathcal{O}_X(t)) \cong [H^{i+1}_P(S/I(X))]_t$ for $i > 0$. Q.E.D.

In [7], p. 82 J. Harris and D. Eisenbud assert for reduced and irreducible subschemes X of $\mathbb{P}^n_K$ ($\mathrm{char}(K) = 0$):

$$r'(X) \leqslant \left\{\frac{\mathrm{degree}(X)}{\mathrm{codim}(X)}\right\}. \qquad (**)$$

The letter [25] gives rise to study this claim again. In this connection, we will prove the following theorem.

THEOREM 3. (i) *We have for nondegenerate, irreducible and reduced curves X:*

$$r'(X) \leqslant \left\{\frac{\text{degree}(X)-1}{\text{codim}(X)}\right\} - 1,$$

that is, the assertion (**) *is true for such curves.*

(ii) *There are nondegenerate, irreducible and reduced subschemes X of* $\mathbb{P}^n$, $n \geqslant 4$, *of dimension d such that the assertion* (**) *is not true for all* $d \geqslant 2$.

Proof. (i) Corollary 2 gives us for $d = 1$

$$r_2(X) - 1 = 2 + e(H^2_{\mathfrak{P}}(S/I(X))) \leqslant \left\{\frac{\text{degree}(X)-1}{\text{codim}(X)}\right\}.$$

Hence (i) follows from Lemma 4(ii).

(ii) We consider the following class of examples: Let $m \geqslant 3$ be an integer. Let $X_m \subseteq \mathbb{P}^4$ be the surface given parametrically by $\{u^m, u^{m-1}v, u^{m-2}vw, uw^{m-1}, w^m\}$. It follows from [8], Proposition 3 that degree$(X_m) = m + 1$. Moreover, Corollary 3.4(ii) of [24] shows $H^1_{\mathfrak{P}}(S/\mathfrak{p}_m) = 0$, that is depth$(S/\mathfrak{p}_m) \geqslant 2$ where $\mathfrak{p}_m \subset S = K[x_0, \ldots, x_4]$ denotes the defining prime ideal of X_m. That is why we can apply Theorem 2(ii.2) and obtain reg$(X_m) \leqslant$ degree$(X_m) -$ codim$(X_m) + 1 = m$. Hence $\mathfrak{p}_m$ is generated by forms of degree $\leqslant m$ according to Theorem 1. Thus we can compute a minimal basis of $\mathfrak{p}_m$ from its parametrization and obtain

$$\mathfrak{p}_m = (x_1x_4 - x_2x_3, x_0x_2^ix_4^{m-1-i} - x_1^ix_3^{m-i}, \quad i = 0, \ldots, m-1).$$

Since $\mathfrak{p}_m$ needs a generator of degree m we get reg$(X_m) = m$. Since $p_{X_m}(t) - h'_{X_m}(t) = \Sigma_{i>0}(-1)^i \text{rank}_K[H^{i+1}_{\mathfrak{P}}(S/\mathfrak{p}_m)]_t$ we obtain from Corollary 2 that

$$p_{X_m}(t) - h'_{X_m}(t) = -\text{rank}_K[H^2_{\mathfrak{P}}(S/\mathfrak{p}_m)]_t$$
$$\text{for } t \geqslant \left\{\frac{\text{degree}(X_m)-1}{\text{codim}(X_m)}\right\} - 2 = \left\{\frac{m}{2}\right\} - 2.$$

Applying Theorem 1 we get from reg$(X_m) = m$ and Corollary 2 that $[H^2_{\mathfrak{P}}(S/\mathfrak{p}_m)]_t \neq 0$ for

$$\left\{\frac{m}{2}\right\} - 2 \leqslant t \leqslant m - 3. \text{ Hence } r'(X_m) = m - 2 > \left\{\frac{m+1}{2}\right\} = \left\{\frac{\text{degree}(X_m)}{\text{codim}(X_m)}\right\}$$

for $m \geqslant 7$. This shows (ii) in case $d = 2$.

Let $j \geqslant 0$ be an integer. We denote by Y_m the projective cone over X_m in $\mathbb{P}^{j+4}$. Then we get $d := \dim(Y_m) = \dim(X_m) + j = 2 + j$, degree$(Y_m) =$ degree(X_m) and

$\operatorname{depth}(S/I(Y_m)) = j + 2$ where $S = K[x_0, \ldots, x_{j+4}]$. Moreover, Lemma 2 gives us $\operatorname{reg}(Y_m) = \operatorname{reg}(X_m) = m$. Therefore we obtain from Corollary 2 as above:

$$r'(Y_m) = m - 2 - j > \left\{\frac{m+1}{2}\right\} = \left\{\frac{\operatorname{degree}(Y_m)}{\operatorname{codim}(Y_m)}\right\}$$

for $m \geqslant 2j + 7$ and for all $d = 2 + j$. Q.E.D.

REMARKS. (i) Using results of [6] the subschemes Y_m show that the conjecture of D. Eisenbud and S. Goto is sharp in the sense that there are nondegenerate, irreducible and reduced varieties X with $\operatorname{reg}(X) = \operatorname{degree}(X) - \operatorname{codim}(X) + 1$ in any dimension $\geqslant 1$ and of any degree $\geqslant 4$.

(ii) If the assertion (**) were true we could deduce $r_2(X) \leqslant \{\operatorname{degree}(X)/\operatorname{codim}(X)\} + 2$. But this is also not true in general as the varieties Y_m show.

(iii) (**) is true in assuming, for example, that the subschemes X are arithmetically Buchsbaum, i.e., that the homogeneous coordinate ring $S/I(X)$ is a graded Buchsbaum K-algebra. In this case we obtain from [21], Theorem 1. $\operatorname{reg}(X) \leqslant \{\operatorname{degree}(X) - 1/\operatorname{codim}(X)\} + 1$. Therefore Lemma 4(ii) yields

$$r'(X) < \left\{\frac{\operatorname{degree}(X) - 1}{\operatorname{codim}(X)}\right\}.$$

(iv) The varieties Y_m are not arithmetically Buchsbaum for $m \geqslant 3$ due to [24], Lemma 4.11 and Corollary 4.7 and even not locally Buchsbaum for $m \geqslant 4$ because $Y_m \subset \mathbb{P}^{j+4}$ has a singularity in the point $\mathfrak{p} = (x_1, \ldots, x_{j+4})$ which is not Buchsbaum for $m \geqslant 4$. Otherwise $(S/I(Y_m))_{\mathfrak{p}}$ and consequently also $(S/I(Y_m) + x_3 S)_{\mathfrak{p}}$ would be Buchsbaum. Since $(I(Y_m) + x_3 S)_{\mathfrak{p}} = (x_3, x_1 x_4, x_2^{m-1}, x_2^{m-2} x_4, \ldots, x_2 x_4^{m-2}, x_4^{m-1})_{\mathfrak{p}} = (x_2^{m-1}, x_3, x_4)_{\mathfrak{p}} \cap (x_1, x_3, x_2^{m-1}, x_2^{m-2} x_4, \ldots, x_4^{m-1})_{\mathfrak{p}}$ we have $x_2 (x_2^{m-1}, x_3, x_4)_{\mathfrak{p}} \nsubseteq (I(Y_m) + x_3 S)_{\mathfrak{p}}$ for $m \geqslant 4$. We immediately obtain a contradiction to a Buchsbaum ring property. (For the facts on Buchsbaum rings used here we refer to [20].)

(v) We can not apply the results of [9] or [5] in order to obtain the bound $\operatorname{reg}(X_m) \leqslant \operatorname{degree}(X_m) - \operatorname{codim}(X_m) + 1$ since the varieties X_m are singular.

(vi) Since our counterexamples are singular varieties it is an open problem if (**) is true in the case of smooth varieties, see [26].

(vii) In [15], p. 370 P. Philippon considers the ideal $\mathfrak{a} := (x_1 x_4 - x_2 x_3, x_0 x_2^2 - x_1^2 x_3, x_0 x_4^2 - x_3^3)$. He asserts that $\mathfrak{a}$ is a prime ideal used in his computations. But it follows from $x_4(x_0 x_2 x_4 - x_1 x_3^2) \in \mathfrak{a}$ that $\mathfrak{a}$ is even not a primary ideal because no power of x_4 and $x_0 x_2 x_4 - x_1 x_3^2$ is contained in $\mathfrak{a}$. (Note that $\mathfrak{p}_3 = \mathfrak{a} + (x_0 x_2 x_4 - x_1 x_3^2) S$.)

References

[1] E. Ballico, On the defining equations of subvarieties in $\mathbb{P}^n$. *Boll. Un. Mat. Ital.* A(6) 5 (1986) 243–246.

[2] D. Bayer and M. Stillman, On the complexity of computing syzygies. *J. Symbolic Comput.* 6 (1988) 135–147.

[3] D. Eisenbud and S. Goto, Linear free resolutions and minimal multiplicity. *J. Algebra* 88 (1984) 89–133.

[4] H. Flenner, Die Sätze von Bertini für lokale Ringe. *Math. Ann.* 229 (1977) 97–111.

[5] V.A. Greenberg, A Castelnuovo bound for projective varieties admitting a stable linear projection onto a hypersurface. Thesis, Columbia university, 1987.

[6] L. Gruson, R. Lazarsfeld and C. Peskine, On a theorem of Castelnuovo, and the equations defining space curves. *Invent. Math.* 72 (1983) 491–506.

[7] J. Harris (with the collaboration of D. Eisenbud), Curves in projective space. Les presses de l'Université, Montréal 1982.

[8] L.T. Hoa, A note on projective monomial surfaces. *Math. Nachr.* (to appear).

[9] R. Lazarsfeld, A sharp Castelnuovo bound for smooth surfaces. *Duke Math. J.* 55 (1987) 423–429.

[10] D. Mumford, Lectures on curves on an algebraic surface. Ann of Math. Studies 59, Princeton Univ. Press, Princeton, N.J., 1966.

[11] U. Nagel, Castelnuovo-Regularität und Hilbertreihen. *Math. Nachr.* 142 (1989) 27–43.

[12] U. Nagel and W. Vogel, Bounds for Castelnuovo's regularity and the genus of projective varieties. In: *Topics in Algebra*. Banach Center Publications, Volume 26. PWN-Polish Scientific Publishers, Warsaw 1989 (to appear).

[13] A. Ooishi, Castelnuovo's regularity of graded rings and modules. *Hiroshima Math. J.* 12 (1982) 627–644.

[14] A. Ooishi, Genera and arithmetic genera of commutative rings. *Hiroshima Math. J.* 17 (1987) 47–66.

[15] P. Philippon, Lemmes de zéros dans les groupes algébriques commutatifs. *Bull. Soc. math.* France 114 (1986) 355–383.

[16] P. Quartararo Jr. and H.S. Butts, Finite unions of ideals and modules. *Proc. Amer. Math. Soc.* 52 (1975) 91–96.

[17] J. Rathmann, The uniform position principle for curves in characteristic p. *Math. Ann.* 276 (1987) 565–579.

[18] J. Rathmann, On the completeness of linear series cut out by hypersurfaces. Preprint, M.I.T., June 1988.

[19] J.-P. Serre, Faisceaus algébriques cohérents. *Ann. of Math.* 61 (1955) 197–278.

[20] J. Stückrad and W. Vogel, Buchsbaum rings and applications. Springer, Berlin 1986.

[21] J. Stückrad and W. Vogel, Castelnuovo bounds for certain subvarieties in $\mathbb{P}^n$. *Math. Ann.* 276 (1987), 341–352.

[22] J. Stückrad and W. Vogel, Castelnuovo bounds for locally Cohen-Macaulay schemes. *Math. Nachr.* 136 (1988), 307–320.

[23] J. Stückrad and W. Vogel, Castelnuovo's regularity and cohomological properties of sets of points in $\mathbb{P}^n$. *Math. Ann.* 284 (1989) 487–501.

[24] N.V. Trung and L.T. Hoa, Affine semigroups and Cohen-Macaulay rings generated by monomials. *Trans. Amer. Math. Soc.* 298 (1986) 145–167.

[25] W. Vogel, Letter to D. Eisenbud dated 5 October, 1987.

[26] U. Nagel and W. Vogel, Castelnuovo's regularity of graded k-algebras and applications. *Stud. Cerc. Mat.* (in preparation).

Compositio Mathematica **76**: 277–294, 1990.

Indecomposable Cohen–Macaulay modules and irreducible maps

DORIN POPESCU[1] & MARKO ROCZEN[2]

[1]*INCREST, Dept. of Math., Bd Pacii 220, Bucharest 79622, Romania;* [2]*Humboldt–Universität, Sekt. Mathematik, Unter d. Linden 6, Berlin 1086, DDR*

Received 22 November 1988; accepted 20 July 1989

Introduction

Let (R, m) be a local CM-ring and M a finitely generated (shortly f.g.) R-module. M is a maximal CM module (shortly MCM R-modules. The isomorphism classes of indecomposable MCM R-modules form the vertices of the Auslander-Reiter quiver $\Gamma(R)$ of R. Section 3 studies the behaviour of $\Gamma(R)$ under base change; best results (cf. (3.10), (3.14)) being partial answers to the conjectures from [Sc] (7.3). Unfortunately, the proofs use the difficult theory of Artin approximation (cf. [Ar], or [Po1]).

A different easier method is to use the so-called CM-reduction ideals as we did in [Po2] (4.9) or have in (3.2), (3.3). This procedure is very powerful in proving results describing how large is the set of those positive integers which are multiplicities of the vertices of $\Gamma(R)$, in fact the first Brauer-Thrall conjecture (cf. [Di], [Yo], [Po2] or here (4.2), (4.3)). However the Corollary (3.3) obtained by this method is much weaker than (3.10) which uses Artin approximation theory. The reason is that the conditions under which we know the existence of CM-reduction ideals are still too complicated. Trying to simplify them we see that the difficulty is just to prove some bound properties on MCM modules (cf. Section 2) which we hope to hold for every excellent henselian local CM-ring. Our Theorem (4.4) and Corollary (4.6) give sufficient conditions when the second Brauer–Thrall conjecture holds, and our Corollary (4.7) is a nice application to rational double points (inspirated by [Yo] (4.1)).

We would like to thank L. Badescu and J. Herzog for helpful conversations on (4.7), (3.10), (3.16) respectively.

1. Bound properties on MCM modules

(1.1) Let (R, m) be a local CM-ring, $k := R/m$, $p := \operatorname{char} k$ and $\operatorname{Reg} R = \{q \in \operatorname{Spec} R \mid R_q \text{ regular}\}$. Suppose that $\operatorname{Reg} R$ is open (this happens for instance when R is quasi-excellent). Then $I_s(R) = \bigcap_{q \notin \mathrm{Reg}\, R} q$ defines the singular locus of R. We say that R has *bound properties on* MCM *modules* if the following conditions hold:

(i) there exists a positive integer r such that $I_s(R)^r \mathrm{Ext}^1_R(M, N) = 0$ for every MCM R-module M and for every f.g. R-module N, i.e. $I_s(R)$ is in the radical of the Dieterich Ext-annihilating ideal of R (cf. [Di] §2),

(ii) for every ideal $a \subset R$ and every element $y \in I_s(R)$ there exists a positive integer e such that $(aM : y^e)_M := \{z \in M \mid y^e z \in aM\} = (aM : y^{e+1})_M$ for every MCM R-module M.

Clearly it is enough to consider in (i), (ii) only indecomposable M. Let M be a MCM R-module and $A_M := \{x \in R \mid x\, \mathrm{Ext}^1_R(M, N) = 0$ for every f.g. R-module $N\}$

(1.2) LEMMA. $I_s(R) \subset \sqrt{A_M}$.

Proof. Let $x \in I_s(R)$. Then R_x is a regular ring and so M_x is projective over R_x. Indeed, if q Spec R with $x \notin q$ then M_q is free over R_q by [He](1.1), M_q being still MCM by [Ma2] (17.3) and R_q is regular. Thus

$$R_x \otimes_R \mathrm{Ext}^1_R(M, N) = \mathrm{Ext}^1_{R_x}(M_x, N_x) = 0$$

and so a certain power of x kills $\mathrm{Ext}^1_{R_x}(M, N)$, i.e. $x \in \sqrt{A_M}$.

(1.3) REMARK. The above Lemma shows that for each MCM R-module M we are able to find a positive integer r_M such that $I_s(R)^{r_M} \mathrm{Ext}^1_R(M, N) = 0$ for every f.g. R-module N. Thus the trouble in (1.1)(i) is just to show that r_M can be bounded when M runs in CM(R). Also for each f.g. R-module M by Noetherianity we can find in (1.1)(ii) a positive integer e_M such that $(aM : y^{e_M})_M = (aM : y^{e_M + 1})_M$. Again the trouble is to show that e_M can be bounded when M runs in CM(R). However, if R has finite CM-type (i.e. $\Gamma(R)$ has just a finite set of vertices) then R has bound properties on MCM modules (compare with [Di] Proposition 8).

(1.4) LEMMA. *Suppose that* (R, m) *is reduced complete with* k *perfect and* $\mathrm{Reg}(R/pR) = \{q/pR \mid q \in \mathrm{Reg}\, R, q \supset pR\}$ *if* $pR \neq 0$ (*i.e. if* $p \neq \mathrm{char}\, R$). *Then* R *has bound properties on* MCM *modules.*

Proof. Clearly either R contains k or R is a flat algebra over a Cohen ring of residue field k, i.e. a complete DVR (T, t) which in an unramified extension of $\mathbb{Z}_{(p)}, p > 0, t = p \cdot 1 \in T$. Let $x = (x_1, \ldots, x_n)$ be a system of elements from m such that (t, x) forms a system of parameters in R. By Cohen's structure Theorems the canonical map $j \colon T[[X]] \to R, X = (X_1, \ldots, X_n) \mapsto x$ is finite. As R is CM we obtain R flat (thus free) over the image S_x of j.

Let I_x be the kernel of the multiplication map $R \otimes_{S_n} R \xrightarrow{\mu} R$ and $\mathcal{N}_x :=$

$\mu(\mathrm{Ann}_{R \otimes_{S_n} R} I_x)$ the Noether different of R over S_n. Then

$$\mathcal{N}_x \cdot \Omega_{R/S} = 0 \quad \text{and} \quad I_s(R) = \sqrt{\Sigma_x \mathcal{N}_x}, \tag{1}$$

where the sum is taken over all systems of elements x such that (t, x) forms a system of parameters of R (see [Po2] (2.8), (2.10), the ideas come in fact from [Yo]).

Using the Hochschild cohomology we get a surjective map

$$\mathrm{Hom}_R(\Omega_{R/S_x}, \mathrm{Hom}_{S_x}(M, N)) \cong \mathrm{Der}_{S_x}(R, \mathrm{Hom}_{S_x}(M, N)) \to \mathrm{Ext}^1_R(M, N)$$

for every MCM R-module M and every f.g. R-module N (see e.g. in [Di] Lemma 5). By (1) follows $\mathcal{N}_x \,\mathrm{Ext}^1_R(M, \mathrm{N}) = 0$ and so there is $r \in \mathbb{N}$ such that

$$I_s(R)^r \,\mathrm{Ext}^1_R(M, N) = 0 \tag{2}$$

for every MCM R-module M and every f.g. R-module N, i.e. (1.1)(i) holds.

Now, let $a \subset R$ be an ideal and $y \in I_s(R)$. We show that there exists a positive integer e such that $(aM : y^e)_M = (aM : y^{e+1})_M$ for every MCM R-module M. If there exists x as above such that $y \in \mathcal{N}_x$ then it is enough to apply [Po2] (3.2) for $S_x \subset R$. Otherwise choose in $I_s(R)$ a system of elements $(n_i)_{1 \leqslant i \leqslant s}$ such that $I_s(R) = \sqrt{\Sigma_i n_i R}$ and for every i there exists $x^{(i)}$ as above such that $n_i \in \mathcal{N}_{x^{(i)}}$. Then $y n_i \in \mathcal{N}_{x^{(i)}}$ and so there exists a positive integer e_i such that

$$(aM : (y n_i)^{e_i})_M = (aM : (y n_i)^{e_i + 1})_M$$

for every MCM R-module M. Choose a positive integer e' such that $I_s(R)^{e'} \subset \sum_{i=1}^s n_i^{e_i} R$. We claim that $e = e' + \max_{1 \leqslant i \leqslant s} e_i$ works. Indeed, let M be a MCM R-module, if $y^v z \in aM$ for a certain $z \in M$ and $v \in \mathbb{N}$ then $(y n_i)^v z \in aM$ and so $(y n_i)^{e_i} z \in aM$. Thus

$$y^e z \in y^{e-e'} \left(\sum_{i=1}^s n_i^{e_i} R \right) z \subset \sum_{i=1}^s (y n_i)^{e_i} R z \subset aM.$$

(1.5) LEMMA. *Suppose that*

(i) (R, m) *is quasi-excellent and reduced,*
(ii) $\mathrm{Reg}(R/pR) = \{q/pR \mid q \in \mathrm{Reg}\, R, q \supset pR\}$ *if* $pR \neq 0$,
(iii) *there exists a flat Noetherian complete local* R*-algebra* (R', m') *such that*
  (iii$_1$) R' *is formally smooth over* R,
  (iii$_2$) $k' := R'/m'$ *is perfect.*

Then R has bound properties on MCM *modules.*

Proof. By André-Radu's Theorem the structural morphism $j\colon R \to R'$ is regular (see [An] or [BR1], [BR2]). Then

$$\operatorname{Reg} R' = \{q \in \operatorname{Spec} R' \mid j^{-1}q \in \operatorname{Reg} R\}$$

by [Mal] (33.B). Hence

$$I_s(R') = \sqrt{I_s(R)R'} \quad \text{and} \quad \operatorname{Reg}(R'/pR') = \{q/pR' \mid q \in \operatorname{Reg} R', q \supset pR'\}.$$

Thus R' has bound properties on MCM modules by Lemma (1.4) and it is enough to show the following

(1.6) *Lemma. Let $j\colon R \to A$ be a flat local morphism of local* CM-*rings such that $I_s(A) \supset I_s(R)A$. If A has bound properties on* MCM *modules then R has too.*

Proof. Let r be the positive integer given for A as in (1.1)(i). We claim that r works also for R. Let M be a MCM R-module and N a f.g. R-module. By flatness we have

$$A \otimes_R \operatorname{Ext}^1_R(M, N) = \operatorname{Ext}^1_A(A \otimes_R M, A \otimes_R N).$$

Since

$$\begin{aligned}\operatorname{depth}_A A \otimes_R M &= \operatorname{depth}_R M + \operatorname{depth}_A A/mA\\ &= \operatorname{depth} R + \operatorname{depth}_A A/mA = \operatorname{depth} A,\end{aligned}$$

the A-module $A \otimes_R M$ is a MCM. Thus $I_s(A)^r$ (so $I_s(R)^r$) kills $\operatorname{Ext}^1_A(A \otimes_R M, A \otimes_R N)$ and it follows $I_s(R)^r(A \otimes_R \operatorname{Ext}^1_R(M, N)) = 0$. By faithful flatness we get $I_s(R)^r \operatorname{Ext}^1_R(M, N) = 0$.

Now let $a \subset R$ be an ideal and $y \in I_s(R)$. Then $j(y) \in I_s(A)$, and by hypothesis there exists an $e \in \mathbb{N}$ such that

$$(aP : j(y)^e)_P = (aP : j(y)^{e+1})_P$$

for every MCM A-module P.

Let M be a MCM R-module. As above, $M' := A \otimes_R M$ is a MCM A-module and by flatness we obtain

$$(aM' : (y)^s)_{M'} \cong A \otimes_R (aM : y^s)_M$$

for every $s \in \mathbb{N}$. Thus the inclusion

$$(aM : y^e)_M \subset (aM : y^{e+1})_M \tag{1}$$

goes by base change into an equality. Then (1) itself is an equality, j being faithfully flat.

(1.7) PROPOSITION: *Suppose that*

(i) (R, m) *is a quasi-excellent reduced and* $[k : k^p] < \infty$ *if* $p > 0$,
(ii) $\operatorname{Reg}(R/pR) = \{q/pR \mid q \in \operatorname{Reg} R, q \supset pR\}$ *if* $pR \neq 0$.

Then R has bound properties on MCM *modules.*

Proof. If k is perfect then apply Lemmas (1.5), (1.6), where A is the completion of (R, m). If k is not perfect, let $K := k^{1/p^\infty}$. We have $\operatorname{rank}_K \Gamma_{K/k} \leqslant \operatorname{rank}_k \Omega_{k/k^p} = [k : k^p]$ (see e.g. in [Po2] (4.4)) and so there exists a formally smooth Noetherian complete local R-algebra (R', m') such that $R'/m' \cong K$ and $\dim R' = \dim R + \operatorname{rank} \Gamma_{K/k}$ (see EGA (22.2.6), or [NP] Corollary (3.6)). Now apply Lemmas (1.5), (1.6).

2. CM-reduction ideals

(2.1) LEMMA. (inspirated by [Po2] (3.1)). *Let A be a Noetherian ring, M, N two f.g. A-modules,* $x \in A$ *an element such that* $x \operatorname{Ext}^1_A(M, P) = 0$ *for every factor A-module P of N, e a positive integer such that* $(0 : x^e)_N = (0 : x^{e+1})_N$ *and* $s \in \mathbb{N}$. *Then for every A-linear map* $\varphi: M \to N/x^{e+s+1}$ *there exists an A-linear map* $\psi: M \to N$ *which makes commutative the following diagram*:

$$\begin{array}{ccc} M & \xrightarrow{\varphi} & N/x^{e+s+1}N \\ {\scriptstyle\Psi}\downarrow & & \downarrow \\ N & \longrightarrow & N/x^{e+s}N \end{array}$$

Proof. Let $N' := (0 : x^e)_N$. We have the following commutative diagram

$$\begin{array}{ccccccc} 0 \to N/N' & \xrightarrow{x^{e+s+1}} & N/N' & \longrightarrow & N/N' + x^{e+s+1}N & \longrightarrow & 0 \\ \downarrow & & \| & & \downarrow & & \\ 0 \to N/N' & \xrightarrow{x^{e+s}} & N/N' & \longrightarrow & N/N' + x^{e+s}N & \longrightarrow & 0 \end{array} \qquad (1)$$

in which the lines are exact. This follows from the elementary

(2.1.1) LEMMA. $x^e N \cap N' = 0$.
Indeed, if $x^{e+s}z \in N'$ *for a certain* $z \in N$ *then* $x^{e+s}z = 0$ *by the above Lemma and so* $z \in N'$. *Applying the functor* $\operatorname{Hom}_A(M, -)$ *to* (1) *we obtain the following com-*

mutative diagram

$$\begin{array}{ccccc} \mathrm{Hom}_A(M, N/N') & \to & \mathrm{Hom}_A(M, N/N' + x^{e+s+1}N) & \to & \mathrm{Ext}^1_A(M, N/N') \\ \| & & \downarrow & & \downarrow x \\ \mathrm{Hom}_A(M, N/N') & \to & \mathrm{Hom}_A(M, N/N' + x^{e+s}N) & \longrightarrow & \mathrm{Ext}^1_A(M, N/N') \end{array} \tag{2}$$

with exact lines. Since the last column is zero by hypothesis we obtain a R-linear map $\alpha\colon M \to N/N'$ *such that the following diagram is commutative:*

$$\begin{array}{ccccc} M & \longrightarrow & N/a^{e+s+1}N & \longrightarrow & N/N' + x^{e+s+1}N \\ \alpha \,\vdots\downarrow & & & & \downarrow \\ N/N' & & \longrightarrow & & N/N' + x^{e+s}N \end{array} \tag{3}$$

Note that in the diagram

$$\begin{array}{ccccc} M & & \longrightarrow & & N/x^{e+s+1}N \\ \| & & & & \downarrow \\ M & \overset{\Psi}{\dashrightarrow} & N/N' \cap x^{e+s}N & \longrightarrow & N/x^{e+s}N \\ \| & & & & \downarrow \\ M & \overset{\alpha}{\longrightarrow} & N/N' & \longrightarrow & N/N' + x^{e+s}N \end{array} \tag{4}$$

the small square is cartesian and so there exists Ψ *which makes* (4) *commutative. Now apply Lemma* (2.1.1).

Proof of Lemma (2.1.1). If $y \in N' \cap x^e N$ and $z \in N$ satisfy $y = x^e z$ then $0 = x^e y = x^{2e} z$ and so zN', i.e. $y = x^e z = 0$.

(2.2) LEMMA. *Let* (R, m) *be a local* CM-*ring which has bound properties on* MCM *modules,* $a \subset R$ *an ideal and* $y \in I_s(R)$. *Suppose that* $\operatorname{Reg} R$ *is open. Then there exists a function* $v\colon \mathbb{N} \to \mathbb{N}$, $v \geqslant 1_{\mathbb{N}}$ *such that for every* $s \in \mathbb{N}$, *all* MCM *R-modules* M, N *and every R-linear map* $\varphi\colon M \to N/(a, x^{v(s)})N$ *there exists a R-linear map* $\Psi\colon M \to N/aN$ *such that the following diagram commutes:*

$$\begin{array}{ccc} M & \overset{\varphi}{\longrightarrow} & N/(a, x^{v(s)})N \\ \Psi \,\vdots\downarrow & & \downarrow \\ N/aN & \longrightarrow & N/(a, x^s)N \end{array}$$

Proof. Let r, e be the positive integers given by (1.1)(i), (1.1)(ii), respectively. Define $v(s) = r(1 + \max\{e, s\})$. Let M, N, φ be given. Since $y^r \mathrm{Ext}^1_R(M, P) = 0$

for every f.g. R-module P and $(aN:y^e)_N = (aN:y^{e+1})_N$ we get Ψ by Lemma (2.1).

(2.3) Let A be a local CM-ring and $a \subset A$ an ideal. As in [Po2], (3.6) the couple (A, a) is a CM-approximation if there exists a function $\nu: \mathbb{N} \to \mathbb{N}$, $\nu \geqslant 1_{\mathbb{N}}$ such that for every $s \in \mathbb{N}$, every two MCM A-modules M, N and every A-linear map $\varphi: M \to N/a^{\nu(s)}N$ there exists an A-linear map $\Psi: M \to N$ such that $A/a^s \otimes_A \varphi \cong A/a^s \otimes_A \Psi$, in other words: the following diagram is commutative

$$\begin{array}{ccc} M & \xrightarrow{\varphi} & N/a^{\nu(s)}N \\ {\scriptstyle\Psi}\downarrow & & \downarrow \\ N & \longrightarrow & N/aN \end{array}$$

(2.4) LEMMA. *Let (R, m) be a local* CM-*ring which has bound properties on* MCM *modules. Suppose that* Reg R *is open. Then for every ideal $a \subset I_s(R)$ the couple (R, a) is a* CM-*approximation.*

Proof. Let $y_1, \ldots, y_t$ be a system of generators of a. Apply induction on t. If $t = 1$ then use Lemma (2.2) for $a = 0$. Suppose $t > 1$ and let ν' be the function given by induction hypothesis for $b := (y_1, \ldots, y_{t-1})$. Let ν''_s be the function given by Lemma (2.2) for $b^{\nu'(s)}$ and y_t, $s \in \mathbb{N}$. Then the function ν given by $\nu(s) = \nu'(s) + \nu''_s(s)$ works. Indeed, let M, N be two MCM R-modules, $s \in \mathbb{N}$ and $\varphi: M \to N/a^{\nu(s)}N$ a R-linear map. Then there exists a R-linear map $\alpha: M \to N/b^{\nu'(s)}N$ such that $(R/y_t^sR) \otimes_R \alpha \cong (R/y_t^sR) \otimes_R \varphi$ (note that $a^{\nu(s)} \subset (b^{\nu'(s)}, y_t^{\nu''_s(s)})$). Moreover, there exists a R-linear map $\Psi: M \to N$ such that $(R/b^s) \otimes_R \psi = (R/b^s) \otimes_R \alpha$. As $b^s \subset a^s$ we get $R/a^s \otimes_R \psi \cong R/a^s \otimes_R \varphi$.

(2.5) Let b be an ideal in a local CM-ring A. Then b is a CM-*reduction ideal* if the following statments hold:

(i) A MCM A-module M is indecomposable iff M/bM is indecomposable over A/b.

(ii) Two indecomposable MCM A-modules M, N are isomorphic iff M/bM and N/bM are isomorphic over A/b.

(2.6) LEMMA. *Let R be a Henselian local* CM-*ring and $a \subset R$ an ideal such that (R, a) is a* CM-*approximation. Then a^r is a* CM-*reduction ideal for a certain positive integer r.*

The proof follows from [Po2] (4.5), (4.6).

(2.7) PROPOSITION. *Let (R, m) be a Henselian local* CM-*ring which has bound properties on* MCM *modules. Suppose that* Reg R *is open. Then for every ideal $a \subset I_s(R)$ there exists a positive integer r such that a^r is a* CM-*reduction ideal.*

The proof follows from Lemmas (2.4), (2.6).

(2.8) COROLLARY ([Po2] (4.8)): *Let (R, m) be a reduced quasi-excellent Henselian local* CM-*ring*, $k := R/m$ *and* $p :=$ char k. *Suppose that*

(i) $[k: k^p] < \infty$ if $p > 0$
(ii) $\mathrm{Reg}(R/pR) = \{q/pR \mid q \in \mathrm{Reg}\, R, q \supset pR\}$ *if* $p \neq$ char R.

Then $I_s(R)^r$ is a CM-*reduction ideal for a certain positive integer r.*

The result follows from Propositions (1.7), (2.7).

3. Stability properties of Auslander–Reiten quivers under base change

(3.1) Let (R, m) be a local CM-ring, $k := R/m, p :=$ char k and M, N two indecomposable MCM R-modules. The R-linear map f is *irreducible* if it is not an isomorphism, and given any factorization $f = gh$ in CM(R), g has a section or k has a retraction. The AR-quiver $\Gamma(R)$ of R is a directed graph which has as vertices the isomorphism classes of indecomposable MCM R-modules, and there is an arrow from the isomorphism class of M to that of N provided there is an irreducible linear map from M to N. Let $\Gamma_0(R)$ be the set of vertices of $\Gamma(R)$. The multiplicity defines a map $e_R: \Gamma_0(R) \to \mathbb{N}$, $M \mapsto e_R(M)$. If R is a domain and K is its fraction field then let $\mathrm{rank}_R: \Gamma_0(R) \to \mathbb{N}$ be the map given by $M \mapsto \dim_K K \otimes_R M$. Clearly $e_R = e(R)\,\mathrm{rank}_R$ by [Ma2] (14.8).

(3.2) PROPOSITION. *Suppose that*
(i) *R is an excellent henselian local ring,*
(ii) *R has bound properties on MCM modules.*

Let A be the completion of R with respect to $I_s(R)$. Then the base change functor $A \otimes_R$-induces a bijection $\Gamma_0(R) \to \Gamma_0(A)$.

Proof. Clearly A is a CM-ring because the canonical map $R \to A$ is regular by (i). First we prove that a f.g. R-module M is an indecomposable MCM module iff $A \otimes_R M$ is an indecomposable MCM A-module. Using the following elementary Lemma, it is enough to show that if M is an indecomposable MCM R-module then $A \otimes_R M$ is indecomposable over A.

(3.2.1) LEMMA. *Let B be a local* CM-*ring, M a* f.g. *B-module and C a flat local B-algebra. Suppose that C is a* CM-*ring. Then*

(i) *M is a* MCM *B-module iff $C \otimes_B M$ is a* MCM *C-module,*
(ii) *if $C \otimes_B M$ is indecomposable then M is so.*

By Proposition (2.7) $I_s(R)^r$ is a CM-reduction ideal for a certain $r \in \mathbb{N}$ and so $R/I_s(R)^r \otimes_R M$ is still indecomposable. As $R/I_s(R)^r \cong A/I_s(R)^r A$ it follows that $A \otimes_R M/I_s(R)^r A \otimes_R M$ is indecomposable and so $A \otimes_R M$ is indecomposable by Nakayama's Lemma. If M, N are two indecomposable MCM R-modules such that $A \otimes_R M \cong A \otimes_R N$ then $M/I_s(R)^r M \cong A \otimes_R M/I_s(R)^r A \otimes_R M \cong A \otimes_R N/I_s(R)^r A \otimes_R N \cong N/I_s(R)^r{}_N$ and so $M \cong N$, because $I_s(R)^r$ is a CM-reduction ideal. Thus $A \otimes_R$-induces an injective map $\alpha: \Gamma_0(R) \to \Gamma_0(A)$. But α is also surjective by [E1] Theorem 3 because a MCM A-module is locally free on Spec $A \setminus V(I_s(A))$ and $I_s(A) = \sqrt{I_s(R)A}$ (see (1.2) and (1.5)), the map $R \to A$ being regular by (i).

Using (1.7) we get

(3.3) COROLLARY. Suppose that

(i) *(R, m) is an excellent reduced Henselian local ring and $k{:}k^p] < \infty$ if $p > 0$,*

(ii) $\mathrm{Reg}(R/pR) = \{q/pR \mid q \in \mathrm{Reg}\, R, q \supset pR\}$ *if* $p \neq \mathrm{char}\, R$.

Let A be the completion of R with respect to $I_s(R)$. Then the base change functor $A \otimes_R$-induces a bijection $\Gamma_0(R) \to \Gamma_0(A)$. In particular $\#\Gamma_0(R) = \#\Gamma_0(A)$.

The above Corollary can be improved if we use Artin approximation theory (see [Ar] or [Po1]).

(3.4) Let $n: B \to C$ be a morphism of rings. We call n *algebraically pure* (see [Po1] §3) if every finite system of polynomial equations over B has a solution in B if it has one in C. Also n is called *strong algebraically pure* (see [Po1] (9.1)) if for every finite system of polynomials f from $B[X]$, $X = (X_1, \ldots, X_s)$ and for every finite set of finite systems of polynomials $(g_i)_{1 \leqslant i \leqslant r}$ in $B[X, Y]$, $Y = (Y_1, \ldots, Y_t)$ the following conditions are equivalent:

(1) f has a solution x in B such that for every i, $1 \leqslant i \leqslant r$ the system $g_i(x, Y)$ has no solution in B,

(2) f has a solution x in C such that for every i, $1 \leqslant i \leqslant r$ the system $g_i(x, Y)$ has no solutions in C.

Suppose that B is Noetherian and let $b \subset B$ be an ideal, C the completion of B with respect to b and n the completion map. Then (B, b) has the *property of Artin approximation* (shortly (B, b) is an AP-couple) if for every finite system of polynomials f in $B[X]$, $X = (X_1, \ldots, X_s)$, every positive integer e and every solution x of f in C, there exists a solution x_e of f in B such that $x_e \equiv x \bmod b^e C$. It is easy to see that (B, b) is an AP-couple iff n is algebraically pure (see [Po1] §1). When B is local and b its maximal ideal then (B, b) is an AP-couple too iff n is strong algebraically pure (see [BNP] (5.1) where these morphisms are called T''-existentially complete). If (B, b) is a Henselian couple and n is regular then (B, b) is an AP-couple by [Po1] (1.3).

(3.5) LEMMA. *Let $A \to B$ be an algebraically pure morphism of Noetherian rings and M, N two* f.g. *A-modules. Then $B \otimes_A M \cong B \otimes_A N$ over B iff $M \cong N$ over A.*

Proof. Let $M \cong A^n/(u)$, $N \cong A^m/(v)$, $u_i = \sum_{j=1}^n u_{ij} e_j$, $i = 1, \ldots, n'$, $v_r = \sum_{s=1}^m v_{rs} e'_s$, $r = 1, \ldots, m'$, where u_{ij}, $v_{rs} \in A$ and (e_j) resp. (e'_j) are canonical bases in A^n resp. A^m. Let $\varphi: A^n \to A^m$ be a linear A-map given by $e_j \mapsto \sum_{s=1}^m x_{js} e'_s$, where

$x_{js} \in A$. Then φ induces a map $f: M \to N$ iff there exist (z_{ir}) in A such that $\varphi(u_i) = \sum_{r=1}^{m'} z_{ir} v_r$, i.e.

$$\sum_{j=1}^{n} u_{ij} x_{js} = \sum_{r=1}^{m'} z_{ir} v_{rs}, \qquad 1 \leqslant i \leqslant n', \quad 1 \leqslant s \leqslant m. \tag{1}$$

Clearly f is an isomorphism iff there exists $\psi: A^m \to A^n$ given by $e'_s \mapsto \sum_{j=1}^{n} y_{sj} e_j$, $y_{sj} \in A$ such that $\Psi(v) \subset (u)$, $(\varphi\Psi - 1)(e') \subset (v)$ and $(\Psi\varphi - 1)(e) \subset (u)$, i.e. there exist $t_{ri}, w_{sr}, w'_{ji} \in A$ such that

$$\begin{cases} \sum_{s=1}^{m} v_{rs} y_{sj} = \sum_{i=1}^{n'} t_{ri} u_{ij}, & 1 \leqslant r \leqslant m', \quad 1 \leqslant j \leqslant n \\ \sum_{s=1}^{m} x_{js} y_{sj'} - \delta_{jj'} = \sum_{i=1}^{n'} w_{ji} u_{ij'}, & 1 \leqslant j, \quad j' \leqslant n \\ \sum_{j=1}^{n} y_{sj} x_{js'} - \delta_{ss'} = \sum_{r=1}^{m'} w'_{sr} v_{rs'}, & 1 \leqslant s, \quad s' \leqslant m \end{cases} \tag{2}$$

where $\delta_{ss'}$ denotes Kronecker's symbol. Thus $M \cong N$ iff the following system of polynomial equations:

$$\begin{cases} \sum_{j=1}^{n} u_{ij} X_{js} = \sum_{r=1}^{m'} Z_{ir} v_{rs}, & 1 \leqslant i \leqslant n', \quad 1 \leqslant s \leqslant m \\ \sum_{s=1}^{m} v_{rs} Y_{sj} = \sum_{i=1}^{n'} T_{ri} u_{ij}, & 1 \leqslant r \leqslant m', \quad 1 \leqslant j \leqslant n \\ \sum_{s=1}^{m} X_{js} Y_{sj'} - \delta_{jj'} = \sum_{i=1}^{n'} W_{ji} u_{ij'}, & 1 \leqslant j, \quad j' \leqslant n \\ \sum_{j=1}^{n} Y_{sj} X_{js'} - \delta_{ss'} = \sum_{r=1}^{m'} W'_{sr} v_{rs'}, & 1 \leqslant s, \quad s' \leqslant m \end{cases} \tag{$*$}$$

has a solution in A. Similarly $B \otimes_A M \cong B \otimes_A N$ as B-modules iff $(*)$ has a solution in B. But $(*)$ has a solution in A iff it has one in B because $A \to B$ is algebraically pure.

(3.6) LEMMA. *Let $h: A \to B$ be a morphism of Noetherian rings and M a* f.g. *A-module. Suppose that either*

(i) *h is strong algebraically pure, or*
(ii) *h is algebraically pure and B is the completion of A with respect to an ideal $a \subset A$ contained in the Jacobson radical of A.*

Then $B \otimes_A M$ is an indecomposable B-module iff *M is an indecomposable A-module.*

Proof. Conserving the notations from the proof of (3.5) for $M = N$ we note that

(1) f is idempotent iff $(\varphi^2 - \varphi)(e) \subset (u)$, i.e. there exist $d_{ji} \in A$ such that

$$\sum_{s=1}^{n} x_{js}x_{sr} - x_{jr} = \sum_{i=1}^{n'} d_{ji}u_{ir}, \qquad 1 \leqslant j, \quad r \leqslant n$$

(2) $f \neq 0, 1$ iff the following two systems of polynomials

$$\sum_{i=1}^{n} Q_{ji}u_{is} = X_{js} \qquad 1 \leqslant j, \quad s \leqslant n \tag{G_1}$$

$$\sum_{i=1}^{n'} Q'_{ji}u_{is} = X_{js} - \delta_{js} \qquad 1 \leqslant j, \quad s \leqslant n \tag{G_2}$$

have no solutions in A with $X = x$. Thus M is decomposable iff the following system

$$\begin{cases} \sum_{j=1}^{n} u_{ij}X_{js} = \sum_{r=1}^{n'} Z_{ir}u_{rs}, & 1 \leqslant i \leqslant n', \quad 1 \leqslant s \leqslant n \\ \sum_{s=1}^{n} X_{js}X_{sr} - X_{jr} = \sum_{i=1}^{n'} \Delta_{ji}u_{ir}, & 1 \leqslant j, \quad r \leqslant n \end{cases} \tag{F}$$

has a solution (x, z, d) for which the systems $G_1(x, Q)$, $G_2(x, Q')$ have no solutions in A. A similar statement is true for $B \otimes_A M$ and they are equivalent if h is strong algebraically pure.

Now suppose that (ii) holds. Then h is faithfully flat and we obtain: M is indecomposable if $B \otimes_A M$ is so (cf. (3.2.1)). If $B \otimes_A M$ is decomposable then it has an idempotent endomorphism $\neq 0, 1$ which gives a solution $(\tilde{x}, \tilde{z}, \tilde{d})$ of F in B. Since (A, a) is an AP-couple, h being algebraically pure (see (3.4)) there exists a solution (x, z, d) of F in A such that $(x, z, d) = (\tilde{x}, \tilde{z}, \tilde{d}) \bmod aB$. Let f be the idempotent endomorphism of M given by x. Then $A/a \otimes_A f \cong B/a \otimes_B f$ because $A/a \cong B/aB$. By Nakayama's Lemma $(B/aB) \otimes_B \tilde{f} \neq 0, 1$ and so $f \neq 0, 1$, i.e. M is decomposable.

(3.7) REMARK. When A is a local ring, a its maximal ideal and B the completion of A with respect to a then h is strong algebraically pure if h is algebraically pure. Thus in the above Lemma (ii) may be a particular case of (i).

(3.8) PROPOSITION. *Let (A, a) be an AP-couple and B the completion of A with* *respect to a. Suppose that A, B are local* CM-*rings. Then the base change functor* $B \otimes_A$*-induces an injection* $\Gamma_0(A) \to \Gamma_0(B)$.

The result follows from Lemmas (3.5), (3.6).

(3.9) LEMMA. *Conserving the hypothesis of the above Proposition, let M, N be two indecomposable* MCM *A-modules and $f: M \to N$ an irreducible A-map. Suppose that the base change functor $B \otimes_A$-induces a bijection $\Gamma_0(A) \to \Gamma_0(B)$. Then $B \otimes_A f$ is an irreducible B-map.*

Proof. By faithfully flatness $B \otimes_A f$ is not bijective. Let $B \otimes_A f = \tilde{g}\tilde{h}$ be a factorization in the category of MCM B-modules, $\tilde{h}: B \otimes_A M \to \tilde{P}$, $\tilde{g}: \tilde{P} \to B \otimes_A N$. By hypothesis $\tilde{P} = B \otimes_A P$ for a MCM A-module P. Suppose that $\tilde{g}$ has no section and $\tilde{h}$ has no retraction. Then $B/a^sB \otimes_A \tilde{g}$ has no section and $B/a^sB \otimes_A \tilde{h}$ has no retraction for a certain $s \in \mathbb{N}$ by the following:

(3.9.1) LEMMA. *Let (B, n) be a Noetherian local ring, $b \subset B$ an ideal and $u: M \to N$ a B-linear map. Then there exists a positive integer $s \in \mathbb{N}$ such that u has a retraction (resp. a section) iff it has one modulo b^s.*

As (A, a) is an AP-couple we can find a factorization $f = gh$, $h: M \to P$, $g: P \to N$ such that $(B/a^sB) \otimes_A \tilde{h} \cong (B/a^sB) \otimes_A h$, $(B/a^sB) \otimes_A \tilde{g} \cong (B/a^sB) \otimes_A g$ (the idea follows the proofs of (3.5), (3.6)). Since $A/a^s \cong B/a^sB$ it follows that $(A/a^s) \otimes h$ has no retraction and $(A/a^s) \otimes g$ has no section. Thus h has no retraction and g has no section. Contradiction (f is irreducible)!

Proof of (3.9.1). As in the proof of (3.5) we see that u has a retraction (resp. a section) iff a certain linear system L of equations over B has a solution in B. Let $\hat{B}$ be the completion of B with respect to n. Then by a strong approximation Theorem (cf. e.g. [Po1] (1.5) there exists a positive integer $s \in \mathbb{N}$ such that L has solutions in $\hat{B}$ iff it has solutions in $\hat{B}/n^s\hat{B}$.

If u has a retraction (resp. a section) modulo b^s then L has a solution in B/b^s. Thus L has a solution in $\hat{B}/n^s\hat{B}$ and so a solution in $\hat{B}$. Then by faithfully flatness L has a solution in B, i.e. u has a retraction (resp. a section).

(3.10) THEOREM. *Suppose that (R, m) is an excellent Henselian local ring and A is the completion of R with respect to $I_s(R)$. Then the base change functor $A \otimes_R$-induces an inclusion $\Gamma(R) \subset \Gamma(A)$ which is surjective on vertices. In particular $\#\Gamma_0(R) = \#\Gamma_0(A)$.*

Proof. By hypothesis $(R, I_s(R))$ is an AP-couple (cf. [Pol] (1.3)) and thus $A \otimes_R$-induces an inclusion $\Gamma_0(R) \subset \Gamma_0(A)$ (cf. (3.8)) which is in fact an equality by [El] Theorem 3 (cf. the proof of ((4.2)). Now it is enough to apply Lemma (3.9).

(3.11) COROLLARY. *Conserving the hypothesis of Theorem* (3.10) *suppose that*

(i) *R is a Gorenstein isolated singularity and $p =$* char R *(i.e. R is of equal characteristic)*

(ii) *k is algebraically closed.*

Then R is of finite CM-*type iff its completion A is a simple hypersurface singularity.*

Proof. Note that R is of finite CM-type iff A is a simple hypersurface singularity by [Kn], [BGS] Theorem A and [GK] (1.4) since $\#\Gamma_0(R) = \#\Gamma_0(A)$.

(3.12) COROLLARY. *Conserving the hypothesis of Theorem* (2.10), *suppose that $k = \mathbb{C}$ and the completion B of R with respect to m is a hypersurface. Then R has countable infinite* CM-*type iff B is a singularity of type A_∞, D_∞.*

The result follows from [BGS] Theorem B and our Theorem (3.10).

(3.13) REMARK. Concerning Theorem (3.10), it would be also nice to know when $\Gamma(R) = \Gamma(A)$. Unfortunately it seems that Artin approximation theory does not help because the definition of irreducible maps involves in fact an infinite set of equations corresponding to all factorizations.

(3.14) PROPOSITION. *Let A be a flat local R-algebra such that mA is the maximal ideal of A. Suppose that*

(i) *R is an excellent Henselian local ring,*
(ii) *A is a* CM-*ring,*
(iii) *the residue field extension of $R \to A$ is strong algebraically pure (e.g. if k is algebraically closed).*

Then the base change functor $A \otimes_R$-induces an injective map $\Gamma_0(R) \to \Gamma_0(A)$. In particular $\#\Gamma_0(R) \leqslant \#\Gamma_0(A)$.

Proof. (R, m) is an AP-couple by [Po1] (1.3) and so the map RA is strong algebraically pure by (iii) (cf. [BNP] (5.6)). Now apply Lemmas (3.5), (3.6).

(3.15) REMARK. If R is not Henselian or (iii) does not hold then our Proposition does not hold in general:

(3.16) EXAMPLE: (i) Let $R = R[X, Y]_{(X,Y)}/(X^2 + Y^2)$, $A := \mathbb{C}[X, Y]_{(X,Y)}/(X^2 + Y^2)$.

Then $M := (X, Y)R$ is an indecomposable MCM R-module but $A \otimes_R M \cong (X + iY)A \oplus (X - iY)A$ is not. Moreover, $\#\Gamma_0(R) = 2$ and $\#\Gamma_0(A) = 3$ by [BEH] (3.1).

(ii) Let $R := \mathbb{C}[X, Y]_{(X,Y)}/(Y^2 - X^2 - X^3)$ and A its henselization. Clearly A contains a unit u such that $u^2 = 1 + X$. Then $M := (X, Y)R$ is an indecomposable MCM R-module but $A \otimes_R M \cong (Y - uX)A \oplus (Y + uX)A$ is not. Also note that $\#\Gamma_0(A) = \#\Gamma_0(\hat{A}) = 3$, $\hat{A}$ being the completion of A (see (3.10)).

4. The Brauer–Thrall conjectures on isolated singularities

Let (R, m) be a Henselian local CM-ring, $k := R/m$, $p :=$ char k. We suppose that R is an isolated singularity, i.e. $I_s(R) = m$.

(4.1) PROPOSITION. *Let* Γ^0 *be a connected component of* $\Gamma(R)$. *Suppose that*

(i) *R has bound properties on* MCM *modules,*

(ii) Γ^0 *is of bounded multiplicity type, i.e. all indecomposable* MCM *R-modules M whose isomorphic classes are vertices in* Γ^0 *have multiplicity* $e(M) \leqslant s$ *for a certain constant integer* $s = s(\Gamma_0)$. *Then* $\Gamma(R) = \Gamma^0$ *and* $\Gamma(R)$ *is a finite graph.*

Proof. By Proposition (2.7) there is a positive integer r such that m^r is a Dieterich ideal, i.e., a CM-reduction ideal which is m-primary. Now it is enough to follow [Po2] (5.4) (in fact the ideas come from [Di] Proposition 2 and [Yo] Theorem (1.1)).

(4.2) COROLLARY. *Suppose that*

(i) *R has bound properties on* MCM *modules,*

(ii) *R has infinite* CM*-type.*

Then there exist MCM *R-modules of arbitrarily high multiplicity* (*or rank if R is a domain*).

(4.3) COROLLARY. ([Po2] (1.2)) *Suppose that*

(i) *R is an excellent ring and* $[k:k^p] < \infty$ *if* $p > 0$,

(ii) $\mathrm{Reg}(R/pR) = \{q/pR \mid q \in \mathrm{Reg}\, R, q \supset pR\}$ *if* $pR \neq 0$.

Then the first Brauer-Thrall conjecture is valid for R, i.e., if R has infinite CM-type then there exist MCM *R-modulas of arbitrarily high multiplicity* (*or rank if R is a domain*).

(4.4) PROPOSITION. Suppose that

(i) (R, m) *is a two dimensional excellent Gorenstein ring*

(ii) *R has bound properties on MCM modules,*

(iii) *the divisor class group Cl(R) of R is infinite.*

Then for all $n \in \mathbb{N}$, $n \geqslant 1$, *there are infinitely many isomorphism classes of indecomposable MCM R-modules of rank n over R. In particular, the second Brauer-Thrall conjecture holds for R, i.e., if R is of infinite CM-type then for arbitrarily high positive integers n, there exist infinitely many vertices in* $\Gamma_0(R)$ *with multiplicity n* (*or rank n if R is a domain*).

Proof. Let K be the fraction field of R and Δ a Weil divisor on Spec R. Then $J_\Delta := \{x \in K \mid \mathrm{div}\, x \geqslant \Delta\}$ is a reflexive R-module of rank one and the correspondence $\Delta \mapsto J$ defines an injective map $u: \mathrm{Cl}(R) \to \Gamma_0(R)$.

Fix $\alpha \in \mathrm{Im}\, u$. Let Γ_α be the connected component of α in $\Gamma(R)$, I the set of vertices of Γ_α and M an indecomposable MCM R-module whose isomorphism class $\hat{M}$ belongs to I. If $M \not\cong R$ then there exists an almost split sequence

$$0 \to P \to E \to M \to 0$$

because R is an isolated singularity (cf. [Au3]). As R is a two dimensional Gorenstein ring we obtain $P \cong M$ by [Au1] III Proposition (1.8) (cf. also [Yo]A14 or [Re]).

Let N be an indecomosable MCM R-module. If there is an irreducible map $N \to M$ (resp. $P \cong M \to N$) then it factorizes by $E \to M$ (resp. $M \to E$) and thus N is a direct summand in E. The converse is also true (cf. e.g. [Re] (3.2)) and in particular,

(1) there exists an irreducible map $N \to M$ iff there exists an irreducible map $M \to N$ (so we can consider instead of Γ_α its undirected graph $|\Gamma_\alpha|$ obtained by removing all the loops and forgetting the direction of the arrows).

(2) $2\,\mathrm{rank}_R M = \mathrm{rank}_R E \geqslant \Sigma_{\tilde{N}}\,\mathrm{rank}_R \tilde{N}$, where $\tilde{N}$ runs through the vertices of $|\Gamma_\alpha|$ which are incident to $\tilde{M}$.

For $M = R$ the fundamental sequence (an exact sequence of this form corresponding to a nonzero element of $\mathrm{Ext}^2_R(k, R) \cong k$, cf. [Au2] Section 6 or [AR] Section 1 or [Re]) shows that (2) is valid also in this case.

We proceed with a combinatorial remark (cf. also [HPR]). Let $\Gamma = (\Gamma_0, \Gamma_1)$ be an undirected graph, Γ_0 the set of vertices, $\Gamma_1 \subseteq \Gamma_0 \times \Gamma_0$ the set of edges ($(i,i) \notin \Gamma_1$ for all $i \in \Gamma_0$ and $(i,j) \in \Gamma_1$ iff $(j,i) \in \Gamma_1$). $\Gamma' = (\Gamma'_0, \Gamma'_1)$ is a subgraph of Γ if $\Gamma'_0 \subset \Gamma_0$ and $\Gamma'_1 = \{(i,j) \in \Gamma_1 \mid i, j \in \Gamma'_0\}$. A function $r\colon \Gamma_0 \to \mathbb{N}$ is subadditive if

$$2r(i) \geqslant \sum_{(i,j)\in\Gamma_1} r(j) \quad \text{for all } i \in \Gamma_0. \qquad (*)$$

(4.5) LEMMA. *Assume that*

(i) Γ *is connected.*
(ii) r *is subadditive,*
(iii) r *is not bounded and has minimal value* 1.

Then Γ *is* A_∞: 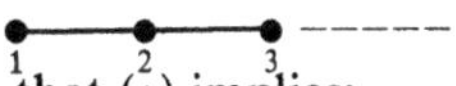 *and* $r(i) = i$ *for all* i.

Proof. Note that $(*)$ implies:

(a) $2r(j) \geqslant d_j := \#\{i \in \Gamma_0 \mid (i,j) \in \Gamma_1\}, \quad j \in \Gamma_0,$

(b) (Monotony) Let  be a subgraph of Γ. Then $r(i) \geqslant r(j)$ implies $r(j) \geqslant r(s)$ (Indeed by $(*)$ we have $2r(j) \geqslant \Sigma_{(t,j)\in\Gamma_1} r(t) \geqslant r(i) + r(s)$, thus $2r(j) \geqslant r(j) + r(s)$).

The assertion of the Lemma immediately follows from

$(**)$ For all $n \geqslant 1$ there is a subgraph

of Γ such that

(c) for $1 \leqslant j \leqslant n-1$, $d_j = 1$ if $j = 1$ and $d_j = 2$ if $j > 1$,

(d) $r(j) = j$ for $1 \leqslant j \leqslant n$.

(Then Γ contains a subgraph A_∞, which must be Γ itself by (c) since Γ is connected.)

We prove (**) by induction on n. If $n = 1$, choose an element $1 \in \Gamma_0$ such that $r(1) = 1$. Now assume (**) is true for $n \geqslant 1$. Then we find a subgraph

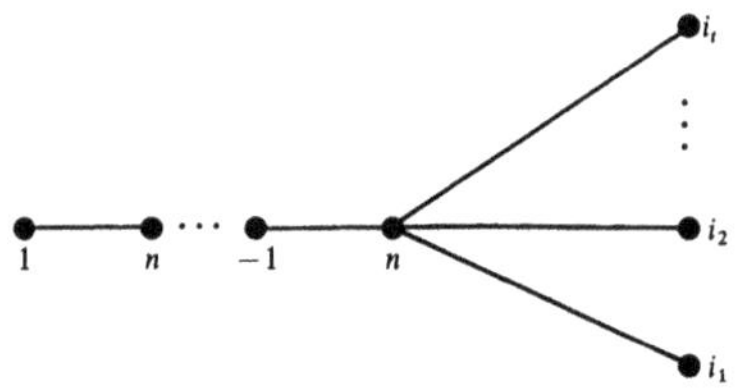

where $t = \begin{cases} d_n - 1 & \text{for } n > 1, \\ d_n & \text{for } n = 1. \end{cases}$

By (*) we have

$$2n \geqslant n - 1 + \sum_{s=1}^{t} r(i_s),$$

and one of the numbers $r(i_s)$ is $>n$ (otherwise r is bounded on Γ_0 by (b)). This implies $n + 1 \geqslant \Sigma_{s=1}^{t} r(i_s) > n$, i.e. $t = 1$, $d_n = 1$ for $n = 1$, respectively, $d_n = 2$ for $n > 1$. If we denote i_1 by $n + 1$, we obtain $r(n + 1) = n + 1$, which completes the proof.

Back to our Proposition, we want to apply (4.5) to $|\Gamma_\alpha|$ for $r := \text{rank}$. By (2) r is subadditive. If r is bounded we obtain $\Gamma_\alpha = \Gamma(R)$ finite (cf. (5.1)) and thus $\text{Cl}(R)$ finite, contradiction! Thus r is unbounded. By (4.5) $|\Gamma_\alpha| = A_\infty$ and $\text{rank}_R(i) = i$ for every i.

Let $\alpha' \in \text{Im}\, u$, $\alpha' \neq \alpha$. Then $\alpha' \notin I$ because Γ_α contains only one vertex of rank one. Thus $\Gamma_\alpha \cap \Gamma_{\alpha'} = \emptyset$ and so for each $n \in \mathbb{N}$ we find $\#\ \text{rank}^{-1}(\{n\}) \geqslant \#\text{Cl}(R)$.

(4.6) COROLLARY. *Suppose that*

(i) (R, m) *is a two dimensional excellent Gorenstein ring,*
(ii) $[k: k^p] < \infty$ *if* $p > 0$,
(iii) $\text{Reg}(R/pR) = \{q/pR \mid q \in \text{Reg}\, R, q \supset pR\}$ *if* $p \neq \text{char}\, R$,
(iv) $\text{Cl}(R)$ *is infinite.*

Then for all $n \in \mathbb{N}$, $n \geqslant 1$ *there are infinitely many isomorphism classes of indecomposable MCM R-modules of rank n. In particular the second Brauer-Thrall conjecture holds.*

The result follows from (1.7) and (4.5).

(4.6.1) REMARK. Dieterich shows in [Di] Theorem 20, that the second Brauer-Thrall conjecture is valid in arbitrary dimension for complete isolated hypersurface singularities over algebraically closed fields of characteristic $\neq 2$. Using our (3.10) we are able to extend it for a large class of excellent henselian local rings. However in the two dimensional case our Corollary (4.6) gives a sharper version.

(4.7) COROLLARY. *Let n be a positive integer. Suppose that*

(i) *(R, m) is a two dimensional excellent Gorenstein ring*
(ii) *R is a R-algebra and k algebraically closed.*

Then the following conditions are equivalent:
(1) *R is not a rational double point,*
(2) *There are infinitely many isomorphism classes of indecomposable MCM R-modules of rank n.*

Proof. By a well-known result (cf. e.g. [Ba]) R is not a rational double point iff R is not rational (R is Gorenstein!) and thus iff $\mathrm{Cl}(R)$ is infinite (cf. [Li] (17.4), (16.2)). Thus (1)$\Rightarrow$(2) follows from the above Corollary. On the other hand, (2) implies: R is of infinite CM-type and thus R is not a rational double point by [AV] (1.11) (see also [EK]) and our (3.3) or (3.10)).

Note added in proof

In (3.9.1), the argument can be simplified by a recent result of V. Nica (*Stud. Cerc. Mat., Vol. 41, Nr. 3* (1989), 179–184), who gives a direct proof for the "linear strong approximation".

References

[An] M. André, Localisation de la lissité formelle, *Manuscripta Math.* 13 (1974) 297–307.

[Ar] M. Artin, Algebraic approximation of structures over complete local rings, *Publ. Math. IHES* 36 (1969) 23–58.

[AV] M. Artin and J. L. Verdier, Reflexive modules over rational double points, *Math. Ann.* 270 (1985) 79–82.

[Au1] M. Auslander, Functors and morphisms determined by objects, *Proc. Conf. on Representation Theory*, Philadelphia 1976, Marcel Dekker, 1978, 1–244.

[Au2] M. Auslander, Rational singularities and almost split sequences, *Trans. Amer. Math. Soc.* 293 (1986) 511–531.

[Au3] M. Auslander, Isolated singularities and existence of almost split sequences, Proc. ICRA IV, Lect. *Notes in Math.* 1178 (1986) 194–241.

[AR] M. Auslander and I. Reiten, Almost split sequences for rational double points, *Trans. Amer. Math. Soc.* 302 (1987) 87–97.

[Ba] L. Badescu, Applications of the Grothendieck duality theory to the study of normal isolated singularities, *Rev. Roum. Math. Pures Appl.* 24 (1979), 673–689.

[BNP] S. Basarab, V. Nica and D. Popescu, Approximation properties and existential completeness for ring morphisms, *Manuscripta Math.* 33 (1981) 227–282.

[BR1] A. Brezuleanu and N. Radu, Sur la localisation de la lissité formelle, C.R. *Acad. Sci.* Paris 276 (1973) 439–441.

[BR2] A. Brezuleanu and N. Radu, Excellent rings and good separation of the module of differentials, *Rev. Roum. Math. Pures Appl.* 23 (1978) 1455–1470.

[BEH] R.-O. Buchweitz, D. Eisenbud and J. Herzog: Cohen-Macaulay modules on quadrics, in Proceedings, Lambrecht 1985, Lect. *Notes in Math.* 1273 (1987) 58–116.

[BGS] R.-O. Buchweitz, G.-M. Greuel and F.-O. Schreyer: Cohen-Macaulay modules on hypersurface singularities II, *Invent. math.* 88 (1987) 165–183.

[Di] E. Dieterich: Reduction of isolated singularities: *Comment. Math.* Helvetici 62 (1987) 654–676.

[El] R. Elkik: Solutions d'equations á coefficients dans un anneau hensélien, *Ann. Sci. Ecole Norm. Sup.* 4e serie 6 (1973) 533–604.

[EK] H. Esnault and H. Knörrer: Reflexive modules over rational double points, *Math. Ann.* 272 (1985) 545–548.

[GK] G.M. Greuel and H. Kröning: Simple singularities in positive characteristic, Preprint, Bonn, *Max-Planck Institut für Math.* 19 (1988).

[EGA] A. Grothendieck and J. Diendonne, Elements de geometrie algebrique, *Publ. Math. IHES* (1964).

[He] J. Herzog: Ringe mit nur endlich vielen Isomorphieklassen von maximalen, unzerlegbaren Cohen-Macaulay Moduln, *Math. Ann.* 233 (1978) 21–34.

[HPR] D. Happel, U. Preiser and C.M. Ringel, Vinberg's characterization of Dynkin diagrams using subadditive functions with application to DTr-periodic modules, Proceedings of the third international conference on representations of algebras, Carleton University, Ottawa 1979, *Springer Lecture Notes in Mathematics* 832, 280–294.

[Kn] H. Knörrer, Cohen Macaulay modules on hypersurface singularities I, *Invent. Math.* 88 (1987) 153–165.

[Li] J. Lipman, Rational singularities with applications to algebraic surfaces and unique factorization, *Publ. Math. IHES* 36 (1969) 195–297.

[Ma1] H. Matsumura, Commutative Algebra, Benjamin, New York, 1970.

[Ma2] H. Matsumura, Commutative Ring Theory, Cambridge University Press, Cambridge, 1986.

[NP] V. Nica and D. Popescu, A structure theorem on formally smooth morphisms in positive characteristic, *J. of Algebra* 100 (1986) 436–455.

[Po1] D. Popescu, General Néron desingularization and approximation, *Nagoya Math. J.* 104 (1986) 85–119.

[Po2] D. Popescu, Indecomposable Cohen Macaulay modules and their multiplicities, to appear in *Trans. AMS.*

[Re] I. Reiten, Finite dimensional algebras and singularities, in Proceedings, Lambrecht 1985, *Springer Lecture Notes in Mathematics*, 1273 (1987) 35–57.

[Sc] F.-O. Schreyer, Finite and countable CM-representation type, in Proceedings, Lambrecht 1985, *Springer Lecture Notes in Mathematics*, 1273 (1987) 9–34.

[Yo] Y. Yoshino, Brauer-Thrall type theorem for maximal Cohen-Macaulay modules, *J. Math. Soc. Japan* 39 (1987) 719–739.

Compositio Mathematica **76**: 295–305, 1990.

Fano bundles of rank 2 on surfaces

MICHAŁ SZUREK & JAROSŁAW A. WIŚNIEWSKI

Department of Mathematics, University of Notre Dame, Notre Dame, Indiana 46556; (on leave from Warsaw University, PkiN IX p., 00-901 Warszawa, Poland)

Received 4 November 1988; accepted 9 October 1989

0. We say that a vector bundle $\mathscr{E}$ on X is Fano if $\mathbb{P}(\mathscr{E})$ is a Fano manifold. In [12], we classified rank-2 Fano vector bundles over the complex projective space $\mathbb{P}^3$ and over a smooth quadric $\mathbb{Q}_3 \subset \mathbb{P}^4$. This paper is thus complementary to [12].

Ruled Fano 3-folds were classified by Demin already in 1980. However, his paper [2] contains gaps and omissions, one of them corrected in the addendum. Here we correct the other. Moreover, we present a vector-bundle approach to the problem, i.e., we try to get our results by pure vector bundle methods. Even if the proofs that utilize Mori-Mukai-Iskovskich list of all Fano threefolds are sometimes shorter and even if we did not succeed, in some cases, to avoid using the classification list mentioned above (namely, to exclude the cases $c_1 = 0$, $c_2 = 6$ over $\mathbb{P}^2$ and $c_1 = (-1, -1)$, $c_2 = 5$ or 6 over $\mathbb{P}^1 \times \mathbb{P}^1$), we believe that our method is worthy of dealing with, also because it works for ruled Fano 4-folds, as well, see [12]. Besides, we are able to draw some conclusions on the geometry of stable vector bundles with $c_1 = 0$, $c_2 = 2$ or $c_2 = 3$, one of them being the 1-ampleness of such bundles.

THEOREM. *There are* 21 *types of ruled Fano* 3*-folds* $V = \mathbb{P}(\mathscr{E})$ *with a* 2*-bundle* $\mathscr{E}$ *on a surface. Up to a twist, they are the following:*

	$\mathscr{E}$	$-K^3$	# in M-M-I list, [9]
over $\mathbb{P}^2$:			
1	$\mathcal{O}(1) \oplus \mathcal{O}(-1)$	62	36, Table 2
2	$\mathcal{O} \oplus \mathcal{O}(1)$	56	35
3	$\mathcal{O} \oplus \mathcal{O}$	54	34
4	$T_{\mathbb{P}^2}(-2)$	48	32
5	$0 \to \mathcal{O} \to \mathscr{E} \to J_x \to 0, \quad x \in \mathbb{P}^2$	46	31
6	$\mathscr{E}$ *stable with* $c_1 = 0$, $c_2 = 2$, *thus:* $0 \to \mathcal{O}(-1)^2 \to \mathcal{O}^4 \to \mathscr{E}(1) \to 0$	38	27
7	$\mathscr{E}$ *stable,* $\mathscr{E}(1)$ *spanned,* $c_1 = 0$, $c_2 = 3$, *thus:* $0 \to \mathcal{O}(-2) \to \mathcal{O}^3 \to \mathscr{E}(1) \to 0$	30	24

over $\mathbb{P}^1 \times \mathbb{P}^1$:			
8	$\mathcal{O}(-1,-1) \oplus \mathcal{O}$	52	31, Table 3
9	$\mathcal{O} \oplus \mathcal{O}$	48	27
10	$\mathcal{O}(0,1) \oplus \mathcal{O}$	48	28
11	$\mathcal{O}(0,-1) \oplus \mathcal{O}(-1,0)$	44	25
12	$\mathcal{E}$ *stable*, $\mathcal{E}(1,1)$ *spanned*, $c_1 = (-1,-1), c_2 = 2$, thus: $0 \to \mathcal{O}(-1,-1) \to \mathcal{O}^3 \to \mathcal{E}(1,1) \to 0$	36	17
over the Hirzebruch surface F_1, with a blow-down map $\beta\colon F_1 \to \mathbb{P}^2$:			
13	$\beta^*(\mathcal{O} \oplus \mathcal{O}(-1))$	50	30, Table 3
10	$\mathcal{O} \oplus \mathcal{O}$	48	28
14	$\beta^*(T_{\mathbb{P}^2}(-2))$	42	24
over a Del Pezzo surface being a blow-up of $k \leqslant 8$ points on $\mathbb{P}^2$:			
15–21	$\mathcal{O} \oplus \mathcal{O}$	$6(9-k)$	10/4.

1. Preliminaries

Throughout the paper $\mathcal{E}$ is a rank-2 vector bundle on a smooth complex projective surface S and $\xi_{\mathcal{E}}$ is the antitautological line bundle on $V = \mathbb{P}(\mathcal{E})$. By p we denote the projection morphism $p\colon \mathbb{P}(\mathcal{E}) \to S$ and by F – the fibre of p. We have the following exact sequence on $V = \mathbb{P}(\mathcal{E})$:

$$0 \to \mathcal{O}_V \to p^*(\mathcal{E})^\vee \otimes \xi_{\mathcal{E}} \to T_{V|S} \to 0. \tag{1.1}$$

with the relative tangent bundle $T_{V|S}$ fitting in

$$0 \to T_{V|S} \to T_V \to p^*(T_S) \to 0. \tag{1.2}$$

We call (1.1) the relative Euler sequence. We then obtain

(1.3) COROLLARY. $c_1(V) = p^*(c_1(S) - c_1(\mathcal{E})) + 2\xi_{\mathcal{E}}$.

The theorem of Leray and Hirsch yields that in the cohomology ring of V the following holds

$$\xi_{\mathcal{E}}^2 - p^*(c_1(\mathcal{E}))\xi_{\mathcal{E}} + p^*(c_2(\mathcal{E})) = 0. \tag{1.4}$$

We infer that S must be a Del Pezzo surface:

(1.5) PROPOSITION. *If $V = \mathbb{P}(\mathcal{E})$ is a ruled Fano threefold, then S is a Del Pezzo surface.*

Proof. See [2]; and [12] for a more general version.

(1.6) PROPOSITION. *In the above notation, we have:*

$$b_2(V) = b_2(S) + 1, \quad \text{and} \quad b_3(V) = 0.$$

Proof. The first statement is obvious. To prove that b_3 vanishes, let us observe first that $h^{3,0}(V) = h^3(\mathcal{O}_V) = h^0(K_V) = 0$ by the Kodaira Vanishing Theorem. To show that $b_3 = 0$, it is then sufficient to prove that $b^{1,2} = b^{2,1} = 0$, i.e., that $h^1(\Omega^2) = h^2(\Omega^1) = 0$. Dualizing (1.2), we get

$$0 \to p^*(\Omega_S) \to \Omega_V \to (T_{V/S})^\vee \to 0.$$

Because $h^2(p^*(\Omega_S) = h^2(\Omega_S) = h^1(K_S) = h^1(\mathcal{O}_S) = 0$ by rationality, we have to show only that $h^2(T^\vee_{V|S}) = 0$. Recall that $T^\vee_{V|S}$ fits in a short exact sequence

$$0 \to (T_{V/S})^\vee \to p^*(\mathcal{E}) \otimes \xi^\vee \to \mathcal{O}_V \to 0.$$

Since $p^*(\mathcal{E})$ restricted to fibres of $p\colon \mathbb{P}(\mathcal{E}) \to \mathbb{P}^2$ is a trivial 2-bundle and $\xi^\vee$ is $\mathcal{O}(-1)$ on fibres, we have $R^i p_*(p^*(\mathcal{E}) \otimes \xi^\vee) = 0$ for $i > 0$ and Leray's spectral sequence gives $h^2(p^*(\mathcal{E}) \otimes \xi^\vee) = 0$. Finally, vanishing of $H^1(\mathcal{O}_V)$ follows directly from the rationality of V.

2. Fano bundles over $\mathbb{P}^2$

This case is the most interesting one. We may assume $\mathcal{E}$ is normalized, i.e., $c_1(\mathcal{E}) = 0$ or -1.

(2.1) PROPOSITION. *If $\mathcal{E}$ is a normalized Fano bundle on $\mathbb{P}^2$, then $\mathcal{E}(2)$ is ample.*

Proof. Let $H = p^*(\mathcal{O}(1))$. By (1.3), $c_1(V) = -H + 2\xi_{\mathcal{E}(2)}$ if $c_1 = 0$ and $c_1(V) = 2\xi_{\mathcal{E}(2)}$ if $c_1 = -1$; but H is nef and we are done.

(2.2) PROPOSITION. *Let $\mathcal{E}$ be a Fano bundle on $\mathbb{F}^2$ with $c_1(\mathcal{E}) = 0$. Then $c_2 \leqslant 3$ and in the cohomology ring of $\mathbb{P}(\mathcal{E})$ we have $H^3 = H\xi^2_{\mathcal{E}} = 0$, $H^2\xi_{\mathcal{E}} = 1$, $\xi^3_{\mathcal{E}} = -c_2(\mathcal{E})$.*

Proof. Vanishing of H^3 and the equality $H^2\xi_{\mathcal{E}} = 1$ is obvious. The relations between the generators of $\mathbb{P}(\mathcal{E})$ are then easy consequences of the Leray-Hirsch formulae. To prove that $c_2 \leqslant 3$, we calculate: $0 > K^3_V = -c^3_1(V) = (3H + 2\xi_{\mathcal{E}})^3 = 8c_2 - 54$, hence $c_2 \leqslant 6$. However, as follows from the Mori-

Mukai-Iskovskich classification, [9], there are no Fano threefolds with $K_V^3 = 6$, 14, 22 and $b_2(V) = 0$, cf. (1.6); see also Remark in Section 3.

(2.3) PROPOSITION. *Let* $\mathscr{E}$ *be a normalized Fano bundle on* $\mathbb{P}^2$. *Then*
(a) *if* $\mathscr{E}$ *is not semistable, then either* $\mathscr{E} = \mathcal{O}(1) \oplus \mathcal{O}(-1)$ *or* $\mathscr{E} = \mathcal{O} \oplus \mathcal{O}(-1)$;
(b) *if* $\mathscr{E}$ *is semistable, but not stable, then either* $\mathscr{E} = \mathcal{O} \oplus \mathcal{O}$ *or* $\mathscr{E}$ *fits in* $0 \to \mathcal{O}_{\mathbb{P}^2} \to \mathscr{E} \to \mathscr{I}_x \to 0$, *where* $\mathscr{I}_x$ *is the sheaf of ideals of a point* $x \in \mathbb{P}^2$;
(c) *if* $c_1(\mathscr{E}) = -1$, *then either* $\mathscr{E} = \mathcal{O} \oplus \mathcal{O}(-1)$ or $\mathscr{E} = T_{\mathbb{P}^2}(-2)$,

Proof. We start from (c). Let us take a line L and let $\mathscr{E}(2)|L$ be $\mathcal{O}_L(a_1) \oplus \mathcal{O}_L(a_2)$. We then have $a_1 + a_2 = c_1(\mathscr{E}(2)) = 3$ and $a_1, a_2 > 0$ by (2.1). Therefore $a_1 = 2$, $a_2 = 1$ up to permutation. By the Van de Ven theorem on uniform bundles, [13], $\mathscr{E}$ is as we claim. This proves (c). To show that (a) holds we may then assume that $c_1(\mathscr{E}) = 0$. If $\mathscr{E}$ is not semistable, there is a non-trivial section $s \in H^0(\mathscr{E}(-1))$. Assuming s does not vanish anywhere we find a trivial subbundle $\mathcal{O} \subset \mathscr{E}(-1)$, hence the quotient of $\mathscr{E}(-1)$ by this trivial bundle is $\mathcal{O}(2)$ and we get (a). Assume now that s vanishes at a point x. Let us take a line that contains a finite number of zeros of s. The bundle $\mathscr{E}(-1)|L$ then splits as $\mathcal{O}(k) \oplus \mathcal{O}(-2-k)$ with $k \geqslant 1$, so that $\mathscr{E}(2)|L = \mathcal{O}(k+3) \oplus \mathcal{O}(1-k)$ in contradiction with (2.1). This concludes the proof of (a). To show (b), assume $\mathscr{E}$ is not stable. Since for $c_1(\mathscr{E}) = -1$ a semistable rank-2 bundle is stable, we infer that $c_1(\mathscr{E}) = 0$. Pick a non-trivial section $s \in H^0(\mathscr{E})$. If the set $\{s = 0\}$ is empty, we get an embedding $\mathcal{O} \subset \mathscr{E}$ whose cokernel is also a trivial bundle and then $\mathscr{E} = \mathcal{O} \oplus \mathcal{O}$. If s vanished at two points (not necessarily distinct), the line through these points would be a jumping one of type $(-2, 2)$, contradicting (2.1). Finally, if s vanishes at a single point x, then $c_2(\mathscr{E}) = 1$ and (b) follows. It is known that any non-trivial extension as in (b) is a bundle, (see [10], ch.1, §1.5).

This proves 1 through 5 of our theorem. To conclude the case of $\mathbb{P}^2$ we must, in view of (2.2) and (2.3), study bundles with $c_1 = 0$, $c_2 = 2$ or 3 in more detail.

Case $c_1 = 0$, $c_2 = 2$. Let $\mathscr{E}$ be a stable bundle with $c_1(\mathscr{E}) = 0$, $c_2(\mathscr{E}) = 2$. The twisted bundle $\mathscr{E}(1)$ is then generated by global sections, though not ample, since there are lines L such that $\mathscr{E}(1)|L = \mathcal{O} \oplus \mathcal{O}(2)$, [1]. Then $\xi_{\mathscr{E}(1)}$ is globally generated, hence nef. Recall that the cone of numerically effective divisors on V is generated by two (classes of) divisors. Because H and $\xi_{\mathscr{E}(1)}$ are not numerically equivalent, their sum must be then in the interior of the cone, i.e., $-K_V = H + 2\xi_{\mathscr{E}(1)}$ is ample. This gives 6 of our theorem.

REMARK. To see that any stable 2-bundle on $\mathbb{P}^2$ with $c_1 = 0$, $c_2 = 2$ fits in the exact sequence as in (6), observe first that $h^2(\mathscr{E}(1)) = h^2(\mathscr{E}) = h^0(\mathscr{E}) = 0$ by stability of $\mathscr{E}$ and of its dual. Then $h^1(\mathscr{E}) = -\chi(\mathscr{E}) = 0$ by Riemann-Roch, so $h^1(\mathscr{E}) = 0$ by the Castelnuovo criterion. By Horrock's criterion of decomposability the kernel of the evaluation $\mathcal{O}^4 \to \mathscr{E}(1) \to 0$ splits. Computing the Chern classes of the kernel then gives the sequence as in 6 of the Theorem.

APPLICATION 1. For $\mathscr{E}$ as above, the resulting Fano threefold arises from blowing up a twisted cubic in $\mathbb{P}^3$, [9]. The generic section s of $\mathscr{E}(1)$ vanishes at three points whose associated lines in $(\mathbb{P}^2)^\vee$ form a triangle inscribed in the non-singular conic of jumping lines of $\mathscr{E}$, [1]. Let $Z = \text{zero}(s)$ and $0 \to \mathcal{O} \xrightarrow{s} \mathscr{E}(1) \to \mathscr{I}_Z(2) \to 0$ be the corresponding exact sequence. It gives rise to an embedding

$$S_3 := \mathbb{P}(\mathscr{I}_Z(2)) \subset \mathbb{P}(\mathscr{E}(1)) = \mathbb{P}(\mathscr{E})$$

(over $\mathbb{P}^2$) of the Del Pezzo surface S_3 (the blow-up of three points in $\mathbb{P}^2$). Since $\xi_{\mathscr{E}(1)}$ is spanned, S_3 is the inverse image of a plane in $\mathbb{P}^3 = \mathbb{P}(\Gamma(\mathscr{E}(1)))$. In other words:

(2.4) The Fano threefold $\mathbb{P}(\mathscr{E})$ with $\mathscr{E}$ as above, admits two projections p: $\mathbb{P}(\mathscr{E}) \to \mathbb{P}^2$, q: $\mathbb{P}(\mathscr{E}) \to \mathbb{P}^3$ such that q^{-1} is the blow-up of a twisted cubic and p is a $\mathbb{P}^1$-bundle. On a generic plane $P \subset \mathbb{P}^3 = \mathbb{P}(\Gamma(\mathscr{E}(1)))$, the rational map pq^{-1}: $P \to \mathbb{P}^2$ is a quadratic map

$$(x_0, x_1, x_2) \to (x_0x_1, x_1x_2, x_2x_0)$$

that blows up the points where P meets the twisted cubic and contracts the three exceptional curves that arise. The rational map pq^{-1} is then a family of such "elementary" quadratic maps.

APPLICATION 2. Stable rank-2 vector bundles on $\mathbb{P}^2$ with $c_1 = 2$, $c_2 = 3$ are 1-ample, see [11] for the definition of 1-ampleness.

Case $c_1 = 0$, $c_2 = 3$.

(2.5) Let $\mathscr{E}$ be a stable rank-2 vector bundle on $\mathbb{P}^2$ with $c_1 = 0, c_2 = 3$. Then $h^0(\mathscr{E}(1)) = 3$.

Proof. As a rank-2 bundle with $c_1 = 0$, $\mathscr{E}$ is autodual, hence $h^0(\mathscr{E}) = h^2(\mathscr{E}) = h^2(\mathscr{E}(1)) = 0$. From the Riemann-Roch formula we obtain $h^1(\mathscr{E}) = -\chi(\mathscr{E}) = 1$. As a stable bundle, $\mathscr{E}$ has the generic splitting type $\mathcal{O} \oplus \mathcal{O}$ (the theorem of Grauert and Mülich) and then an easy lemma of LePotier, [3], prop. 2.17, gives $h^1(\mathscr{E}(k)) = 0$ for $k \geqslant 1$. Hence $h^0(\mathscr{E}(k)) = \chi(\mathscr{E}(k))$ if only $k \geqslant 1$; in particular $h^0(\mathscr{E}(1))$ equals 3.

(2.6) PROPOSITION. *Let $\mathscr{E}$ be a stable, rank-2 vector bundle on $\mathbb{P}^2$ with $c_1 = 0$, $c_2 = 3$. Then $V := \mathbb{P}(\mathscr{E})$ is a Fano threefold if and only if $\mathscr{E}(1)$ is spanned.*

REMARK. For a general bundle $\mathscr{E} \in \mathscr{M}(0, 3)$, $\mathscr{E}(1)$ is spanned and from Barth's description of $\mathscr{M}(0.3)$ it follows that there are stable rank-2 vector bundles $\mathscr{E}$ with $c_1 = 0$, $c_2 = 3$ and $\mathscr{E}(1)$ not spanned (namely, type "a" in Section 7 of [1]).

To prove (2.6), assume first that $\mathscr{E}(1)$ is spanned. Then by the same arguments as we used in the case of bundles with $c_1 = 0, c_2 = 2$ (namely, that the sum $H + \xi_{\mathscr{E}(1)}$ lies in the interior of the cone of numerically effective divisors and hence is ample) we conclude that $-K_V$ is ample. Assume then that $\mathscr{E}(1)$ is not spanned, so that the base point locus $Bs|H + \xi|$ is not empty.

(2.7) *Claim.* $|\xi + H|$ has no base components.

Proof. Assume then that $|H + \xi|$ has a fixed component B_0 and consider a divisor D in the system $|H + \xi|$. Let $D = B_0 + U$, with U in the variable part of $|H + \xi|$. Because the fibres of $p\colon \mathbb{P}(\mathscr{E}) \to \mathbb{P}^2$ are curves, $p(B_0)$ and $p(U)$ are at least one-dimensional. If $p(B_0)$ and $p(U)$ were curves, they would give rise to sections of $\mathscr{E}(1) \otimes \mathscr{I}_{P(B_0)}$ and of $\mathscr{E}(1) \otimes \mathscr{I}_{P(U)}$, contradicting stability. Hence $P(B_0)$ and $P(U)$ are the whole $\mathbb{P}^2$, so that $B_0F \geqslant 1$, $UF \geqslant 1$, where F is the fibre of p. Then $1 = 0 + 1 = HF + \xi F \geqslant B_0F + UF \geqslant 2$, a contradiction.

Since $(H + \xi)^3 = 0$, $Bs|\xi + H|$ is not zero-dimensional thus in view of (2.7) it contains one-dimensional components. Let B the sum of them counted with multiplicities so that we can write $D_1 \cdot D_2 = B + C$ where 1-cycle C does not contain one-dimensional components of the base point locus $Bs|\xi + H|$ and D_1, D_2 are general divisors from $|\xi + H|$.

(2.8) *Claim.* The cycle C contains at least one fibre F of the projection $p\colon \mathbb{P}(\mathscr{E}) \to \mathbb{P}^2$.

Proof. Let $x \in p(B)$. We show that $h^0(\mathscr{E}(1) \otimes \mathscr{I}_x) = 2$. Indeed, if $h^0(\mathscr{E}(1) \otimes \mathscr{I}_x)$ were 1, then at x there would exist two independent sections of $\mathscr{E}(1)$ (cf. (2.5)), hence it would be generated by global sections at x which is not the case. Let us now take a line L through x such that $\mathscr{E}(1)|L = \mathcal{O}(1) \oplus \mathcal{O}(-1)$. The inclusion $H^0(\mathscr{E}(1)) \to H^0(\mathscr{E}(1)|L)$ induces the embedding $H^0(\mathscr{E}(1) \otimes \mathscr{I}_x) \to H^0(\mathscr{E}(1) \otimes \mathscr{I}_x|L)$ and from $h^0(\mathscr{E}(1) \otimes \mathscr{I}_x|L) = 2$ we infer that $h^0(\mathscr{E}(1) \otimes \mathscr{I}_x) \leqslant 2$ and therefore equals 2. To conclude the proof of (2.8). let us take two sections of $\mathscr{E}(1)$ that vanish at x. The corresponding divisors of $|\xi + H|$ then vanish along the fibres over x.

(2.9). In the above notation, $(H + \xi)\cdot C \geqslant 1$.

Proof. In general, if a curve C is not contained in the base point set of a linear system Λ, then $\Lambda \cdot C \geqslant 0$. This shows that in our situation $(H + \xi)\cdot C \geqslant 0$. As C contains the fibre, then $(H + \xi)\cdot C \geqslant 1$.

(2.10). $HB \leqslant 2$. Indeed, we know already that $(H + \xi)^2 = B + C$. Since $(H + \xi)^2 H = 2$ and H is nef, we have $H \cdot C \geqslant 0$, so that $HB \leqslant 2$.

(2.11). $(H + \xi)B \leqslant -1$. Indeed, let us take two divisors as in (2.8). Since $(H + \xi)^3 = 0$, then by inequality (2.9) we get (2.11).

To conclude the proof of (2.6), let us notice that, by (2.11) and (2.10), $c_1(V)B = 2(H + \xi)B + HB \leqslant 0$, i.e., $c_1(V)$ cannot be ample. □

To study the structure of $\mathbb{P}(\mathscr{E})$ more closely, we consider the evaluation morphism $\mathcal{O}^3 \to \mathscr{E}(1)$. Its kernel is a line bundle with $c_1 = -2$, i.e., we have an exact sequence

$$0 \to \mathcal{O}(-2) \to \mathcal{O}^3 \to \mathscr{E}(1) \to 0. \tag{2.12}$$

Now, $\mathscr{E}(1)$ admits a section s with four ordinary zeros at points x_1, x_2, x_3, x_4; see e.g. [5], proposition 1.4b. No three of these points are collinear, since otherwise $\mathscr{E}(1)$ would have a jumping line of type $(3, -1)$ or $(4, -2)$ which contradicts (2.1), Hence, in the terminology of [1], $\mathscr{E}$ is a Hulsbergen bundle. By standard arguments (see e.g. [1], §5.2), every such $\mathscr{E}$ is obtained as an extension $0 \to \mathcal{O} \to \mathscr{E}(1) \to \mathscr{I}(2) \to 0$ with $\mathscr{I} \subset \mathcal{O}$ the ideal sheaf of $Z = \{x_1, x_2, x_3, x_4\}$.

Recall that $\Upsilon := \mathbb{P}(\mathscr{I}_Z(2))$ is the blow-up of $\mathbb{P}^2$ at x_1, x_2, x_3, x_4 and is then a Del Pezzo surface. The above extension gives rise to a $\mathbb{P}^2$-embedding of Υ into $\mathbb{P}(\mathscr{E}(1))$. Let C_1, C_2, C_3 and C_4 be the blow-ups of x_1, x_2, x_3, x_4 and H' be the pull-back of the divisor of a line in $\mathbb{P}^2$. We have:

(a) $-K_\Upsilon = 3H' - \sum C_i$, [7], Proposition 25.1(*i*);

(b) $\xi_{\mathscr{E}(1)} | \Upsilon = 2H' - \sum C_i$ and $(\xi_{\mathscr{E}(1)} | \Upsilon)^2 = \xi^3_{\mathscr{E}(1)} = (H + \xi_{\mathscr{E}})^3 = 0$.

(c) because $-K_\Upsilon$ is ample, by Mori's Cone Theorem the cone of curves is spanned by the extremal ones, [8], Theorem 1.2. Since Υ is neither $\mathbb{P}^2$, nor a $\mathbb{P}^1$-bundle over a curve, all extremal rational curves are exceptional, [8], Theorem 2.1., hence they are of the form C_i or $H' - C_i - C_j$, as follows directly from [7], Proposition 26.2. From the above discussion, the following geometrical interpretation of (2.12) emerges:

(2.13). Let s be a generic section of $\mathscr{E}(1)$ such that zero(s) $= \{x_1, x_2, x_3, x_4\}$. Let $L \subset \mathbb{P}^2 = \mathbb{P}(\Gamma(\mathscr{E}(1)))$ be a line corresponding to the section s viewed at as an element of $H^0(\mathscr{E}(1))$. Then $\Upsilon \to L$ is a conic bundle whose fibres are (strict transforms of) conics through x_1, x_2, x_3, x_4. In particular, three fibres are reducible – they correspond to pairs of lines through x_i, x_j and x_k, x_l, (i, j, k, l) being a permutation of the indices $(1, 2, 3, 4)$. It follows that the map p^{-1} composed with $\Upsilon \to L$ contracts all conics through our points.

COROLLARY. *$\mathscr{E}(1)$ is 1-ample.*

(2.14) REMARK. Globally generated bundles are dense in the moduli space $\mathscr{M}_{\mathbb{P}^2}(0, 3)$. Indeed, by standard cohomological algebra (cf. e.g. [10], ch. II, Lemma 4.1.3) the bundle $\mathscr{E}(1)$ is uniquely determined by choosing an embedding $\mathcal{O}(-2) \subset \mathcal{O}^3$. Giving such an embedding is, in turn, equivalent to picking a three-dimensional linear system in $|\mathcal{O}(2)|$ at each point $x \in \mathbb{P}^2$. Such systems are in an 1–1 correspondence with an open set of 2-planes in $\mathbb{P}^5$, i.e., with points of an open set in Grass $(3, 6)$. Since dim Grass $(3, 6) = 9 = \dim \mathscr{M}_{\mathbb{P}^2}(0, 3)$ and the latter space is irreducible, globally generated bundle form a dense subspace. From the (proof of) Lemma (5.4) in [1] we also infer that a general bundle in $\mathcal{M}_{\mathbb{P}^2}(0, 3)$ has a non-singular cubic as its curve of jumping lines.

3. Fano bundles over $\mathbb{P}^1 \times \mathbb{P}^1$

Let D_1 and D_2 be the divisors corresponding to the two rulings of $\pi_i: \mathbb{P}^1 \times \mathbb{P}^1 \to \mathbb{P}^1$ and let $D = D_1 + D_2$. It is easy to derive the Riemann-Roch formula for rank-2 bundles on $\mathbb{P}^1 \times \mathbb{P}^1$:

$$\chi(\mathscr{E}) = 2 + \tfrac{1}{2}c_1^2 - c_2 + c_1 \cdot D.$$

To study whether a bundle $\mathscr{E}$ on $\mathbb{P}^1 \times \mathbb{P}^1$ is Fano, we may assume $\mathscr{E}$ to be normalized, i.e., $c_1(\mathscr{E}) = (a_1, a_2)$ with $-1 \leqslant a_i \leqslant 0$, $i = 1, 2$. Let us denote $H = p^*(D)$, $H_i = p^*(D_i)$, $i = 1, 2$. As in Section 1, we obtain a formula for $c_1(V)$, where $V = \mathbb{P}(\mathscr{E})$: if $c_1(\mathscr{E}) = (0, 0)$, then $c_1(V) = 2H + 2\xi$; if $c_1(\mathscr{E}) = -H_i$, then $c_1(V) = 2H + H_i + 2\xi$ and if $c_1(\mathscr{E}) = (-1, -1)$, then $c_1(V) = 3H + 2\xi$. Because H is numerically effective, we easily derive an analogue of (2.1):

(3.1). If $\mathscr{E}$ is normalized 2-bundle on $\mathbb{P}^1 \times \mathbb{P}^1$ such that $\mathbb{P}(\mathscr{E})$ is a Fano manifold, then $\mathscr{E}(2, 2) := \mathscr{E} \otimes \mathcal{O}(2,2)$ is ample. If $c_1(\mathscr{E}) = 0$ then already $\mathscr{E}(1.1)$ is ample.

Because a bundle $\mathscr{F} = \bigoplus \mathcal{O}(a_i)$ on a line is ample iff all a_i's are positive, we easily obtain the following corollaries:

(3.3). If $c_1(\mathscr{E}) = 0$, then $\mathscr{E} \mid D_i = \mathcal{O} \oplus \mathcal{O}$ for $i = 1, 2$;

(3.4). If $c_1(\mathscr{E}) = (-1, 0)$, then $\mathscr{E} \mid D_2 = \mathcal{O} \oplus \mathcal{O}(-1)$ and $\mathscr{E} \mid D_1 = \mathcal{O} \oplus \mathcal{O}$ with the obvious symmetry when $c_1(\mathscr{E}) = (0, 1)$;

(3.5). If $c_1(\mathscr{E}) = (-1, -1)$, then $\mathscr{E} \mid D_1 = \mathcal{O} \oplus \mathcal{O}(-1)$ and $\mathscr{E} \mid D_2 = \mathcal{O} \oplus \mathcal{O}(-1)$.

Let us notice that in cases (3.3) and (3.4) the push-forward $\pi_{i*}(\mathscr{E})$ is a rank-2 vector bundle on $\mathbb{P}^1$ (for i chosen such that $\mathscr{E} \mid D_i = \mathcal{O} \oplus \mathcal{O}$). Moreover, the natural morphism $\pi_i^* \pi_{i*}(\mathscr{E}) \to \mathscr{E}$ is an isomorphism and hence we have

(3.6) If $c_1(\mathscr{E}) = (0, 0)$, then $\mathscr{E} = \mathcal{O} \oplus \mathcal{O}$;

(3.7) If $c_1(\mathscr{E}) = (-1, 0)$ or $c_1(\mathscr{E}) = (0, -1)$, then either $\mathscr{E} = \mathcal{O}(-1, 0) \oplus \mathcal{O}$ or $\mathscr{E} = \mathcal{O}(0, -1) \oplus \mathcal{O}$, respectively.

It remains then to study the cases $c_1(\mathscr{E}) = (-1, -1)$. If this is the case, the (1.3) reads as $c_1(V) = 3H + 2\zeta_{\mathscr{E}}$ and in the cohomology ring of $\mathbb{P}^1 \times \mathbb{P}^1$ the following relations hold

$$H_1^3 = H_2^3 = 0, \ H_1 H_2 \xi = 1, \ H^2 \xi = 2, \ H_1 \xi^2 = -1, \ H\xi^2 = 2, \ \xi^3 = 2 - c_2(\mathscr{E}),$$

and thus $0 < (c_1(V))^3 = (3H + 2\xi)^3 = 56 - 8c_2$, i.e., $c_2 \leqslant 6$. Combining this with (1.7) and Table 3 in [9], we get $c_2 \leqslant 4$. Then we show that cases $c_2 = 3$ or 4 cannot occur:

(3.8). If $\mathscr{E}$ is a Fano 2-bundle on $\mathbb{P}^1 \times \mathbb{P}^1$ with $c_1(\mathscr{E}) = (-1, -1)$, then $c_2(\mathscr{E}) \leqslant 2$.

Proof. First we show that $H^0(\mathscr{E}(-1, -1)) = 0$. Indeed, assume that s is a non-zero section of $\mathscr{E}(-1, -1)$ and $\mathcal{O} \to \mathscr{E}(-1, -1)$ is the corresponding inclusion. On a line L we have then the exact sequence

$$0 \to \mathcal{O}_L(3) \to \mathscr{E}(2, 2)|_L \to Q \to 0,$$

where rank-1 quotient sheaf Q has degree 0 in contrary with the ampleness of $\mathscr{E}(2,2)$.

In general, for a 2-bundle $\mathscr{F}$ we have $\mathscr{F} = \mathscr{F}^{\vee} \otimes \det \mathscr{F}$, hence in our situation $H^2(\mathscr{E}(1, 1)) = H^0\mathscr{E}((-1, -1)) = 0$ by Serre duality and then the Riemann-Roch formula gives

$$\chi(\mathscr{E}(1, 1)) = 5 - c_2.$$

We see that if $c_2 \leqslant 4$, then $H^0(\mathscr{E}(1, 1)) \neq 0$. Let s be a non-zero section.

Claim. If c_2 were 3 or 4, then it would exist a curve $C \in |\mathcal{O}(1, 1)|$ such that the multiplicity of the zero set Z of s on C was at least three.

Indeed, let us consider the Segre embedding of $\mathbb{P}^1 \times \mathbb{P}^1$ into $\mathbb{P}^3$ determined by the linear system $|\mathcal{O}(1, 1)|$. Let us pick a plane in P^3 meeting Z at three points, counted with multiplicities. The intersection of the plane and the cubic surface $\mathbb{P}^1 \times \mathbb{P}^1$ in $\mathbb{P}^3$ is the curve C, therefore C is a conic. If C were reducible, e.g. $C = C_1 \cup C_2$, the zero set of s would meet one of C_i's with multiplicity at least two. It would give rise to an embedding $\mathcal{O}_{\mathbb{P}^2}(2) \to \mathscr{E}(1, 1)$, in a contrary with $\mathscr{E}|D_i = \mathcal{O} \oplus \mathcal{O}(-1)$. Hence C must be a smooth conic. Assume then that $\mathscr{E}(1, 1)|C = \mathcal{O}(a_1) \oplus \mathcal{O}(a_2)$, $a_1 \geqslant a_2$. As $c_1(\mathscr{E}(1, 1)) = (1, 1)$, we have $a_1 + a_2 = 2$, but $s|C$ gives an embedding $\mathcal{O}(3)|C \to \mathscr{E}(1, 1)|C$, so that $a_1 \geqslant 3$ and then $a_1 - a_2 \geqslant 4$, which contradicts Lemma 1.5 in [12] and hence proves (3.8).

REMARK. A similar method may be used to exclude (without using the M.-M.-I. classification) the case $c_1 = 0, c_2 = 4, 5$ oń $\mathbb{P}^2$.

(3.9) PROPOSITION *A Fano bundle $\mathscr{E}$ on $\mathbb{P}^1 \times \mathbb{P}^1$ with $c_1(\mathscr{E}) = (-1, -1)$ fits into the exact sequence*

$$0 \to \mathcal{O}(0, -c_2) \to \mathscr{E} \to \mathcal{O}(-1, c_2 - 1) \to 0 \tag{3.10}$$

and for $c_2 \leqslant 1$ the sequence splits.

Proof. We know that $\mathscr{E}|D_i = \dot{\mathcal{O}} \oplus \mathcal{O}(-1)$, $i = 1, 2$, hence the push-forward

$\pi_{i*}(\mathscr{E})$ is a line bundle on $\mathbb{P}^1$, say $\mathcal{O}(k)$. The natural morphism $\pi_i^* \pi_{i*}(\mathscr{E}) \to \mathscr{E}$ is an evaluation on every fibre and because the sections of $\mathcal{O} \oplus \mathcal{O}(-1)$ are constant (in particular, they do not vanish), we have an exact sequence

$$0 \to \pi_i^* \pi_{i*}(\mathscr{E}) \to \mathscr{E} \to Q \to 0$$

with a line bundle Q as a cokernel. Calculating the Chern classes we obtain (3.10). Finally, the fact that for $c_2 \leqslant 1$ this sequence splits follows immediately from the vanishing of first cohomology groups of appropriate bundles on $\mathbb{P}^1 \times \mathbb{P}^1$. This proves (3.9).

COROLLARY. *For $\mathscr{E}$ as above, $c_2(\mathscr{E}) \geqslant 0$.*

Proof. (3.10) gives the exact sequence

$$0 \to \mathcal{O}(2, .2 - c_2) \to \mathscr{E}(2, 2) \to \mathcal{O}(1, c_2 + 1) \to 0$$

and $c_2 < 0$ would contradict the ampleness of $\mathscr{E}(2, 2)$. This proves (8) and (11) of the Theorem.

(3.11) PROPOSITION. *If $\mathscr{E}$ is a Fano bundle and $c_1(\mathscr{E}) = (-1, -1)$, $c_2(\mathscr{E}) = 2$, then $\mathscr{E}(1, 1)$ is globally generated and fits in an exact sequence*

$$0 \to \mathcal{O}(-1, -1) \to \mathcal{O}^{\oplus 3} \to \mathscr{E}(1, 1) \to 0.$$

Proof. By (3.9) we have $h^i(\mathscr{E}(0, 1)) = h^i(\mathscr{E}(1, 0)) = \mathcal{O}$, all i, and $h^i(\mathscr{E}(1, 1)) = 3$ if $i = 0$ and 0 otherwise. Restricting $\mathscr{E}$ to the ruling D_1 gives

$$0 \to \mathscr{E}(0, 1) \to \mathscr{E}(1, 1) \to \mathscr{E}(1, 1)|_{D_1} \to 0.$$

The induced evaluation morphism $H^0(\mathscr{E}(1, 1)) \to H^0(\mathscr{E}(1, 1)|_{D_1})$ is then an isomorphism. But $\mathscr{E}(1, 1)|_{D_i}$ is globally generated, so is $\mathscr{E}(1, 1)$. Since $h^0(\mathscr{E}(1, 1)) = 3$, computing the Chern classes of the kernel of the evaluation $\mathcal{O}^3 \to \mathscr{E}(1, 1)$ gives (3.11). Conversely, if the inclusion $\mathcal{O} \to \mathcal{O}(1, 1)^3$ corresponds to a non-vanishing section of $\mathcal{O}(1, 1)^3$, the quotient is a 2-bundle. To complete our discussion of the case $c_2 = 2$, we must show that $\mathbb{P}(\mathscr{E})$ is Fano. Because $\mathscr{E}(1, 1)$ is globally generated, it is nef and $H + 2\xi_{\mathscr{E}(1,1)}$ is nef, as well. Therefore, to prove that it is ample, it is sufficient (by the theorem of Moishezon and Nakai) to check that $H + 2\xi_{\mathscr{E}(1,1)}$ has positive intersections with curves in $\mathbb{P}(\mathscr{E})$. However, if $H \cdot C = 0$, then C is contained in a fibre and then $\xi_{\mathscr{E}(1,1)} \cdot C > 0$.

4. Fano bundles over non-minimal Del Pezzo surfaces

Let us recall that any non-minimal Del Pezzo surface S_k is a blow-up of k points $x_i (1 \leqslant i \leqslant k \leqslant 8)$ on the plane, no three on one line and no six of them on a conic.

The canonical divisor of S_k has the self-intersection number equal to $9-k$. S_1 is the same as the Hirzebruch surface F_1. Let $\beta: S_k \to \mathbb{P}^2$ be the blow-down morphism, C_i be the exceptional divisors of β and H be the inverse image of the divisor of a line of $\mathbb{P}^2$. Let $\mathscr{E}$ be a Fano bundle on S_k. As in the preceeding sections, we may assume $\mathscr{E}$ to be normalized, i.e.,

$$-1 \leqslant c_1(\mathscr{E})\cdot C_i \leqslant 0 \quad \text{and} \quad -1 \leqslant c_1(\mathscr{E})\cdot H \leqslant 0.$$

Since $K\cdot C_i = 1$, we may apply the same methods as in Section 3 (using Lemma 1.5 from [12]) to conclude easily that $\mathscr{E}|_{C_i} = \mathcal{O} \oplus \mathcal{O}$ and consequently $\mathscr{E} = \beta^*(\mathscr{E}')$ with a 2-bundle $\mathscr{E}'$ on $\mathbb{P}^2$. Moreover, if $c_1(\mathscr{E}')\cdot H = 0$, then $\mathscr{E}$ is trivial. Indeed, let $\bar{L}$ be the strict transform of a line $L \subset \mathbb{P}^3$ that passes through one of the points x_i. Then $K_{S_k}\cdot\bar{L} \leqslant 2$ and in virtue of Lemma 1.5 in [12] we have $\mathscr{E}|\bar{L} = \mathcal{O} \oplus \mathcal{O}$, therefore $\mathscr{E}/L = \mathcal{O} \oplus \mathcal{O}$ and Van de Ven's theorem shows that $\mathscr{E}'$ is trivial, so is $\mathscr{E}$.

Let us notice that for $k \geqslant 2$ we can always choose a line L that passes through two of the points x_i, so that $-K_{S_k}\cdot\bar{L} = -1$ and, as above, $c_1(\mathscr{E})\cdot\bar{L} = 0$, implying $c_1(\mathscr{E}')\cdot H = 0$. In other words, we have proved that for $k \geqslant 2$, the only ruled Fano 3-fold over a Del Pezzo surface S_k is $\mathbb{P}^1 \times S_k$. Finally, on the Hirzebruch surface F_1 we have

(a) if $c_1(\mathscr{E}') = 0$, then, as above, $\mathscr{E} = \mathcal{O} \oplus \mathcal{O}$,

(b) if $c_1(\mathscr{E}') = -1$, then, as in (2.3), we infer that $\mathscr{E}' = \mathcal{O} \oplus \mathcal{O}(-1)$ or $\mathscr{E}' = T_{\mathbb{P}^2}(-2)$.

References

1. Barth, W., Moduli of Vector Bundles on the Projective Plane. *Inv. Math.* 42, (1977), 63–91.
2. Demin, I.V., Three-dimensional Fano manifolds representable as line fiberings (Russian). *Izv. Acad. Nauk SSSR*, 44, no. 4 (1980). English translation in *Math. USSR Izv.* 17. Addendum to this paper in *Izv. Acad. Nauk SSSR*. 46, no. 3. English translation in *Math. USSR Izv.* 20.
3. Elencwajg, F. and Forster, O., Bounding Cohomology Groups of Vector Bundles on $\mathbb{P}^n$, *Math. Ann.* 246 (1980) 251–270.
4. Hartshorne, R., Ample Subvarieties of Algebraic Varieties. *Lecture Notes* 156 (1970).
5. Hartshorne, R., Stable Vector Bundles of Rank 2 on $\mathbb{P}^3$. *Math. Ann.* 238 (1978) 229–280.
6. Kawamata, Y., The cone of curves of algebraic varieties. *Ann. Math.* 119, 603–633 (1984).
7. Manin, Yu.I Cubic forms, Algebra, Geometry, Arithmetic. North Holland 1974.
8. Mori, Sh., Threefolds Whose Canonical Bundle is not Numerically Effective. *Ann. Math.* 116, 133–176 (1982).
9. Mori, Sh. and Mukai, Sh.: Classification of Fano 3-folds with $B_2 \geqslant 2$. *Manuscripta Math.* 36, 147–162 (1981).
10. Okonek, Ch., Schneider, M. and Spindler, H.: Vector Bundles on Complex Projective Spaces, Birkhäuser, 1981.
11. Schiffman, B, and Sommese, A. J., Vanishing theorems on complex manifolds. *Birkhäuser* 1985.
12. Szurek, M., Wiśniewski, J.A., Fano Bundles on $\mathbb{P}^3$ and $\mathbb{Q}_3$. *Pacific Journ. Math.* 140, no. 2. (1989).
13. Van de Ven, A., On uniform vector bundles. *Math. Ann.* 195 (1972) 245–248.

GPSR Compliance

The European Union's (EU) General Product Safety Regulation (GPSR) is a set of rules that requires consumer products to be safe and our obligations to ensure this.

If you have any concerns about our products, you can contact us on ProductSafety@springernature.com

In case Publisher is established outside the EU, the EU authorized representative is:

Springer Nature Customer Service Center GmbH
Europaplatz 3
69115 Heidelberg, Germany

Batch number: 10584501

Printed by Printforce, the Netherlands